Leading Personalities in Statistical Sciences

WILEY SERIES IN PROBABILITY AND STATISTICS

Established by WALTER A. SHEWHART and SAMUEL S. WILKS

A complete list of the titles in this series appears at the end of this volume.

Leading Personalities in Statistical Sciences

FROM THE SEVENTEENTH CENTURY TO THE PRESENT

Edited by

NORMAN L. JOHNSON
University of North Carolina

and

SAMUEL KOTZ
University of Maryland

A Wiley-Interscience Publication
JOHN WILEY & SONS, INC.
New York • Chichester • Weinheim • Brisbane • Singapore • Toronto

Cover photograph depicts Lambert Aldolphe Quetelet, 1796–1874

Some material in this work may have originally appeared in or been adapted from *Encyclopedia of Statistical Sciences, Volumes 1 to 9*, and *Encyclopedia of Statistical Sciences: Supplement Volume*, all edited by Norman L. Johnson and Samuel Kotz or *Encyclopedia of Statistical Sciences: Update Volumes 1 and 2*, edited by Samuel Kotz, Campbell B. Read, and David L. Banks.

Printed on acid-free paper.

Library of Congress Cataloging in Publication Data:

Leading personalities in statistical sciences: from the 17th century to the present/[edited by] Norman L. Johnson, Samuel Kotz.
p. cm.—(Wiley series in probability and mathematical statistics. Probability and mathematical statistics)
"A Wiley-Interscience publication."
Includes bibliographical references (p. –) and index.
ISBN 0-471-16381-3 (pbk. : alk. paper)
1. Statisticians–Biography. I. Johnson, Norman Lloyd.
II. Kotz, Samuel. III. Series.
QA276.156.L43 1997
510'92'2—dc21 97-9152
[B] CIP

Printed in the United States of America
10 9 8 7 6 5 4 3 2 1

"If I have seen further, it is by standing on the shoulders of giants."

Sir Isaac Newton
in a letter written on February 5, 1676
to Robert Hooke

Preamble

Statistical personalities in this volume have been classified into seven groups called "Sections" according to their status or the field of specialization. The Sections are as follows:

1. Forerunners
2. Statistical Inference
3. Statistical Theory
4. Probability Theory
5. Government and Economic Statistics
6. Applications in Medicine and Agriculture
7. Applications in Science and Engineering

It may happen that a certain person may also be assigned to some other section, in addition to the main section. The numbers of these additional sections are indicated in parenthesis to the right of the name in the Contents listing. The name of the author of an entry is indicated at the end. Unsigned entries have been composed by the editors, either jointly or separately.

Contents

SECTION 3 STATISTICAL THEORY

SECTION 4 PROBABILITY THEORY

SECTION 5 GOVERNMENT AND ECONOMIC STATISTICS

SECTION 6 APPLICATIONS IN MEDICINE AND AGRICULTURE

Preface

This book contains biographical sketches of persons who have contributed to the growth and acceptance of statistical methods since the early Seventeenth Century. It is intended to provide both serious students and interested laypersons with some appreciation of the social and intellectual backgrounds of those to whom statistical science owes much of its current flourishing state. The diversity of these personalities reflects the rich variety of sources and endeavors—seemingly only slightly related—which merged together in the first half of the present century to form the body of ideas and techniques known as "modern statistical methodology." This consolidation was achieved—perhaps fortuitously—just in time to prevent statistical procedures from being abosorbed piecemeal into small segments of "computer technology." As it is, we now possess a recognized discipline, providing bases for deciding *what* to compute, as distinct from directions *how* to compute.

The main developments leading to the merging of diverse statistical techniques into a *bona fide* scientific discipline took place in the Nineteenth Century—more particularly in the second half of that century. These developments were, however, based on the pioneering work of many authors during the preceding 200 years. Many aspects of this phenomenon have been described—notably in three remarkable scholarly treatises—by S. M. Stigler (1986), T. M. Porter (1986), and I. Hacking (1992)—together with valuable contributions from several authors in the last 30 years. Much of this work is described in the books of M. G. Kendall, E. S. Pearson and R. L. Plackett (1970, 1977). These works, however, do not cover developments over the last 25 years, *or* from the earliest periods. The two volumes of *Breakthroughs in Statistics* (1992, published by Springer-Verlag New York and edited by S. Kotz and N. L. Johnson) and a third volume in preparation at the time when this introduction is being written, may serve as a partial effort at such a study.

However, all of the contributions just described are, perhaps, too formal and "scholarly" for many present-day users of statistical methods. We have, therefore, chosen biographical studies in a way designed to throw light on the interplay of currents of thought in private and public affairs which has led to much of the present-day outlook in statistical matters. For each personality we have tried to provide, at least, cursory information on the milieu in which his/her activities were conducted, keeping in mind the dictum of Karl Pearson (a major architect in the founding of modern statistics): "It is impossible to understand a man's work unless you understand something of his environment."

Even casual browsing through the pages of this book will demonstrate the wide range of scientific interests among those whose lives and work are described. We note that F. Y. Edgeworth was an economist, J. C. Maxwell was a physicist and P. S. Laplace was a distinguished mathematician and physicist. Even R. A. Fisher—regarded by many as the most influential statistician of all ages—was known in wider scientific circles as a geneticist. T. M. Porter (1986) has presented, in a brilliant and convincing manner, the thesis that all the forerunners of statistics (as we know the subject at present) were primarily mathematicians, physicists, or biologists. We accept this evaluation, adding astronomers to the list. We venture to suggest that the first world-renowned "pure" statistician—with no other label attached—was M. G. Kendall (1907–1983). Only gradually has the profession of "statistician" without qualifying adjective(s) become widely recognized. There is, of course, a considerable range of types of statisticians defined, more or less, by the work in which they are engaged. In particular, biostatisticians, who are concerned with application of statistical methods to problems associated with living organisms, especially with medical matters, have become a well-defined group. Nevertheless, there is an awareness of a basic corpus of concepts and techniques underlying the various specific fields of statistical effort.

Statistics as a science, indeed has deep roots in political economy, astronomy, geodesy, engineering, medicine, and social and biological sciences. It uses probability theory—more specifically, mathematical techniques for analyzing data, derived from probability theory on the basis of various philosophical principles of inductive inference.

Although this volume concentrates on "statistical personalities," we feel it is appropriate to provide a broad picture of milestone events in the history of statistics over the last four centuries. A summary of events up to 1917 is provided in the following table, which is based on a translation, with some corrections and emendations, of a table in *Manuale di Statistica* by Felice Vinci (pp. 284–286 of Volume 1, 3rd edition, published in 1939 by Nicola Zanichelli, Editore, Bologna, Italy). In the table, the reference *Hist. Stat.* refers to *The History of Statistics, their Development and Progress in Many Countries* published in 1918 by Macmillan, New York.

No.	Year	Events	Author(s) or Bibliographic Origin
1.	1532	First weekly data on deaths in London	From C. H. Hill, *Econ. Writ. of Sir W. Petty,* page LXXI.
2.	1539	Beginning of official data collection on baptisms, marriages, and deaths in France	From F. Favre, *Hist. Stat.,* page 242.
3.	1608	Beginning of the Parish Registry in Sweden	From E. Arosonius, *Hist. Stat.,* page 537.
4.	1662	Publicaion of *Natural and political observations mentioned in a following index and made upon the bills of mortality.*	J. Graunt (1620–1764)
5.	1666	First census in Canada (the first modern demographic census)	From E. H. Godfrey, *Hist. Stat.*, page 179.
6.	1693	Publication of *Estimate of the degrees of mortality in mankind,* in Philosophical Transactions of the Royal Society	
7.	1713	Publication of *Ars Conjectandi*	J. Bernoulli (1654–1705)
8.	1718	Publication of *The Doctrine of Chances*	A. De Moivre (1667–1754)
9.	1735	Beginning of demographic data collection in Norway	From A. N. Kiaer, *Hist. Stat.,* page 447.
10.	1741	Publication of *Die göttliche Ordnung in den Veränderung en des menschlichen Geschlects aus der Geburt, dem Tode und der Fortpflanzung desselben erwiesen*	J. P. Süssmilch (1707–1767)
11.	1746	Publication in France of particular tables on mortality data under the title *Essai sur les probabilités de la durée de la vie humaine*	A. Déparcieux (1703–1768)
12.	1748	Beginning of official demographic statistics in Sweden	From E. Arosonius, *Hist. Stat.*, page 540.
13.	1749	First complete demographic census in Sweden	From W. S. Rossiter, *A Century Popul. Growth,* page 2.
14.	1753	First demographic census in Austria	From R. Meyer, *Hist. Stat.,* page 85.
15.	1766	Publication of the first national table on mortality in Sweden: *Mortaliteten i Sverige, i andelning of Tabel-Verket*	W. Wargentin (1717–1783)
16.	1769	First demographic census in Denmark and Norway	From A. Jensen, *Hist. Stat.*, page 201.

No.	Year	Events	Author(s) or Bibliographic Origin
17.	1790	Publication of the *Riflessioni sulla popolazione delle nazioni per rapporto all'economia nazionale*	G. Ortes (1713–1790)
18.	1790	First federal demographic census in U.S.A.	
19.	1798	Anonymous publication of *An Essay on the principle of population*	T. R. Malthus (1766–1834)
20.	1801	First complete demographic census in Great Britain	From W. S. Rossiter, *A Century Pop. Growth*, l.c.
21.	1801	First complete demographic census in France	From W. S. Rossiter, *A Century for Growth,*
22.	1812	Publication of *Théorie analytique des probabilités*	P. S. Laplace (1740–1827)
23.	1825	Publication of *Mémoire sur les lois des naissances et de la mortalité a Bruxelles*	A. Quetelet (1796–1874)
24.	1825	Publication of *On the nature of the function expressive of the law of human mortalitiy* in Philosophical Transactions of the Royal Society.	B. Gompertz (1779–1865)
25.	1829	First demographic census in Belgium	From A. Julin, *Hist. Stat.*, page 128.
26.	1832	Publication of *Recherches sur la reproduction et sur la mortalité de l'homme au differents ages et sur la population de la Belgique*	A. Quetelet and E. Smith.
27.	1834	Establishment of the Statistical Society of London (later, Royal Statistical Society)	
28.	1835	Publication of *Sur l'homme et le developpement de ses facultés, ou essai de physique sociale*	A. Quetelet (1796–1874)
29.	1837	Publication of *Recherches sur la probabilité des jugements en matiére criminelle et en matiére civile, precedées des regles genérales du calcul des probabilités*	S. D. Poisson (1781–1840)
30.	1837	Public data collection of the demographic statistics in England. Establishment of the Registrar General Office	From A. Baines, *Hist. Stat.*, page 370.

No.	Year	Events	Author(s) or Bibliographic Origin
31.	1839	Organization of the American Statistical Association (Boston)	
32.	1844	Publication of *Recherches mathematiques sur la loi d'accroissement de la population* in "Mémoires de l'Académie Royal de Bruxelles"	P. F. Verhulst (1804–1849).
33.	1861	First complete demographic census in Italy	
34.	1869	Publication of *Hereditary Genius*	F. Galton (1822–1907).
35.	1869	Establishment of the "Société de Statistique de Paris"	
36.	1877	Publication of *Zur theorie der Massenerscheinungen der menschlichen Gesellschaft*	W. Lexis (1837–1914).
37.	1885	Establishment in Den Haag (Netherlands) of the International Statistical Institute	
38.	1889	Publication of *Natural Inheritance*	F. Galton (1822–1907).
39.	1890	Introduction of the automated screening of census data (John S. Billings and Herman Hollerith) (U.S.A.)	
40.	1901	Publication of the first issue of *Biometrika*.	F. Galton, K. Pearson, W. F. R. Weldon, C. B. Davenport.
41.	1910	Publication of *An Introduction to the Theory of Statistics*	G. Udny Yule.
42.	1911	Publication of *A Problem in Age-Distribution* in The London, Edinburgh and Dublin Philosophical Magazine	A. J. Lotka (and F. R. Sharpe).
43.	1917	Publication of *The Mathematical Theory of Population*	G. H. Knibbs.

It is a complex task to provide an adequate representative continuation of this table. Here we add a few "points of light" that have illuminated subsequent progress. We note:

Year	Event
1930	Establishment of the Institute of Mathematical Statistics, and appearance of *Annals of Mathematical Statistics*.
1931	Establishment of the Indian Statistical Institute
1933	Appearance of *Sankhyā*.
1935	R. A. Fisher's *The Design of Experiments*; published by Oliver and Boyd, London, the first of 12 editions.
1937	G. W. Snedecor and W. G. Cochran's *Statistical Methods* (published by Iowa State University Press, Ames, IA). First of 8 editions.
1943–1946	M. G. Kendall's *The Advanced Theory of Statistics*; 2 volumes, published by Charles Griffin, London. Later editions with A. Stuart and then A. Stuart and J. K. Ord have increasing number of volumes.
1946	H. Cramér's *Mathematical Methods of Statistics*; published by Almqvist and Wiksell, Uppsala and Princeton University Press, Princeton, NJ.
1968	Publication of *International Encyclopedia of Statistics*; 2 volumes, edited by J. M. Tanur and W. H. Kruskal, published by Crowell-Collier/Macmillan: New York and London.
1974–1982	World Fertility Survey, conducted by the International Statistical Institute, funded by the United Nations Fund for Population Activities and the United States Agency for International Development, under the leadership of Sir Maurice Kendall.
1982–1989	Publication of *Encyclopedia of Statistical Sciences*; edited by S. Kotz, N. L. Johnson and C. B. Read, 9 volumes plus supplemental volume, John Wiley and Sons, New York.

To assist initial reading of the contents of this book, each personality has been assigned a number reflecting the primary interest of the subject according to the following broad scheme of classification:

1. Forerunners
2. Statistical Inference
3. Statistical Theory
4. Probability Theory
5. Government and Economic Statistics
6. Applications in Medicine and Agriculture
7. Applications in Science and Engineering

For several personalities, their interests could not be limited to a single one of these categories. In the Contents, secondary interests of sufficient importance are indicated by numbers in parentheses. Thus the biography for R. A. Fisher appears with numbers 2 (3,6,7).

We have made substantial efforts to obtain photographs, or other likenesses, of the personalities included in this volume. We thank many statisticians (in many parts of the world) who replied positively to our requests; and also the publishers who graciously granted permission to reproduce. The biographies in this volume are, in the majority of cases, adapted from entries in the *Encyclopedia of Statistical Sciences*, or will appear in the *Updates* to those volumes, expected to be published in 1997–1998. Several entries were written especially for the present publication.

It should be borne in mind that we have tried to present accounts of the "personalities" of the chosen subjects. For this reason, there is generally more discussion of their lives than of technical details of their work. One should not expect, for example, details of the way John Graunt constructed the so-called "first life table," or how to use the sequential probability ratio test of Abraham Wald.

As always, we expect and hope for comments from readers. We welcome criticism on matters of opinion and fact, and correction on matters of record. After a combined total of some 80 years in statistical work, we realize that exchanges of opionions and knowledge are essential to stimulate awareness, among statisticians and others, of the (mostly) glorious past and (hopefully) bright future of this often maligned and misunderstood discipline.

NORMAN L. JOHNSON SAMUEL KOTZ

REFERENCES

Hacking, I. (1990) *The Taming of Chance*. New York: Cambridge University Press.

Kendall, M. G. and Plackett, R. L. (eds.) (1977) *Studies in the History of Statistics and Probability*, **2**, New York: Macmillan.

Pearson, E. S. and Kendall, M. G. (eds.) (1970) *Studies in the History of Statistics and Probability*, **1**, New York: Macmillan.

Porter, T. M. (1986) *The Rise of Statistical Thinking, 1820–1900*. Princeton, NJ: Princeton University Press.

Stigler, S. M. (1986) *History of Statistics: The Measurement of Uncertainty Before 1900*, Cambridge, MA: Belknap Press.

Leading Personalities in Statistical Sciences

SECTION **1**

Forerunners

Abbe, Ernst

Born: January 23, 1840, in Eisenach, Germany.
Died: January 14, 1905, in Jena, Germany.
Contributed to: theoretical and applied optics, astronomy, mathematical statistics.

Ernst Abbe overcame a childhood of privation and a financially precarious situation early in his academic career when he completed his inaugural dissertation, "On the Law of Distribution of Errors in Observation Series." With this, at the age of 23, he attained a lectureship at Jena University. The dissertation, partly motivated by the work of C. F. Gauss,* seems to contain his only contributions to the probability analysis of observations subject to error. These contributions anticipate his later work in distribution theory and time-series analysis, but they were overlooked until the late 1960s. With the notable exception of Wölffing [10], almost none of the early bibliographies on probability and statistics mention this work. In 1866, Abbe was approached by Carl Zeiss, who asked him to establish a scientific basis for the construction of microscopes; this was the beginning of a relationship that lasted the rest of his life. From this period on, his main field of activity was optics [9] and astronomy.

Abbe shows, first, that the quantity $\Delta = \sum_{i=1}^{n} Z_i^2$, where Z_i, $i = 1, \ldots, n$, are n independently and identically distributed $N(0,1)$ random variables, is described by a chi-square density with n degrees of freedom [5, 8], although this discovery should perhaps be attributed to I. J. Bienaymé* [4]. Second, again initially by means of a "discontinuity factor" and then by complex variable methods, Abbe obtains the distribution of $\Theta = \sum_{j=1}^{n} (Z_j - Z_{j+1})^2$, where $Z_{n+1} = Z_1$, and ultimately that of Θ/Δ, a ratio of quadratic forms in $Z_1, \ldots, Z_n$ very close in nature to the definition of what is now called the first circular serial correlation coefficient, and whose distribution under the present conditions is essentially that used to test the null hypothesis of Gaussian white noise against a first-order autoregression alternative, in time-series analysis [3]. (The distribution under such a null hypothesis was obtained by R. L. Anderson in 1942.) Knopf [6] expresses Abbe's intention in his dissertation as being to seek a numerically expressible criterion to determine when differences between observed and sought values in a series of observations are due to chance alone.

REFERENCES

[1] Abbe, E. (1863). *Über die Gesetzmässigkeit der Vertheilung der Fehler bei Beobachtungsreihen.* Hab. schrift., Jena. (Reprinted as pp. 55–81 of ref. [2].)

[2] Abbe, E. (1906). *Gesammelte Abhandlungen,* Vol. 2. G. Fischer, Jena.

[3] Hannan, E. J. (1960). *Time Series Analysis.* Methuen, London, pp. 84–86.

[4] Heyde, C. C. and Seneta, E. (1977). *I. J. Bienaymé: Statistical Theory Anticipated.* Springer-Verlag, New York, p. 69.

[5] Kendall, M. G. (1971). *Biometrika,* **58,** 369–373. (Sketches Abbe's mathematical reasoning in relation to the contributions to mathematical statistics.)

[6] Knopf, O. (1905). *Jahresber. Dtsch. Math.-Ver.,* **14,** 217–230. [One of several obituaries by his associates; nonmathematical, with a photograph. Another is by S. Czapski (1905), in *Verh. Dtsch. Phys. Ges.,* **7,** 89–121.]

[7] Rohr, L. O. M. von (1940). *Ernst Abbe.* G. Fischer, Jena. (Not seen.)

[8] Sheynin, O. B. (1966). *Nature (Lond.),* **211,** 1003–1004. (Notes Abbe's derivation of the chi-square density.)

[9] Volkman, H. (1966). *Appl. Opt.,* **5,** 1720–1731. (An English-language account of Abbe's life and contributions to pure and applied optics; contains two photographs of Abbe, and further bibliography.)

[10] Wölffing, E. (1899). *Math. Naturwiss. Ver. Württemberg* [Stuttgart], *Mitt.,* (2) **1,** 76–84. [Supplements the comprehensive bibliography given by E. Czuber (1899), in *Jahresber. Dtsch. Math.-Ver.,* **7** (2nd part), 1–279.]

E. SENETA

Achenwall, Gottfried

Born: October 20, 1719, in Elbing, Germany.
Died: May 1, 1772, in Göttingen, Germany.
Contributed to: *Staatswissenschaft* ("university statistics").

Achenwall was born into the family of a merchant. In 1738–1740 he acquired a knowledge of philosophy, mathematics, physics, and history at Jena; then he moved to Halle, where, without abandoning history, he studied the law and *Staatswissenschaft* (the science of the state; also known as "university statistics"). Apparently in 1742 Achenwall returned for a short time to Jena, then continued his education in Leipzig. In 1746 he became Docent at Marburg, and in 1748, extraordinary professor at Göttingen (ordinary professor of law and of philosophy from 1753), creating there the Göttingen school of statistics. Its most eminent member was A. L. Schlözer (1735–1809). Achenwall married in 1752, but his wife died in 1754, and he had no children.

Achenwall followed up the work of Hermann Conring (1606–1681), the founder of *Staatswissenschaft,* and was the first to present systematically, and in German rather than in Latin, the Conring tradition. According to both Conring and Achenwall, the aim of statistics was to describe the climate, geographical postition, political structure, and economics of a given state, to provide an estimate of its population, and to give information about its history; but discovering relations between quantitative variables was out of the question. For Achenwall [1, p. 1], "the so-called statistics" was the *Staatswissenschaft* of a given country.

Since 1741, "statisticians" have begun to describe states in a tabular form, which facilitate the use of numbers, a practice opposed by Achenwall. Even in 1806 and 1811 [5, p. 670] the use of tabular statistics was condemned because numbers were unable to describe the spirit of a nation.

Nevertheless, Achenwall [4, Chap. 12] referred to Süssmilch,* advised state measures fostering the multiplication of the population, recommended censuses, and even [4, p. 187] noted that its "probable estimate" can be gotten by means of "yearly lists of deaths, births, and marriages." The gulf between statistics (in the modern sense) and Staatswissenschaft was not as wide as it is usually supposed to have been. Leibniz's manuscripts, written in the 1680s, present a related case. First published in 1866 and reprinted in 1977, they testify that he was both a political arithmetician and an early advocate of tabular description (both with and without the use of numbers) of a given state; see [10, 222–227, 255].

REFERENCES

[1] Achenwall, G. (1748). *Vorbereitung zur Staatswissenchaft.* Göttingen. (An abridged version of this was included in his next contribution.)

[2] Achenwall, G. (1749). *Abriß der neuesten Staatswissenschaft der vornehmsten europäischen Reiche und Republicken zum Gebrauch in seinen academischen Vorlesungen.* Schmidt, Göttingen.

[3] Achenwall, G. (1752). *Staatsverfassung der europäischen Reiche im Grundrisse.* Schmidt, Göttingen. This is the second edition of the *Abriß*. Later editions: 1756, 1762, 1767, 1768. 1781–1785, 1790–1798. By 1768 the title had changed to *Staatsverfassung der heutigen vornehmsten europäischen Reiche and Völker,* and the publisher was Witwe Wanderhoeck.

[4] Achenwall, G. (1763). *Staatsklugheit nach ihren Grundsätzen.* Göttingen. (Fourth ed., 1779).

[5] John, V. (1883). The term "Statistics." *J. R. Statist. Soc.* **46,** 656 – 679. (Originally published in German, also in 1883.)

[6] John, V. (1884). *Geschichte der Statistik.* Encke, Stuttgart.

[7] Lazarsfeld, P. (1961). Notes on the history of quantification in sociology—trends, sources and problems. *Isis,* **52,** 277– 333. Reprinted (1977) in *Studies in the History of Statistics and Probability,* Sir Maurice Kendall and R. L. Plackett. eds. Vol. 2, pp. 213–269. Griffin, London and High Wycombe

[8] Leibniz, G. W. (1886). *Sämmtliche Schriften und Briefe,* Reihe 4, Bd. 3. Deutsche Akad. Wiss., Berlin.

[9] Schiefer, P. (1916). *Achenwall und seine Schule.* München: Schrödl. (A Dissertation.)

[10] Sheynin, O. B. (1977). Early history of the theory of probability. *Arch. Hist. Ex. Sci.,* **17,** 201–259.

[11] Solf, H. H. (1938). *G. Achenwall. Sein Leben und sein Werk, ein Beitrag zur Göttinger Gelehrtengeschichte.* Mauser, Forchheim, Oberfranken. (A dissertation.)

[12] Westergaard, H. (1932). *Contributions to the History of Statistics.* King, London. (Reprinted, New York, 1968, and The Hague, 1969.)

[13] Zahn, F. and Meier, F. (1953). Achenwall. *Neue deutsche Biogr.* **1,** 32–33. Duncker and Humblot, Berlin.

O. SHEYNIN

Arbuthnot, John

Born: April 29, 1667 in Arbuthnot, Kincardineshire, Scotland.
Died: February 27, 1735 in London, England.
Contributed to: demography, probability.

John Arbuthnot was a Scot with a wide-ranging *curriculum vitae*. A capability in mathematics, and an eagerness to apply mathematics to the real world, were by no means his only or his most celebrated talents. He became a Fellow of both the Royal Society (1704) and the Royal College of Physicians (1710). He was said to be at one time the favorite among Queen Anne's physicians. A noted wit and satirist, he was the creator of the figure John Bull (who has become a symbol of the English character), and was a close colleague of Jonathan Swift, Alexander Pope, and John Gay. Swift said of Arbuthnot in a letter to Pope: "Our doctor hath every quality in the world that can make a man amiable and useful; but alas! he hath a sort of slouch in his walk." An oil painting of Arbuthnot hangs in the Scottish National Portrait Gallery.

Two of Arbuthnot's published works have merited him a place in the history of statistics and probability. He translated, from Latin into English, Huygens' *De ratiociniis in ludo aleae,* the first probability text [6]. Arbuthnot's English edition [2] was not simply a translation. He began with an introduction written in his usual witty and robust style, gave solutions to the problems Huygens had posed, and added further sections of his own about gaming with dice and cards.

It was, however, the paper Arbuthnot presented to the Royal Society of London on April 19, 1711 [1] that has attracted most attention from historians of statistics and probability (it appeared in a volume of the *Philosophical Transactions* dated 1710, but published late). Arbuthnot's paper was "An Argument for Divine Providence, taken from the constant Regularity observ'd in the Births of both Sexes." In it, he maintained that the guiding hand of a divine being was to be discerned in the nearly constant ratio of male to female christenings recorded annually in London over the years 1629 to 1710. Part of his reasoning is recognizable as what we would now call a "sign" test, so Arbuthnot has gone down in statistical history as a progenitor of significance testing.

The data he presented showed that in each of the 82 years 1629–1710, the annual number of male christenings had been consistently higher than the number of female christenings, but never *very much* higher. Arbuthnot argued that this remarkable regularity could not be attributed to chance, and must therefore be an indication of divine providence. It was an example of the "argument from design," a thesis of considerable theological and scientific influence during the closing decades of the seventeenth century and much of the next century. Its supporters held that natural phenomena of many kinds showed evidence of careful and beneficent design, and were therefore indicative of the existence of a supreme being.

Arbuthnot's representation of "chance" determination of sex at birth was the toss of a fair two-sided die, with one face marked M and the other marked F. From there, he advanced his argument on two fronts. First, "chance" could not explain the very close limits within which the annual ratios of male to female christenings had been observed to fall. Secondly, neither could it explain the numerical dominance, year after year, of male over female christenings.

He pursued the first line of argument by indicating how the middle term of the binomial expansion, for even values of the size parameter *n,* becomes very small as n gets large. Though he acknowledged that in practice the balance between male and female births in any one year was not exact, he regarded his mathematical demonstration as evidence that, "if mere Chance govern'd," there would be years when the balance was not well maintained.

The second strand of his argument, concerning the persistent yearly excess of male over female christenings, was the one which ultimately caught the attention of historians of statistics. He set out a calculation showing that the probability of 82 consecutive years in which male exceeded female christenings in number, under the supposition that "chance" determined sex, was very small indeed. This he took as weighty evidence against the hypothesis of chance, and in favor of his alternative of divine providence. He argued that if births were generated according to his representation of chance, as a fair two-sided die, the probability of observing an excess of male over female birthes in any one year would be no higher than one-half. Therefore the probability of observing 82 successive "male years" was no higher than $(\frac{1}{2})^{82}$ (a number of the order of 10^{-25} or 10^{-26}). The probability of observing the data given the "model," as we might now say, was very small indeed, casting severe doubt on the notion that chance determined sex at birth. Arbuthnot proceeded to a number of conclusions of a religious or philosophical nature, including the observation that his arguments vindicated the undesirability of polygamy in a civilized society.

We can see in Arbuthnot's probabilistic reasoning some of the features of the modern hypothesis test. He defined a null hypothesis ("chance" determination of sex at birth) and an alternative (divine providence). He calculated, under the assumption that the null hypothesis was true, a probability defined by reference to the observed data. Finally, he argued that the extremely low probability he obtained cast doubt on the null hypothesis, and offered support for his alternative.

Arbuthnot's reasoning has been given a thorough examination by modern statisticians and logicians, most notably by Hacking [4,5]. We have, of course, the benefit of more than 250 years of hindsight and statistical development. We can see that the probability of $(\frac{1}{2})^{82}$, on which hinged Arbuthnot's dismissal of the "chance" hypothesis, was one of a well-defined reference set, the binomial distribution with parameters 82 and one-half. It was the lowest probability in this reference set, and because it was the most extreme was also in effect a tail-area probability. And it was an *extremely low* probability. Arbuthnot made only the last of these points explicit.

Arbuthnot's advancement of an "argument from design" did not single him out from his contemporaries. Nor were his observations on the relative constancy of the male to female birth ratio radical. What was novel was his attempt to provide a statistical "proof" of his assertions, based on a quantitative concept of chance, and with the argument explicitly expressed and concluded in numerical terms.

An unpublished manuscript in the Gregory collection, held at the University of

Edinburgh, indicates that Arbuthnot had been flirting with ideas of probabilistic proof well before 1711, possibly as early as 1694 [3] (David Gregory, 1661–1708, was Savilian Professor of Astronomy at Oxford). In Arbuthnot's 10-page "treatise on chance," there is an anticipation of his 1711 argument concerning the middle term of the binomial as n gets large, as well as two other statistical "proto-tests" concerning the lengths of reign of the Roman and Scottish kings. The chronology of the first seven kings of Rome was suspect, he suggested, because they appeared to have survived far longer on average than might reasonably be expected from Edmund Halley's life table, based on the mortality bills of Breslau. In the case of the Scottish kings, on the other hand, the evidence seemed to indicate that mortality amongst Scottish kings was higher than might be expected from Halley's table. However, neither of the calculations Arbuthnot outlined had the clarity of statistical "model" evident in his 1711 paper, nor did they culminate in a specific probability level quantifying the evidence.

Arbuthnot's 1711 paper sparked off a debate which involved, at various times, William 'sGravesande.* (a Dutch scientist who became professor of Mathematics at the University of Leiden), Bernard Nieuwentijt (a Dutch physician and mathematician), Nicholas Bernoulli,* and Abraham De Moivre.* 'sGravesande developed Arbuthnot's test further, attempting to take into account the close limits within which the male-to-female birth ratio fell year after year. Bernoulli, on the other hand, questioned Arbuthnot's interpretation of "chance." He proposed that the fair two-sided die could be replaced by a multifaceted die, with 18 sides marked M and 17 marked F. If tossed a large number of times, such a die would, Bernoulli maintained, yield ratios of M's to F's with similar variability to the London christenings data. Certain aspects of the exchanges which took place between the participants in the debate can be seen as attempts to emulate and develop Arbuthnot's mode of statistical reasoning, though they have not proved as amenable to reinterpretation within modern frameworks of statistical logic.

Though Arbuthnot's 1711 argument tends now to be regarded as the first explicitly set out and recognizable statistical significance test, it is doubtful whether his contribution, and the debate it provoked, provided any immediate stimulus to ideas of statistical significance testing. The obvious effect was to fuel interest in the argument from design, in the stability of statistical ratios, and in the interplay of one with the other.

REFERENCES

[1] Arbuthnot, J. (1710). An argument for divine providence, taken from the constant regularity observ'd in the births of both sexes. *Phil. Trans. R. Soc., London,* **27,** 186–190. Reprinted in M. G. Kendall and R. L. Plackett, eds. *Studies in the History of Statistics and Probability,* Griffin, London, 1977, Vol. 2, pp. 30–34.

[2] Arbuthnot, J. (1692). *Huygens' de ratiociniis in ludo aleae.* London.

[3] Bellhouse, D. R. (1989). A manuscript on chance written by John Arbuthnot. *Int. Statist. Rev.,* **57,** 249–259.

[4] Hacking, I. (1965). *Logic of Statistical Inference.* Cambridge University Press, pp. 75–81.

[5] Hacking, I. (1975). *The Emergence of Probability.* Cambridge University Press, pp. 166–171.

[6] Huygens, C. (1657). De ratiociniis in ludo aleae. In *Exercitationum mathematicarum libri quinque,* F. van Schooten, ed. Amsterdam.

BIBLIOGRAPHY

Aitken, G. A. (1892). *The Life and Works of John Arbuthnot.* Clarendon Press, Oxford.

Arbuthnot, J. (1751). *Miscellaneous works of the late Dr Arbuthnot, with Supplement.* Glasgow.

Beattie, L. M. (1935). *John Arbuthnot: Mathematician and Satirist.* Harvard Studies in English, Vol. XVI. Harvard University Press, Cambridge, Mass.

Hald, A. (1990). *A History of Probability and Statistics and their Applications before 1750.* Wiley, New York.

Shoesmith, E. (1987). The continental controversy over Arbuthnot's argument for divine providence. *Historia Math.* **14,** 133–146.

Stephen, L. and Lee, S., eds. (1917). John Arbuthnot. In *Dictionary of National Biography,* Oxford University Press, Vol. I.

E. SHOESMITH

Bayes, Thomas

Born: 1701, in London, England.
Died: April 7, 1761, in Tunbridge Wells, England.
Contributed to: statistical inference, probability.

Rather little is known about Thomas Bayes, for whom the Bayesian school of inference and decision is named. He was a Nonconformist minister, a Fellow of the Royal Society, and according to the certificate proposing him for election to that society, was "well skilled in Geometry and all parts of Mathematical and Philosophical Learning" (see ref. [6], p. 357).

Nonconformist faiths played a major role in the scientific ideas of the eighteenth century, a time when religion and philosophy of science were inextricably linked. The growth of Nonconformism was influenced by the rise of natural philosophy, which encouraged a scientific examination of the works of the Deity so that men could come to understand His character. The Nonconformists included Deists and Arians (forerunners of modern Unitarians) among them, all sharing a rejection of the Trinity and a skepticism about the divinity of Christ. The Royal Society is known to have had a strong antitrinitarian component [6, p. 356].

Thomas was born to Joshua and Ann Bayes in 1702 in London. Joshua was the first Nonconformist minister to be publicly ordained (1694) after the Act of Uniformity was passed. He became minister of the Presbyterian meeting house at Leather Lane in 1723; Thomas served as his assistant.

Thomas was educated privately, as was the custom with Nonconformists at that time. After serving as an assistant to his father, he went to Tunbridge Wells as minister of a dissenting Presbyterian congregation. It is not known precisely when he began that post, but he was there by 1731 when his religious tract, *Divine Benevolence, or an attempt to prove that the Principal End of the Divine Providence and Government is the Happiness of his Creatures,* was printed by John Noon at the White Hart in Cheapside, London [1].

Bayes was elected a Fellow of the Royal Society in 1742. The signers of his certificate for election included officers of the Royal Society; John Eames, mathematician and personal friend of Newton; and Dr. Isaac Watts, Nonconformist minister and well-known composer of hymns. Unfortunately, there is little indication of published work by Bayes meriting his election to the Royal Society. In fact, there is no concrete evidence of any published paper, mathematical or theological, by Bayes from his election in 1742 until his death in 1761 [6, p. 358]. There is however, an anonymous 1736 tract, published by John Noon, which has been ascribed to Bayes, entitled *An Introduction to the Doctrine of Fluxions, and a Defence of the Mathematicians against the objections of the Author of the Analyst, in so far as they are designed to affect the general method of reasoning.* This work was a defense of the logical foundations of Newtonian calculus against attacks made by Bishop Berkeley, the noted philosopher. Augustus DeMorgan, in an 1860 published query about Bayes, noted the practice among Bayes' contemporaries of writing in the author's name on an anonymous tract, and DeMorgan (and the British Museum) accepted the ascription.

DeMorgan gives some sense of Bayes' early importance by writing, "In the last century there were three Unitarian divines, each of whom has established himself firmly among the foremost promoters of a branch of science. Of Dr. Price and Dr. Priestley . . . there is no occasion to speak: their results are well known, and their biographies are sufficiently accessible. The third is Thomas Bayes . . ." [5]. De Morgan recognized Bayes as one of the major leaders in the development of the mathematical theory of probability, claiming that he was "of the calibre of DeMoivre and Laplace in his power over the subject," and noting in addition that Laplace was highly indebted to Bayes, although Laplace made only slight mention of him [5].

Bayes is also known to have contributed a paper in 1761, published in the *Philosophical Transactions* in 1763 (see ref. [2]), on semiconvergent asymptotic series. Although a minor piece, it may have been the first published work to deal with semiconvergence [6, p. 358].

Bayes was succeeded in the ministry at Tunbridge Wells in 1752, but continued to live there until his death on April 17, 1761. He was buried in the Bayes family vault in Bunhill Fields, a Nonconformist burial ground. The bulk of his considerable estate was divided among his family, but small bequests were made to friends, including a £200 bequest to Richard Price.

Bayes greatest legacy, however, was intellectual. His famous work, "An Essay towards Solving a Problem in the Doctrine of Chance," was communicated by Richard Price to the Royal Society in a letter dated November 10, 1763, more than two years after Bayes' death, and it was read at the December 23 meeting of the society. Although Bayes' original introduction was apparently lost, Price claims in his own introduction (see ref. [3]) that Bayes' purpose was "to find out a method by which we might judge concerning the probability that an event has to happen in given circumstances, upon supposition that we know nothing concerning it but that under the same circumstances it has happened a certain number of times and failed a certain other number of times." Price wrote that the solution of this problem was necessary to lay a foundation for reasoning about the past, forecasting the future, and understanding the importance of inductive reasoning. Price also believed that the existence of the Deity could be deduced from the statistical regularity governing the recurrency of events.

Bayes' work begins with a brief discussion of the elementary laws of chance. It is the second portion of the paper in which Bayes actually addresses the problem described by Price, which may be restated as: Find $\Pr[a < \theta < b \mid X = x]$, where $X \sim$ Binomial (n, θ). His solution relies on clever geometric arguments (using quadrature of curves instead of the more modern integral calculus), based on a physical analog of the binomial model—relative placements of balls thrown upon a flat and levelled table. Bayes' findings can be summarized as follows [7]:

$$\Pr[a < \theta < b, X = x] = \int_a^b \binom{n}{x} \theta^x (1 - \theta)^{n-x} \, d\theta,$$

$$\Pr[X = x] = \int_0^1 \binom{n}{x} \theta^x (1 - \theta)^{n-x} \, d\theta,$$

and finally,

$$\Pr[a < \theta < b, 1X = x] = \frac{\int_a^b \binom{n}{x} \theta^x (1-\theta)^{n-x} \, d\theta}{\int_0^1 \binom{n}{x} \theta^x (1-\theta)^{n-x} \, d\theta}.$$

The uniform prior distribution for the unknown θ followed naturally by construction in Bayes' analogy. According to Price, Bayes originally obtained his results by assuming that θ was uniformly distributed, but later decided that the assumption may not always be tenable and resorted to the example of the table. Bayes added a *Scholium* to his paper in which he argued for an *a priori* uniform distribution for unknown probabilities.

The use of a uniform prior distribution to represent relative lack of information (sometimes called the *Principle of Insufficient Reason*) has long been controversial, partly because such a distribution is not invariant under monotone transformation of the unknown parameter. Many statisticians and philosophers have interpreted the *Scholium* in terms of the principal of insufficient reason, but more recently, Stigler [7] has argued that Bayes was describing a uniform predictive distribution for the unknown X, and *not* a prior distribution for the parameter θ. Such an interpretation seems plausible, particularly due to its emphasis on probabilities for ultimately observable quantities and due to Bayes' view of probability as an expectation (see also ref. [4] for a view of probability as expectation). The assumption of a uniform predictive distribution is more restrictive than that of a uniform prior distribution; it does not follow, in general, that the implied distribution of θ would be uniform. One advantage of the uniform predictive distribution is that the problem of invariance no longer arises: If $\Pr[X = x]$ is constant, then so is $\Pr[f(X) = f(x)]$, for monotone functions f.

Thomas Bayes did not extend his results beyond the binomial model, but his views of probability and inductive inference have been widely adopted and applied to a variety of problems in statistical inference and decision theory.

REFERENCES

[1] Barnard, G. A. (1958). *Biometrika,* **45,** 293–295. (A biographical note that introduces a "reprinting" [2] of Bayes' famous paper.)

[2] Bayes, T. (1763). *Phil. Trans. R. Soc.,* London, **53,** 269–271.

[3] Bayes, T. (1763). *Phil. Trans. R. Soc.,* London, **53,** 370–418. (Also reprinted in *Biometrika,* **45,** 296–315. The reprinted version has been edited and modern notation has been introduced.)

[4] De Finetti, B. (1974–1975). *Theory of Probability,* Vols. 1 and 2. Wiley, New York.

[5] DeMorgan, A. (1860). *Notes and Queries,* January 7, 9–10.

[6] Pearson, K. (1978). *The History of Statistics in the 17th and 18th Centuries,* E. S. Pearson, ed. Macmillan, New York, pp. 355–370.

[7] Stigler, S. M. (1982). *J. R. Statist. Soc. A,* **145,** 250–258. (Argues that uniform distribution is intended for the predictive or marginal distribution of the observable events; discusses reasons behind common "misinterpretations" of Bayes' *Scholium.*)

BIBLIOGRAPHY

Holland, J. D. (1962). *J. R. Statist. Soc. A,* **125,** 451–461. (Contains many references to primary sources concerning Bayes' life.)

Laplace, P. S. (1951). *A Philosophical Essay on Probabilities.* Dover, New York. (Laplace's indebtedness to Bayes is apparent in his general principles.)

Stigler, S. M. (1983). *Amer. Statist.,* **37,** 290–296. (Evidence is presented that Nicholas Saunderson, an eighteenth-century mathematician, may have first discovered the result attributed to Bayes.)

Todhunter, I. (1865). *A History of the Mathematical Theory of Probability.* Chelsea, New York (reprint 1965).

R. L. TRADER

Bernoullis, The

Bernoulli is one of the most illustrious names in the history of mathematics. At least eight distinguished mathematicians bore the name (see Fig. 1), and three of these, the brothers James and John and John's son Daniel, were in the first rank of the mathematicians of their time. All the mathematicians Bernoulli were descendants of James' and John's father Nicholas Bernoulli, a prominent merchant in Basel, Switzerland. And all considered Basel their home.

JAMES AND JOHN

James Bernoulli, the eldest of four brothers, was meant by his father to be a theologian. John, the third of the brothers, was meant to be a merchant. Both studied mathematics secretly and against their father's will.

James completed a degree in theology at the University of Basel in 1676. He then left Basel for several years of travel, during which he tutored and taught in several parts of France and Switzerland and further pursued his studies, especially Cartesian philosophy. After his return to Basel in 1680, he turned more to mathematics. In 1681, at a time when his fellow theologians were proclaiming a recent spectacular comet a sign of God's anger, he published a mathematical theory of the motion of comets. And he devoted a second educational journey, to Holland and England in 1681–1682, to visits with mathematicians. He returned to the University of Basel to lecture in experimental physics in 1683, and he obtained the university's chair of mathematics in 1687.

John, too, disappointed his father. He proved unsuited for a business career and obtained permission to enroll at the university when James began teaching there. His father now wanted him to study medicine, and he did so after completing his master of arts degree in 1685. But James secretly tutored him in mathematics, and about the time that James obtained his chair in mathematics, the two began to puzzle out the cryptic version of the infinitesimal calculus that G. W. Leibniz (1646–1716) published in 1684. They were the first to master Leibniz' method, and

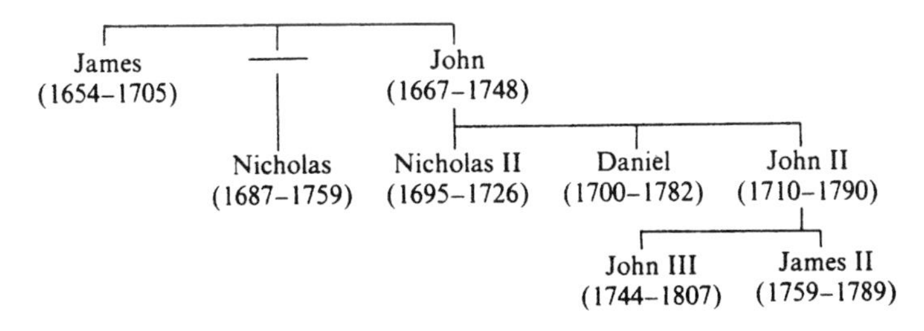

Figure 1. The mathematicians Bernoulli. The Roman numerals have been added to their names by historians. *James, John,* and *Nicholas* are English names. They become *Jakob, Johann,* and *Nikolaus* in German, and *Jacques, Jean,* and *Nicolas* in French.

it was through the brilliant papers that they and Leibniz published from 1690 on that the calculus became known to the learned world.

As their mathematical powers developed, the two pugnacious brothers became bitter rivals. By the late 1690s, after John had obtained his own position in Groningen, the Netherlands, they were publicly trading insults and challenging each other with mathematical problems. It is unfortunate, perhaps, that these problems were so interesting, for memory of them has kept the brothers' quarrels alive even after their own lifetimes.

It is difficult to compare the accomplishments of the two brothers. The slower and more careful James bested his brother in their public exchanges, but John ultimately made the greater contribution to the calculus. John inherited his brother's chair at Basel after his death in 1705, and remained active for 30 years thereafter. After Newton's death in 1727, he was unchallenged as the leading mathematician of all Europe.

In the field of probability, it is certainly James who was the giant. His philosophical training, together with his deep mathematical intuition, uniquely fitted him to attack the conceptual problems involved in applying the theory of games of chance, as it had been developed by his predecessors Pascal* (1623–1662), Fermat (1601–1665), and Huygens* (1629–1695), to problems of probability and evidence. And his struggle with these conceptual problems led him to the first limit theorem of probability—the law of large numbers.

James' great treatise on probability, *Ars Conjectandi* or the "Art of Conjecture," has four parts. The first three parts are in the tradition of his predecessors; he studies Huygens' pioneering book (*De ratiociniis in ludo aleae,* 1657), develops new combinatorial tools, and applies these to new problems. But the fourth part develops a new theory of probability—a theory that uses Huygens' ideas together with the law of large numbers to go beyond games of chance to problems of evidence and practical decision. (See below for a brief explanation of James' theory.)

James' mathematical diary shows that he had begun to work on probability in the 1680s and that he had proven the law of large numbers by 1689. Thus his completion of *Ars conjectandi* was impeded by ill health, and after his death its publication was delayed by his wife and son's fears that an editor might treat it unfairly. It was finally published in 1713.

NICHOLAS

Nicholas Bernoulli studied mathematics under both his uncles. In 1709 he obtained a law degree with a dissertation that applied some the ideas of *Ars conjectandi* to problems of law. Later, he was the editor who finally published *Ars conjectandi.* His first academic position was at Padua, Italy, but from 1722 on he held chairs at Basel, first in logic and then in law.

Nicholas was a gifted mathematician, but he published very little. Most of his achievements are hidden in the some 500 letters he wrote to other mathematicians.

His correspondents included Leibniz, Euler (1707–1783), de Moivre* (1667–1754), and Montmort (1678–1719).

He is especially remembered for his letters to Montmort,* where he generalized many of the results of Montmort's 1708 book Essay d'analyse sur les jeux de hazard. Montmort published most of his correspondence in the second edition of the book, in 1713; he and Nicholas had become so close that Nicholas stayed in Montmort's home for several months helping him prepare that edition.

One of the topics in the correspondence between Nicholas and Montmort was the excess of boys over girls in the christening records in London. This excess had been pointed out around 1711 by John Arbuthnot*, who observed that it was too persistent to attribute to chance. Arbuthnot saw this as evidence of divine intervention, but Nicholas, mindful of James' law of large numbers, explained the excess by saying that the true odds for a boy are closer to 18:17 than to 50:50.

DANIEL

Daniel Bernoulli was born during his father John's stay in Groningen, but he was educated at Basel. John made Daniel repeat his own youthful experience—first he tried to make him a merchant, and then he directed him toward medicine. Daniel's doctoral dissertation, completed in 1721, was on the mechanics of breathing.

In 1725, Daniel and his beloved older brother Nicholas II, who had tutored Daniel in mathematics, both obtained academic posts in St. Petersburg, Russia. Daniel was saddened by his brother's sudden death the following year, but he remained in St. Petersburg for eight years, during which he was very productive. In 1733, he returned to Basel, where he successively held chairs in anatomy and botany, in physiology, and in physics.

Daniel was a prolific and original mathematician, especially in physics. His greatest work was his treatise on hydrodynamics, written while he was in St. Petersburg. He also made major contributions to the mechanics of flexible and elastic bodies. It is an indication of his stature, perhaps, that his competitive father was often antagonistic toward him.

Daniel made several contributions to probability and statistics. The first and most influential was his *Specimen theoriae novae de mensura sortis* (1738) in which he developed the idea of utility. His inspiration for this paper was a problem posed to him by his cousin Nicholas: how to determine the fair price of a game with infinite mathematical expectation. We usually think of the expectation as the game's fair price, but as Nicholas pointed out, it is hardly fair to pay an infinite amount when one is certain to get only a finite amount back. Daniel's solution was to replace the monetary value of the payoff by its utility to the player, which may have a finite expectation. Since Daniel's paper was published in the St. Petersburg Academy's journal, Nicholas' problem has become known as the St. Petersburg Paradox.

Daniel is also remembered for his spirited statistical defense of inoculation for smallpox and for his statistical test of randomness for the planetary orbits. And in several papers written around 1770, he participated in the effort to develop a probabilistic theory of errors that could be applied to astronomical observations. One of these papers is notable, in retrospect, for its use of what we now call the method of maximum likelihood. Daniel did not develop his idea extensively, and his work soon superseded by Laplace's* Bayesian theory.

JAMES' THEORY OF PROBABILITY

In the long tradition of philosophical and theological thought to which James Bernoulli was heir, the idea of probability was not closely tied to the idea of chance. Pascal, Fermat, and Huygens did not even use the word *probability* in their writings on chance; probability, as these scholars knew, was an attribute of opinion, a product of argument or authority. The theory that James set forth in Part IV of Ars conjectandi was an attempt to bridge this theory of games of chance to probability while preserving the idea that probability is based on argument.

James' theory had two parts: his law of large numbers, which he thought could be used to assess the strenght of an argument, and his rules for combining arguments.

Here is one one of the examples of practical reasoning that made James see the need for his law of large numbers. Titius is marrying Caja, and an agreement is to be made concerning the division of her estate between him and their children in case she dies before him. But the size of the estate will depend on whether their own fathers are still living. Titius considers two possible agreements: one specifies that two-thirds of the estate will fall to him in any case; the other varies the proportion according to which fathers are still living. Which agreement is most advantageous to Titius? As James saw, the answer depends on how probable it is that Caja will outlive one or both of the fathers. But we cannot evaluate the probabilities *a priori,* for we cannot count the equally possible ways in which Caja might or might not outlive the fathers. James thought there must be some number r of ways in which she might outlive them and some number s in which she might not. But since we do not fully understand the possible causes of death for Caja or the fathers, we cannot find r and s so as to calculate the probability $r/(r + s)$.

But it should be possible, James reasoned, to evaluate this probability *a posteriori,* from outcomes observed in many similar cases. The numbers r and s should be the same for other young women in similar circumstances. So in large number n of such cases, there will be a total of $(r + s)^n$ equally possible ways things might turn out: r^n ways for all the young women to outlive their fathers and fathers-in-law, $nr^{n-1}s$ ways for all but one to, and so on. Studying this binomial expansion, James realized that most of the $(r + s)^n$ equally possible ways involve a number k of the n young women outliving their fathers and fathers-in-law such that k/n is close to $r/(r + s)$. So if we actually observe n such young women, the observed frequency k/n will probably be close to $r/(r + s)$. As James proved, we can always choose n large enough for it to be morally certain (he thought we should decide on some high value of probability to call "moral certainty") that k/n will be close enough to

$r/(r + s)$ that the two are identical for our practical purposes. This is the theorem we now call "the weak law of large numbers."

There remains the problem of combination. In the preceding example there may be no need to combine the probabilities based on mortality statistics with probabilities based on other arguments. But in general, James believed, probability will depend on the combination of many arguments. Among the examples he used to make his point was the following very traditional one, which goes back to Cicero (106–43 B.C.) and Quintilian (ca. A.D. 35–96). Titius is found dead on the road, and Maevius is accused of committing murder. There are several arguments against him. Maevius hated Titius, and they had been quarreling the day before; Maevius traveled over the road that day; a blood-stained sword is found in his house; and he turns pale and answers apprehensively when he is questioned. Perhaps none of these arguments alone would make Maevius' guilt probable. But altogether they make it almost certain.

As James saw it, his theorem should enable us to assess probabilities based on each argument. By observing, for example, other instances where a blood-stained weapon is found in a suspect's house, we should be able to assess the probability of Maevius' guilt based on that argument alone. But after thus assessing probabilities based on each argument alone, we should use his rules, which were derived from the theory of games of chance, to combine these probabilities. Only then do we have a probability based on the total evidence. The rules of combination James gave were imperfect, but they are interesting even today. Some of them explicitly recognize the possible nonadditivity of probability.

James' theory was not, of course, an adequate theory of statistical inference. He suggested we find out how many observations are required for moral certainty and then make that many. He did not recognize, perhaps because he had no opportunity to apply his theory to actual numerical data, that we cannot always make as many observations as we would like. Statistical methods for dealing with limited observations did not emerge until more than 50 years after James' death, in response to the need to combine discrepant astronomical observations.

Literature

The entries on the Bernoullis in the *Dictionary of Scientific Biography* include excellent descriptions of their contributions to mathematics and mechanics and extensive bibliographies. Detailed accounts of their contributions to probability can be found in Issac Todhunter's *History of the Theory of Probability* (1865; reprinted by Chelsea, New York, in 1965) and in L. E. Maistrov's *Probability Theory; A Historical Sketch* (1974, Academic Press, New York). James' work is also discussed by F. N. David in *Games, Gods and Gambling* (1962, Hafner, New York) and by Ian Hacking in *The Emergence of Probability* (1975, Cambridge University Press, Cambridge).

In 1935, the Bernoulli Commission was created in Basel for the purpose of publishing a new edition of the works of the Bernoullis. The plan is to publish everything left by the three great Bernoullis and a selection of the work of the five lesser ones. So far there have appeared four volumes of James Bernoulli's works, three each of John Bernoulli's works, Daniel's works, and John's letters, one volume of James' letters and one volume of polemical correspondence between James and John. (All published by Birkhauser in Basel, Switzerland.) Among these the third volume of James' works (edited by B. L. van der Waerden and published in 1975) and the second volume of Daniel's works (edited by L. P. Bouckaert and B. L. van der Waerden, published in 1982) are particularly concerned with probability calculus. The first of these includes most of James' work on probability, including the parts of his previously unpublished diary bearing on probability. (It is in this diary that he discusses the problem of Caja's marriage contract.) It also includes Nicholas' dissertation, John de Witt's treatise on annuities, and extensive commentary, in German, on James' work and on most of the Bernoullis' correspondence on probability. [For further information, see Clifford Truesdell's reviews in *Isis:* 1958 (49, 54–62) and 1973 (64, 112–114).]

James' *Ars Conjectandi* was reprinted in the original Latin in 1968 by the Belgian publishing house Culture et Civilisation. A German translation was published in 1899: *Wahrscheinlichkeitsrechnung* (2 vols., Engelman, Leipzig). Unfortunately, there is still no definitive and widely available English translation, but a rough translation of Part IV is available from the National Technical Information Service, 5285 Port Royal Road, Springfield, VA 22161. (*Translations from James Bernoulli,* by Bing Sung, Harvard University Technical Report, 1966; the order number is AD-631 452.) For a faithful account in English of James' proof of his theorem, see pp. 96–101 of J. V. Uspensky's Introduction to *Mathematical Probability* (1937, McGraw-Hill, New York). For a discussion of his ideas on the combination of arguments, see G. Shafer's "Non-additive probabilities in the Work of Bernoulli and Lambert," *Archive for History of Exact Sciences,* 1978 (19, 309–370), and for a discussion of the relation of his ideas to classical rhetoric, see D. Garber and S. Zabell's "On the Emergence of Probability" in the same journal, 1979 (21, 33–54).

An English translation of Daniel's *De mensura sortis* was published in *Econometrica* in 1954 (22, 23–36). In 1967, Gregg Press (Farnborough, Hampshire, England) reprinted this translation, together with an earlier German translation and commentary, under the original Latin title. L. J. Savage discussed the relation between Daniel's concept of utility and contemporary concepts in §5.6 of *The Foundations of Statistics* (1972, Dover, New York). Daniel's essays on inoculation appeared in French in the *Mémoires* of the Paris Academy (1760, pp. 1–45) and in *Mercure de France* (June 1760). An English translation of the paper in which he used maximum likelihood was published in *Biometrika* in 1958 (45, 293–315) and reprinted in Vol. 1 of *Studies in the History of Statistics and Probability* (E. S. Pearson and M. G. Kendall, eds.; Charles Griffin, London, 1970). Articles by O. B. Sheynin on his work are included in Vol. 2 of these *Studies* (Kendall and Plackett, eds.; 1977).

G. SHAFER

Bienaymé, Irenée-Jules

Born: August 28, 1796, in Paris, France.
Died: October 19, 1878, in Paris, France.
Contributed to: probability, mathematical statistics, demography, social statistics.

Bienaymé's life initially followed a somewhat erratic course with regard to scientific pursuits, partly because of the times through which he lived. The École Polytechnique, in which he enrolled in 1815, was dissolved in 1816, because of the Napoleonic sympathies of its students. Subsequently, he worked as a translator for journals, and in 1818 became lecturer in mathematics at the military academy at St. Cyr, leaving in 1820 to enter the Administration of Finances. Here he became Inspector, and, in 1834, Inspector General, but because of the revolution of 1848, retired to devote all his energies to scientific work.

Up to that time he had been active in the affairs of the Société Philomatique de Paris, and his contributions to its meetings were reported in the now obscure scientific newspaper–journal *L'Institut, Paris,* being reprinted at the end of the year of their appearance in the collections *Procés-Verbaux de la Société de Paris—Extraits.* The most startling of Bienaymé's contributions to probability occurs in this context, where he gives a completely correct statement of the criticality theorem for simple branching processes. His communication [2], which may have been stimulated by work of L. F. B. de Châteauneuf (1776–1856), precedes the partly correct statement of F. Galton* and H. W. Watson by some 30 years, and the first subsequent completely correct one by over 80 years [5].

Bienaymé began to publish only in 1829, and his early (Civil Service period) writings lean to demography and actuarial matters. For example, a major one [1] discusses the then used *life tables* of A. Déparcieux* and E. E. Duvillard, with the object of presenting overwhelming evidence against continued use of the Duvillard table, since its predicted mortality rates were heavier than appropriate. These interests persisted in Bienaymé's work, for despite his retirement, Bienaymé had considerable influence as a statistical expert in the government of Napoleon III, being praised in a report to the Senate in 1864 for his actuarial work in connection with the creation of a retirement fund. He was elected to the Academy of Sciences in 1852.

Even those of his papers with primarily sociological or demographic intent sometimes involve significant methodological contributions to probability and mathematical statistics. Bienaymé was active in three general areas: the stability and dispersion theory of statistical trials (before his retirement), theory associated with linear least squares (after his retirement), and limit theorems.

In dispersion theory (which later came to be associated with the names of W. Lexis,* L. Bortkiewicz,* and A. A. Chuprov*) Bienaymé introduced a physically motivated principal of *durée des causes,* under the operation of which the proportion of successes in a sequence of trials exhibits more variability than in the homogeneous case of all Bernoulli trials, and which might therefore be used to explain such observed variability. Bienaymé also manifested an understanding of the concept of a sufficient statistic.

Laplace's treatise [9] acted as a basis for some of Bienaymé's best work. In the area of least squares in particular, he is concerned with generalizing and defending Laplacian positions [3, 4]. Reference [4] contains, *inter alia,* the general Bienaymé–Chebyshev inequality $\Pr[|\bar{X} - EX| \geq \epsilon] \leq \operatorname{Var} X/(\epsilon^2 n)$ (proved by the simple reasoning still used today), which is used to deduce a weak law of large numbers. P. L. Chebyshev* obtained the inequality in 1867 (in a much more restricted setting and with a more difficult proof), in a paper published simultaneously in Russian and in French, juxtaposed to a reprinting of Bienaymé's paper. Later, Chebyshev gave Bienaymé credit for arriving at the inequality via the "method of moments," whose discovery he ascribes to Bienaymé. The slightly earlier paper [3], partly on the basis of which Bienaymé was elected to the Academy, contains, again as an incidental results, the deduction of an almost final form of the continuous chi-square density, with n degrees of freedom, for the sum of squares of n independently and identically distributed normal $N(0, 1)$ random variables.

A limit theorem reobtained in 1919 by R. von Mises [10] (who regarded it as a *Fundamentalsatz* of the same significance as the central limit theorem), was actually proved by Bienaymé in 1838. It asserts the following: If the random variables W_i, $i = 1, \ldots, m$ ($\sum W_i = 1$) have a joint Dirichlet distribution with parameters $x_1, x_2, \ldots, x_m$, then, as $n = \sum x_i \to \infty$ with $r = x_i/n =$ constant $(i) > 0$, the limiting standardized distribution of $V = \sum \gamma_i W_i$, is $N(0, 1)$. Late in his life, he constructed a simple combinatorial "runs up and down" test of randomness of a series of observations on a continuously varying quantity, based on the number of local maxima and minima in the series. In particular he stated that the number of intervals between extrema ("runs up and down") in a sequence of n observations is, under assumption of randomness of sample from a continuous distribution, approximately normally distributed (for large n) about a mean of $(2n - 1)/3$ with variance $(16n - 29)/90$.

Bienaymé was far ahead of his time in the depth of his statistical ideas and, as a result, has been largely ignored in the literature, although he was an important figure in nineteenth-century statistics. He was a correspondent of L. A. J. Quetelet* and a friend of A. A. Cournot and of Chebyshev. However, his papers were verbose, his mathematics was laconic, and he had a penchant for controversy. He invalidly criticized the law of large numbers of S. D. Poisson, and no sooner was he in the Academy than he engaged in a furious controversy with A. L. Cauchy.*

Bienaymé's life and scientific contributions, within a general framework of nineteenth-century statistics, have been extensively described and documented by Heyde and Seneta [6]. Reference [8] provides information on Bienaymé's contributions to linear least squares, and ref. [7] expands on ref. [5] in relation to the discovery of the criticality theorem.

REFERENCES

[1] Bienaymé, I. J. (1837). *Ann. Hyg. Paris,* **18,** 177–218.

[2] Bienaymé, I. J. (1845). *Soc. Philom. Paris Extraits, Ser. 5,* 37–39. (Also in *L'Institut, Paris,* **589,** 131–132; and reprinted in ref. [7].)

[3] Bienaymé, I. J. (1852). *Liouville's J. Math. Pures Appl.,* **17**(1), 33–78. [Also in *Mém. Pres. Acad. Sci. Inst. Fr.,* **15**(2), 615–663 (1858).]

[4] Bienaymé, I. J. (1853). *C. R. Acad. Sci. Paris,* **37,** 309–324. [Also in *Liouville's J. Math. Pures Appl.,* **12**(2), 158–176 (1867).]

[5] Heyde, C. C. and Seneta. E. (1972). *Biometrika,* **59,** 680–683.

[6] Heyde, C. C. and Seneta, E. (1977). *I. J. Bienaymé: Statistical Theory Anticipated.* Springer-Verlag, New York. (The basic modern source on Bienaymé.)

[7] Kendall, D. G. (1975). *Bull. Lond. Math. Soc.,* **7,** 225–253.

[8] Lancaster, H. O. (1966). *Aust. J. Statist.,* **8,** 117–126.

[9] Laplace, P. S. de (1812). *Théorie analytique des probabilitiés.* V. Courcier, Paris.

[10] Mises, R. von (1964). *Mathematical Theory of Probability and Statistics,* Hilda Geiringer, ed. Academic Press, New York, pp. 352–357.

E. SENETA

Boscovich, Ruggiero Giuseppe

Born: May 18, 1711, in Ragusa (now Dubrovnik), Dalmatia (now in Croatia).
Died: February 13, 1787, in Milan, Italy.
Contributed to: astronomy, geodesy, and physics.

Boscovich was a son of a Serb who settled in Ragusa (we use above the Italian version of his name). His early education (in his native city) was at a Jesuit school, and subsequently he attended the Collegium Romanum. Ordained a priest within the Society of Jesus in Rome in 1744, he propounded a very early version of the atomic theory of matter and was recognized by the foremost scientific bodies in Europe; he was made a Fellow of the Royal Society of London in 1760. After the suppression of the Jesuits in 1773, Boscovich, then in his sixties, went to Paris at the invitation of the King of France and spent 9 years as Director of Optics for the Marine before returning to Italy.

In 1750, Pope Benedict XIV commissioned Boscovich and his fellow Jesuit, C. LeMaire, to carry out several meridian measurements in Italy; part of the report of this journey, reprinted as ref. [1], contains his presently known contribution to statistical methodology. (Several other considerations on the effects of chance are reported in ref. [3].) Boscovich considered the simple version $Z_i = \alpha + \beta w_i + \epsilon_i$, $i = 1, \ldots, n$, of the linear model, where the data set is (Z_i, w_i), $i = 1, \ldots, n$, the ϵ_i are residuals due to errors in measurement, and α and β are to be determined from the conditions

$$\sum_{i=1}^{n} |\epsilon_i| = \min, \tag{1}$$

$$\sum_{i=1}^{n} \epsilon_i = 0. \tag{2}$$

The second of the conditions became familiar in the eighteenth and nineteenth centuries. Boscovich noted that it implies that $\overline{Z} = \alpha + \beta\overline{w}$, which yields by substitution in the model that $Y_i = Z_i - \overline{Z} = \beta(w_i - \overline{w}) + \epsilon_i$, $i = 1, \ldots, n$, which is automatically satisfied; thus it can be ignored. Putting $x_i = w_i - \overline{w}$, Boscovich therefore proposes to determine β in this model according to the criterion (1), i. e., by minimizing the sum of absolute deviations (the l_i-criterion). Boscovich's method for achieving this was geometric; Laplace* gave an analytical solution, acknowledging Boscovich, and used it in his own astronomical writings as early as 1789. In the second supplement (1818) to the *Théorie Analytique des Probabilités,* he calls it the "method of situation" to distinguish it from the l_2 (least-squares) procedures, which he calls the "most advantageous method." This robust method of estimation (in principle, tantamount to the use of the sample median as a measure of location of a sample rather than the least-squares measure, the sample mean), whose efficient general application requires linear programming methodology, therefore predates least-squares in the linear model framework. After Laplace, it was taken up [3] by Gauss,* who indicated the form of analytic solution in the general case of the linear model with rank $r \geq 1$.

Additional information on Boscovich's life and work can be found in refs. [2] and [4].

REFERENCES

[1] Bosković, R. J. (1757). In *Geodetski rad R. Boskovića,* N. Cubranić, ed., Zagreb, 1961. (Printed with Serbo-Croatian translation. See especially pp. 90–91.)

[2] Eisenhart, C. (1961). In *R. J. Boscovich: Studies of His Life and Work,* L. L. Whyte, ed. London, pp. 200–213.

[3] Sheynin, O. B. (1973). *Arch. History Exact Sci.,* **9,** 306–324. (The most comprehensive overview of Boscovich's statistical concepts.)

[4] Stigler, S. M. (1973). *Biometrika,* **60,** 439–445. (Section 3 discusses Laplace's treatment of the "method of situation.")

E. Seneta

De Moivre, Abraham

Born: May 26, 1667, in Vitry (in Champagne), France.
Died: November 27, 1754, in London, England.
Contributed to: mathematical analysis, actuarial science, probability.

De Moivre's early education was in the humanities, at the Protestant University of Sedan where he was sent at age 11, and the University of Saumur, but he soon showed a flair for mathematics which was encouraged by his father, a poor surgeon, and at an early age he read, *inter alia,* the work of C. Huygens* on games of chance (*De Ratiociniis in Ludo Aleae*), which was the kernel of his own later work on chance. At the Sorbonne he studied mathematics and physics with the famous Jacques Ozanam. In 1688 he emigrated to London to avoid further religious persecution as a Protestant, after the repeal in 1685 of the Edict of Nantes. Here he was forced to earn his living first as a traveling teacher of mathematics, and then by coffeehouse advice to gamblers, underwriters, and annuity brokers, in spite of his eminence as a mathematician. He became acquainted with Halley in 1692, was elected Fellow of the Royal Society of London in 1697, and to the Berlin and Paris Academies. His contact with the work of Newton at an early stage in England had important consequences for his mathematical growth; the two men became very close friends. Todhunter [9] writes of De Moivre: "In the long list of men ennobled by genius, virtue and misfortune, who have found an asylum in England, it would be difficult to name one who has conferred more honour on his adopted country than De Moivre."

His principal contributions to probability theory are contained in the book [4] *The Doctrine of Chances* (dedicated to Newton), the two later editions of which contain English versions of the rare second supplement (as bound with his *Miscellanea Analytica de Seriebus et Quadraturis*), which is actually a seven-page privately printed pamphlet [3] dated November 12, 1733, discovered by K. Pearson* [6], in which the density function of the normal distribution first appears. In the supplement of the *Miscellanea Analytica,* De Moivre obtained the result $n! \sim Bn^{n+1/2}e^{-n}$, now called Stirling's formula (and of extensive use in the asymptotics of combinatorial probability)—the contribution of James Stirling was the determination of B as $\sqrt{2\pi}$. Using this formula, De Moivre [3] initially investigated the behavior of the modal term of the symmetric binomial distribution $\binom{n}{k}2^{-n}$, $k = 0, \ldots, n,$ and the term t terms distant, for large n. If we call these α_0 and α_t, respectively, he concluded that $\alpha_0 \simeq 2/\sqrt{2\pi n}$, $\alpha_0 \simeq \alpha_t \exp(-2t^2/n)$, and determined a series approximation to $\alpha_0 + \alpha_1 + \ldots + \alpha_t$, which is a series expansion of the integral

$$(2/\sqrt{2\pi n})\int_0^t \exp(-t^2/n)\, dt.$$

He also deduced similar results for the general case of a binomial distribution with probability of success p $(0 < p < 1)$, giving $\alpha_0 \simeq 1/\sqrt{2\pi np(1-p)}$, $\alpha_t \simeq \alpha_0 \exp\{-t^2/[2np(1-p)]\}$, although again in different notation. We may therefore attribute to him the local and global central limit theorem in the case of sums of (0, 1) random variables, now known in probability collectively as "De Moivre's theorem." He also had a very clear notion of the significance of independence.

De Moivre's work is well explored [1]. A detailed account of his other probabilistic writings is given in refs. [2] and [9]; it is, however, worthwhile to comment on his treatment of the simple random walk on $\{0, 1, 2, \ldots, a + b\}$ with starting point a, probability of a step to the right p, and 0 and $a + b$ absorbing barriers. It is desired to find the probabilities of ultimate absorption into each of the two absorbing barriers, the expected time to absorption, and the probability of absorption into a specific barrier in n steps. De Moivre treats it in the framework of "gambler's ruin." The problem and its solution in the simple case $a = b = 12$ originates with Pascal* and Fermat [5], whence it is to be found in Huygens. De Moivre's method of solution in the general case is ingenious and shorter than most modern demonstrations [8], although there is some doubt on *priority* because of solutions by De Montmort and by N. Bernoulli (possibly due to James Bernoulli) in the same year (1711).

REFERENCES

[1] Adams, W. J. (1974). *The Life and Times of the Central Limit Theorem.* Kaedmon, New York. (Chapter 2 contains a careful bibliographical analysis of the origins of De Moivre's theorem, and a picture of De Moivre. There is also a useful bibliography of secondary sources.)

[2] Czuber, E. (1899). *Jahresber. Dtsch. Math.-Ver.,* **7** (2nd part), 1–279.

[3] De Moivre, A. (1733). Approximatio ad Summam terminorum Binomii $\overline{a + b}\backslash^n$ in Seriem expansi. [A facsimile may be found in R. C. Archibald, *Isis,* **8,** 671–683 (1926).]

[4] De Moivre, A. (1738). *The Doctrine of Chances; or a Method of Calculating the Probability of Events in Play,* 2nd ed. H. Woodfall, London. (First ed., 1718; 3rd ed., 1756. The first edition is an enlarged English version of De Moivre's first published work on probability: De Mensura Sortis, seu de Probabilitate Eventuum in Ludis a Casu Fortuito Pendentibus. *Phil. Trans. R. Soc.* (London), **27** (1711) 213–226. Reprinted by Kraus, New York, 1963.)

[5] Ore, O. (1960). *Amer. Math. Monthly,* **67,** 409–419.

[6] Pearson, K. (1978). *The History of Statistics in the 17th and 18th Centuries.* Charles Griffin, London. (Lectures by Karl Pearson 1921–1933, edited by E. S. Pearson. Pages 141–166 contain details of De Moivre's life and work on annuities as well as on De Moivre's theorem.)

[7] Schneider, I. (1968). *Arch. History Exact Sci.,* **5,** 177–317.

[8] Thatcher, A. R. (1957). *Biometrika,* **44,** 515–518.

[9] Todhunter, I. (1865). *A History of the Mathematical Theory of Probability.* Macmillan, London. (Reprinted by Chelsea, New York, 1949 and 1965.)

E. Seneta

De Montmort, Pierre Rémond

Born: October 27, 1678, in Paris, France.
Died: October 7, 1719 (of smallpox) in Paris, France.
Contributed to: probability theory.

The second of three sons of François and Marguerite Rémond, who were of the nobility, Montmort traveled widely in Europe in his youth. He came under the influence of Father Nicholas de Malebranche, with whom he studied religion, philosophy, and mathematics. He succeeded his elder brother as canon of Nôtre-Dame but resigned in 1706 to marry and settle down at the country estate of Montmort, which he had bought with the fortune his father had left him. His marriage was a happy one, and during this simple and retired life he set to work on the theory of probability. In 1708, the first edition of ref. [3] appeared. "... where with the courage of Columbus he revealed a new world to mathematicians ..." according to Todhunter [5]. At the time, Montmort was aware of, and partly motivated by, the work by the Bernoullis* (reviewed in 1705 and 1706) on the book that was to be published posthumously in 1713 as Jacob (James) Bernoulli's *Ars Conjectandi,* under the editorship of Jacob's nephew Nicolaus (Nicholas). This publication in turn was motivated by the appearance of Montmort's work. Nicolaus and Montmort evolved an extensive and fruitful technical correspondence, some of which is included [together with a single letter from Jean (John) Bernoulli] as the fifth part of the substantially expanded second edition of Montmort's treatise [3], also published in 1713. It is clear from the correspondence that the mathematical influence of the Bernoullis (not to mention their contributions) on the second edition, was substantial. Montmort was piqued by De Moivre's* *De Mensura Sortis* (the Latin precursor of the Doctrine of Chances), which appeared in 1711 and which he regarded as plagiaristic [5]. It was, in fact, quite scathing in attacking his own first edition [1]; Montmort retaliated with an *Avertissement* in his second edition. Contrary to popular opinion, the breach was never properly healed [5, p. 102; 2].

The value of Montmort's work is partly in his scholarship. He was well versed in the work on chance of his predecessors (Pascal*, Fermat, Huygens*); he met Newton* on one of a number of visits to England, and corresponded with Leibnitz, but remained on good terms with both sides during the strife between their followers. His friendship with Brook Taylor led to the latter's facilitating the publication of ref. [4] on the summation of infinite series, an element of Montmort's mathematical interests which enters into his probability work and distinguishes it from the earlier purely combinatorial problems arising out of enumeration of equiprobable sample points. Although ref. [3] to a large extent deals with the analysis of popular gambling games, it focuses on the mathematical properties and is thus written for mathematicians rather than gamblers. The Royal Society elected Montmort a Fellow in 1715, and the *Académie Royale des Sciences* made him an associate member (as he was not a resident of Paris) the following year.

An extensive summary of the whole of ref. [3] is given in ref. [5]; Todhunter thus accords him a very substantial place in the history of the subject. Montmort's best-known contribution to elementary probability is a result connected with the problem of matching pairs (in connection with the card games *rencontre, treize,* and *snap*), in which n distinct objects are assigned a specific order, while n matching objects are assigned random order. The probability u_n of at least one match, where A_i is the event that there is a match in the ith position, is, in modern textbook analysis,

$$\Pr(\bigcup_{i=1}^{n} A_i) = \sum \Pr(A_i) - \sum \Pr(A_i \cup A_j) + \sum \Pr(A_i \cup A_j \cup A_k) \cdots$$
$$= \sum_{j=1}^{n} \frac{(-1)^{j-1}}{j!},$$

of which the limit as $n \to \infty$ is $1 - e^{-1}$. This is thought to be the first occurrence of an exponential limit in the history of probability. Montmort's general iterative procedure for calculating u_n from $nu_n = (n-1)u_{n-1} + u_{n-2}$ is, on the other hand, based on a conditional-probability argument (given in a commentary by Nicolaus Bernoulli) according to the outcome at the first position.

Montmort also worked with Jean Bernoulli on the problem of points considered by Pascal and Fermat (for players of unequal skill). He worked with Nicolaus on the problem of duration of play in the gambler's ruin problem, possibly prior to De Moivre; it was at the time the most difficult problem solved in the subject. Finally, in a letter of September 9, 1713, Nicolaus proposed the following problems to Montmort:

> *Quatrième Problème: A* promises to give an *écu* to *B,* if with an ordinary die he obtains a six with the first toss, two *écus* if he obtains a six with the second toss. . . . What is the expectation of *B*?
>
> *Cinquième Problème:* The same thing if *A* promises *B écus* in the progression 1, 2, 4, 8, 16, ...

It is clear that the St. Petersburg paradox, as subsequently treated by Daniel Bernoulli in 1738, is but an insignificant step away. In his reply Montmort indicates the solutions to Nicolaus' problems and describes them as being of no difficulty.

REFERENCES

[1] David, F. N. (1962). *Games, Gods, and Gambling: The Origins and History of Probability and Statistical Ideas from the Earliest Times to the Newtonian Era.* Griffin, London, Chap. 14. (Strong on biographical detail.)

[2] Hacking, I. (1974). "Montmort, Pierre Rémond de." *Dictionary of Scientific Biography.* Vol. 9, C. C. Gillispie, ed. pp. 499–500.

[3] Montmort, P. R. de (1713). *Essay d'Analyse sur les Jeux de Hazard,* 2nd ed. Quillau, Paris (Reprinted by Chelsea, New York, 1980. Some copies of the second edition bear the date 1714. The first edition appeared in 1708.)

[4] Montmort, P. R. de (1720). "De seriebus infinitis tractatus." *Trans. R. Soc.,* **30,** 633–675.

[5] Todhunter, I. (1865). *A History of the Mathematical Theory of Probability.* Macmillan, London, Chap. VIII. (Reprinted by Chelsea, New York, 1949 and 1965.)

BIBLIOGRAPHY

Fontenelle, (1721). Eloge de M. de Montmort. *Histoire de l'Académie Royale des Sciences pour l'Année 1719,* pp. 83–93. (The chief biographical source cited in refs. [1] and [5].)

Maistrov, L. E. (1974). *Probability Theory: A Historical Sketch.* Academic Press, New York. [Translated and edited from the Russian edition (1967) by S. Kotz. Chapter III, §2, alludes to Montmort's work.]

Taylor, B. (1793). *Contemplatio Philosophica,* W. Young, ed. London. (Contains part of Montmort's correspondence with Taylor.)

E. SENETA

Déparcieux, Antoine

Born: October 18, 1703, in Clotet-de-Cessous, near Nîmes, France.
Died: September 2, 1768, in Paris, France.
Contributed to: life-table construction, mathematical and statistical tables.

Antoine Déparcieux was the son of a poor agricultural worker. His precocious early development induced his brother Pierre [possibly with assistance of one or more other patrons] to make possible his study at the Jesuit College in Alés. Here he made rapid progress in the exact sciences, after which he went to Paris in 1730 to study higher mathematics, again with some financial assistance from a patron (Montcarville). This assistance, however, became inadequate, and Déparcieux had to seek work to support himself. He began by constructing sundials, but soon turned to invention of industrial machinery—for example, in tobacco manufacture, and pumps for controlling water level.

It was only somewhat later that he turned his attention to activities forming the basis for his inclusion in this book. Following publication, in 1741 and 1742 respectively, of tables of astronomical and trigonometric functions, there appeared ref. [1] in 1746. This was ostensibly concerned with the computation of prices of annuities, but also included many examples of life-table construction from survival data for a variety of populations. It was in the choice of these populations that Déparcieux contributed most notably to the principles of life-table construction. While the tables of Graunt [5] were based on records of deaths and births (or christenings) in cities (notably London), Déparcieux was well aware of difficulties introduced by migration, rendering assessment of numbers "exposed to risk" at specified ages speculative at best. He therefore tried to use closed groups, in which individual ages at entry and death were available. To this end he constructed separate life tables for members of various specified religious orders, and for members of each of several tontines, for which such records were routinely kept. He also used data from the carefully maintained records of the parish of St. Sulpice in Paris, and somewhat venturesomely, of births and deaths in the city of Breslau (already used by Edward Halley for similar purposes), chosen as being least likely, among a number of cities selected for possible inclusion, to be affected by migration. Of course, the results obtained for restricted populations, although accurately derived—even according to current standards—could not reasonably be ascribed to surrounding areas, a point which caused some criticism. In particular, life expectancies corresponding to his life tables appeared to be unduly low.

A letter sent by one Mr. Thomas to the Editor of the *Journal de Trévoux* [3] made detailed criticisms of ref. [1]. Déparcieux made a corresponding detailed reply, and there was a further pair of letters in the *Journal de Verdun* (also in ref. [3]). K. Pearson [7, pp. 198–199] gives an interesting account of this controversy. It is well worth reading, but note that Pearson assumed that Thomas was a Jesuit priest. This may have been so, but there is no evidence for it in the correspondence, although the editor of the *Journal de Trévoux* (Rev. Fr. Berthier) *was* a Jesuit priest.

Possibly as a consequence of ref. [1], in 1746, Déparcieux was elected a member of the Royal Academy of Science in Paris. Similarly to the situation with Graunt, whose fame is based on a single book [5]—see also the entry on Graunt and Petty, elsewhere in this book—Déparcieux's reputation, in the present context, derives, almost entirely, from the single book [1]. A further account [2] was published in 1760, dealing with other sets of data—for example, from Sweden—but it adds little to the ideas in ref. [1].

Nicolas [6] gave a favorable assessment of Déparcieux's character. He was modest and not overambitious, always keeping in mind his humble origins. In his will he remembered the schools of Porte and Saint-Florent, where he had learned to read and write.

There was another Antoine Déparcieux—known as "Déparcieux the Younger" (1753–1799), who was a nephew of the subject of this entry. Evidently he was equally precocious. His uncle brought him to Paris, and, at the very early age (even in the eighteenth century) of 20, he became a professor. He planned to be a chemist, but said that to be a good chemist it was necessary to be expert in mathematics, physics, and astronomy. His early death, at age 46, was ascribed to his "bad or sad habit of working immediately after his meals" (Pearson [7, p. 198]).

REFERENCES

[1] Déparcieux, A. (1746). *Essai sur les Probabilités de la Durée de la Vie Humaine.* Frères Guerin, Paris.

[2] Déparcieux, A. (1760). *Addition á 1'Essai sur . . . Humaine.* H. L. Guerin and L. F. Delatour, Paris.

[3] Déparcieux, A. and Thomas (1746). *Objection faites á M. Déparcieux . . . sur son Livre des Probabilités de le durée de la vie Humaine; avec les responses á ces objections.* Académie Royale des Sciences, Paris. (Includes letters to the Editors of the *Journal de Trévoux* and the *Journal de Verdun.*)

References [1–3] are included in a limited reprint (250, plus 30 review copies) published by EDHIS, Paris, in 1973.

[4] Déparcieux, A. ("The Younger") (1781). *Traité des Annuités, Accompagné de Plusieurs Tables Trés Utiles.* Chez L'Auteur, Paris.

[5] Graunt, A. (1662). *Natural and Political Observations. Mentioned in a Following Index, and Made Upon the Bills of Mortality.* Martin, Allestry and Dorcas, London.

[6] Nicolas, M. (1855). Déparcieux, Antoine. In *Nouvelle Biographie Générale,* Vol. 13, pp. 694–696.

[7] Pearson, K. (1978). *History of Statistics in the 17th and 18th Centuries.* E. S. Pearson, ed. Macmillan, New York and London.

De Witt, Johan

Born: September 25, 1625 (birthplace unknown, probably outside Dordrecht, Holland).
Died: August 20, 1672, in The Hague, Holland.
Contributed to: actuarial science, mathematics, economic statistics.

The place of Johan de Witt in the history of the statistical sciences is due to his pioneering calculations of the value of annuities, based on information about mortality rates and probabilistic considerations. In Roman times no theory of probabilities existed, but a table had been produced for evaluating annuities, the basis of which is unknown. The newly developed calculus of probabilities had been given publicity by Huygens' (1629–1695) tract of 1656 on Calculations in Games of Chance, included as *De Ratiociniis in Ludo Aleae* in Van Schooten's (1615–1660) *Exercitationes Mathematicae* (1657) and three years later appearing in the original Dutch version (*Van Rekenigh in Spelen van Geluck*). No doubt de Witt was acquainted with this publication when he started his work on annuities. He collected data from the annuity registers of Holland, but these were insufficient as a basis for a mortality table of potential annuitants, and so he decided to postulate the following simple model or mortality as a basis of his calculation: In each of four age groups considered (4–53, 54–63, 64–73, and 74–79), the rate of mortality was taken to be constant, being one and one-half times as big in the second group as in the first, twice as big in the third group as in the first, and three times as big in the third. The calculations were set out in a tract called *Waerdye van Lyf-Renten Naer Proportie van Los-Renten,* included in Resolutions of the States of Holland for 1671. The title may be translated as "The Worth of Life Annuities in Proportion to Redemption Bonds." The tract demonstrated that the issue of life annuities with a rate of some 7% would be more favorable to issuer and purchasers alike than the issue of perpetual or redemption bonds (which had no termination date, but could be called in by the issuer at any time) with a rate of 4%. The conclusion was so novel that it was not widely accepted by the public at the time in spite of the simple way the argument was presented (mostly with the aid of a numerical example) in the tract. A more fundamental argument is found in the author's correspondence with Hudde (1628–1704), in which he also calculated the value of annuities based on several lives.

De Witt also supplied Pieter de la Court (1618–1685) with economic statistics for the latter's publication *Interest van Holland* which appeared in various versions from 1662 on, in some of which de Witt is mentioned incorrectly as author.

De Witt was the most prominent Dutch statesman of the third quarter of the seventeenth century, a time when the Dutch Republic had just gained its independence from Spain and, in spite of its population of only a little above one million, was the second richest country (after France) and was a major world power. At the age of 16, he and his elder brother Cornelis entered Leyden University for the study of law, finishing in 1645. Then they went to "finish their education" in the manner then customary for sons of prominent families, namely by a grand tour of France lasting one and one-half year, seeing the sights and meeting prominent people, and also gaining a doctor of law degree during a three months' stay in Angers. The brothers then spent a few months in England and returned to Holland. Johan became an apprentice with a prominent lawyer in The Hague, and in December 1647 was nominated as a representative of his home city, Dordrecht, to the governing body ("States") of the Province of Holland, whose seat was in The Hague. Because of the seniority of Dordrecht among the cities of the Province (whose representatives together with one representative of the nobility made up the "States"), its deputies were always called upon first to give their opinions in the debates. Moreover, the resident representative of Dordrecht had to take the place of the Raadspensionaris (chief executive officer) during the latter's absence. Because of these facts, together with de Witt's extraordinary abilities and devotion to his task, as well as his great integrity (a quality not common among politicians at the time), he soon became a leading figure in the "States." He was appointed Raadspensionaris in 1652, and reappointed every five years, the last time in 1668. Since the Dutch Republic was a loose federation of seven Provinces, of which the Province of Holland was by far the richest and most influential, and because of de Witt's personal qualities, his position became, in fact, one of chief executive officer of the Republic. More and more, he gave direction to its foreign, fiscal, and defense policies and their execution, even though his formal powers were very restricted and he had to operate largely by persuasion and political means. During his time in office the rulers of England several times attacked the Dutch Republic, driven largely by envy of its leading position in commerce and navigation. Moreover, the country was threatened by Louis XIV of France, who was intent on expanding his domains with the help of a powerful army. The year 1672 is known as the "Year of Disaster" in Dutch history. The English fleet attacked and Louis XIV invaded the country (which in the end was saved only by flooding the polders in the Western part of the country and by depletion of the supplies of the French army). Internally, passions ran high and many accused the de Witt brothers of treason. There was an attempt on the life of Johan on June 21, which put him out of action for some time. He submitted his resignation on August 4, which was granted with reluctance. Meanwhile, Cornelis was arrested because of a (false) accusation of plotting to assassinate the Prince of Orange. Johan came to visit his brother at the prison; and later a mob forced its way into the prison and killed both. A committee of the States of Holland, upon examining the papers of the late Raadspensionaris, found no evidence of traitorous behavior; one of the members, who had not been a political friend, when asked what had been found, replied: "Nothing but honour and virtue."

Already during his studies at Leyden, de Witt developed a strong interest in mathematics. Before his appointment as Raadspensionaris, he had drafted a treatise on conic sections, which he sent about 1657 to his former fellow student Van Schooten, then Professor of Mathematics at Leyden University. The latter was then preparing a second edition of a translation into Latin of Descartes's (1596–1650) *Géometrie* with commentaries, which was brought out by Elsevier in two volumes (1659, 1661) under the title *Geometria a Renate Des Cartes.* He proposed that de Witt's treatise should be included in this publication, but reformulated in closer accordance with Descarte's new approach and notation. Since he knew that de Witt was a very busy man, he offered to take upon himself the reformulation, the checking of calculations, and the drafting of figures. De Witt gladly accepted, but went carefully over the reformulated version and made important further amendments. The result was the treatise named *Elementa Curvarum Linearum.* It was much praised by Huygens*; Newton* (1642–1727), when asked which books would be helpful to the study of his *Principia,* recommended it for the geometrical background it supplied. It was widely used during the seventeenth century.

On the other hand, there are no signs that de Witt's work on annuities influenced the development of actuarial science or practice to any large event. James (Jacob) Bernoulli (1654–1705) requested from Leibniz(1646–1716) a loan of his copy of de Witt's tract, but the latter could not locate it and asserted that it was not of much value. The *Waerdye* was rediscovered more or less accidentally by Bierens De Haan and reissued in 1879 by the Mathematical Society in Amsterdam. Different interpretations and evaluations of this work have appeared; the interested reader may consult the appended bibliography as well as the references quoted in these publications.

BIBLIOGRAPHY

Bernoulli, J. (1975). *Die Werke,* Vol. 3. Birkhäuser Verlag, Basel. (Contains a facsimile copy of the *Waerdye,* and commentaries.)

Brakels, J. van (1976). Some remarks on the prehistory of the concept of statistical probability. *Arch. History Exact Sci.,* **16,** 119–136. (The footnote on pp. 130–131 corrects a number of mistakes in Hacking and others.)

Fenaroli, G., Garibaldi, U., and Penco, M. A. (1981). Giochi, scommesse sulle vita, tabelli di mortalità: nascita del calcolo probabilastico, statistica e teoria delle populazioni. *Arch. History Exact Sci.,* **25,** 329–341.

van Geer, P. (1915). Johan de Witt als Wiskundige. *Nieuw Arch. Wiskunde, 2,* **11,** 98–126.

Hacking, I. (1975). *The Emergence of Probability.* Cambridge University Press, Cambridge.

Rowen, H. H. (1986). *John de Witt, Statesmen of the "True Freedom."* Cambridge University Press, Cambridge. (This is a short biography containing a brief bibliographical essay at the end.)

Rowen, H. H. (1975). *John de Witt, Grand Pensionary of Holland, 1625–1672.* Princeton University Press, Princeton, NJ. (This work is much more extensive and fully documented.)

Hendrik S. Konijn

Graunt, John

Born: April 24, 1620, in London, England.
Died: April 18, 1674, in London, England.

Petty, William

Born: May 26, 1623, in Romsey (Hampshire), England.
Died: December 16, 1687, in London, England.
Contributed to: collection and analysis of demographic data, life-table construction.

We have taken the unusual step of combining our accounts of the lives and labors of these two seventeenth century pioneers in life-table construction, because there are so many contacts between the two. In particular, we note the ongoing controversy on the authorship of *Natural and Political Observations Mentioned in a Following Index and Made Upon the Bills of Mortality* (Martin, Allestry and Dicas, London, 1662). This book (referred to as *Observations* below) is generally regarded as initiating the idea that vital statistics (in this case, records of christenings and burials in London, mostly in the years 1629–1636 and 1647–1653, selected as periods relatively unaffected by plague) could be used to construct life tables for the relevant population.

Much of the present article is based on material in refs. [3] and [4]. The latter was in celebration of the tercentenary of the publication of *Observations*—not of the birth or death of John Graunt.

Although work on the material in *Observations* represents only a small part of the preoccupations of Graunt and Petty, it is a very considerable part of the intersection of their lives, as well as being the reason for inclusion of an entry dedicated to them in the present volume. However, we are here concerned with *personalities,* as well as work, of our subjects, and these provide a sharp contrast in the present case, as do their careers. In regard to the latter, while Graunt started adult life as a solid merchant, he later fell on hard times. Petty made his way up from a disadvantaged youth to wealth and acceptance in high social circles. We now summarize some details.

Both Graunt and Petty were of Hampshire stock,

> John Graunt was born in London on April 24th, 1620, to Henry and Mary Graunt, and was christened a week later at St. Michael, Cornhill. His father was a draper who had moved to London from Hampshire, and carried on his business in Birchin Lane. John was the eldest of a large family and, after serving an apprenticeship, took up and eventually succeeded to his father's business, as a "Haberdasher of small-wares," which he carried on in the family house for the greater part of his life. He became a Freeman of the Drapers' Company ("by Patrimony") at the age of 21, was granted the Livery of the Company when he was 38, and rose to the distinguished position of Renter Warden three years before his death at the age of 53. In this and other ways Graunt became a respected London citizen. He passed through the various ward offices of the city and was eventually elected to the common council for two years. He was a captain in the (military) trained band for several years and a major for two or three more. Even be-

> fore he was 30 his influence was sufficient to procure the professorship of music at Gresham College for his friend Dr. William Petty [in 1650]. [4, p. 538]

We will refer to the last sentence again, below.

William Petty was born in Hampshire. Although he showed promise of his later originality and versatility, he

> did not find a patron in Romsey and was shipped for a cabin boy at the age of fourteen. His short sight earned him a taste of the rope's end, and after rather less than a year at sea he broke his leg and was set ashore in Caen to shift for himself. "Le petit matelot anglois qui parle latin et grec" attracted sympathy and obtained instruction in Caen. [3, p. 3]

Petty returned to England in 1642, having sufficient knowledge of "arithmetic, geometry, astronomy conducing to navigation, etc." to join the navy.

> His naval career was short, for in 1643 he was again on the continent. Here he wandered in the Netherlands and France and studied medicine or at least anatomy. He frequented the company of more eminent refugees, such as Pell and Hobbes, as well as that of the French mathematician Mersen. [3, p. 3]

On returning to England:

> At first Petty seems to have tried to make a living out of his father's business, but he soon went to London with a patented manifold letter writer and sundry other schemes of an educational character. These occupied him between 1643 and 1649 and made him acquainted with various men of science, among other Wallis and Wilkins, but were not remunerative, and in 1649 he migrated to Oxford.
>
> Petty was created Doctor of Medicine on 7 March 1649 by virtue of a dispensation from the delegates (no doudt the parliamentary equivalent of the Royal Mandate of later and earlier times). He was also made a Fellow of Brasenose and had already been appointed deputy to the Professor of Anatomy. He was admitted a candidate of the College of Physicians in June 1650 (he was not elected a Fellow until 1655 and was admitted on 25 June 1658). At Oxford he became something of a popular hero by resuscitating (on 14 December 1651) an inefficiently hanged criminal, who, condemned for the murder of an illegitimate child, is said to have survived to be the mother of lawfully begotten offspring.
>
> Academically Petty rose to be full Professor of Anatomy and Vice-Principal of Brasenose. It is at this point (as usual the precise dates are dubious) that he became a candidate for a Gresham professorship and made contact with John Graunt. [3, p. 3]
>
> . . .
>
> Why the anatomy professor who had resuscitated half-hanged Ann Green should be made a professor of music is not obvious, and if the Gresham appointments were jobs, why should the job be done for Petty? The modern imaginative historian might suggest various reasons. For instance, that Petty made a conquest of Graunt, perhaps had Hampshire friends who were friends of the Graunt family, perhaps talked about political arithmetic. [3, p. 4]

However, Petty obtained a leave of absence from Brasenose in 1652 to go to Ireland as an agent of the Commonwealth government, as John Evelyn relates in his diary for March 22, 1675:

> Sir William came from Oxon, to be a tutor to a neighbor of mine; thence when the rebells were dividing their conquests in Ireland, he was employ'd by them to measure and set out the land, which he did on an easy contract, so much per acre. This he effected so exactly, that it not only furnish'd him with a greate sum of money, but enabled him to purchase an estate worth £.4000 a yeare. He afterwards married the daughter of Sir Hardresse Waller; she was an extraordinary witt as well as beauty, and a prudent woman. [1, p. 382]

Note the "*Sir* William." Despite Petty's association with the Commonwealth government, he was able to attain royal favor after the Restoration of the monarchy. There is evidence that he was an entertaining companion, and this, combined with the wealth derived from his activities in Ireland, presumably procured his entrance into high levels of society. There are some details in ref. [3, p. 5], and also in Evelyn's diary:

> The Map of Ireland made by Sir William Petty is believ'd to be the most exact that ever yet was made of any country. He did promise to publish it; and I am told it has cost him neare £.1000 to have it engrav'd at Amsterdam. There is not a better Latine poet living when he gives himselfe that diversion; nor is his excellence less in Council and prudent matters of state; but he is so exceeding nice in sifting and examining all possible contingencies, that he adventures at nothing which is not demonstration. There were not in the whole world his equal for a superintendant of manufacture and improvement of trade, or to govern a Plantation. If I were a Prince, I should make him my second Counsellor at least. [1, p. 382]
>
> . . .
>
> Having never known such another genius, I cannot but mention these particulars amongst a multitude of others which I could produce. When I who knew him in mean circumstances have ben in his splendid palace, he would himselfe be in admiration how he ariv'd at it; nor was it his value or inclination for splendid furniture and the curiosities of the age, but his elegant lady could endure nothing meane, or that was not magnificent. He was very negligent himselfe, and rather so of his person, and of a philosophic temper. "What a to-do is here!" would he say, "I can lie in straw with as much satisfaction." [1, p. 383]

This was in 1675, just under twelve months after Graunt had died in poverty.

Up to the time (1662) of the publication of the *Observations,* Graunt's career as a "prosperous city tradesman" [3, p. 7] had moved smoothly. In the Preface to the *Observations* he explains how he came to be interested in making use of the data contained in the weekly Bills of Mortality:

> Having been born, and bred in the City of *London,* and having always observed, that most of them who constantly took in the weekly Bills of *Mortality,* made little other use of them, then to look at the foot, how the *Burials* increased, or decreased; and,

among the Casualties, what had happened rare, and extraordinary in the week currant: so as they might take the same as a Text to talk upon, in the next Company: and withall, in the Plague-time, how the Sickness increased, or decreased, that so the Rich might judge of the necessity of their removall, and Trades-men might conjecture what doings there were like to have in their respective dealings:
2. Now, I thought that the Wisdom of our City had certainly designed the laudable practice of takeing, and distributing these Accompts, for other, and greater uses then those above-mentioned, or at least, that some other uses might be made of them: And thereupon I casting mine Eye upon so many of the General Bills, as next came to hand, I found encouragement from them, to look out all the Bills I could, and (to be short) to furnish my self with as much matter of that kind, even as the Hall of the Parish-Clerks could afford me; the which, when I had reduced into Tables (the Copies whereof are here inserted) so as to have a view of the whole together, in order to the more ready comparing on one Year, Season, Parish, or other Division of the City, with another, in respect of all the Burials, and Christnings, and of all the Diseases, and Casualties happening in each of them respectively; I did then begin, not onely to examine the Conceits, Opinions, and Conjectures, which upon view of a few scattered Bills I had taken up; but did also admit new ones, as I found reason, and occasion from my Tables.
3. Moreover, finding some Truths, and not commonly-believed Opinions, to arise from my Meditations upon these neglected Papers, I proceeded further, to consider what benefit the knowledge of the same would bring to the World; that I might not engage my self in idle, and useless Speculations, but like those Noble Virtuosi of Gresham-Colledge (who reduce their subtile Disquisitions upon Nature into downright Mechanical uses) present the World with some real fruit from those ayrie Blossoms.
4. How far I have succeeded in the Premisses, I now offer to the World's censure. Who, I hope, will not expect from me, not professing Letters, things demonstrated with the same certainty, wherewith Learned men determine in their Scholes; but will take it well, that I should offer at a new thing, and could forbear presuming to meddle where any of the Learned Pens have ever touched before, and that I have taken the pains, and been at the charge, of setting out those Tables, whereby all men may both correct my Positions, and raise others of their own; For herein I have, like a silly Schole-boy, coming to say my Lesson to the World (that Peevish, and Tetchie Master) brought a bundle of Rods wherewith to be whipt, for every mistake I have committed. [4, pp. 541–543]

In fact the *Observations* were well received:

. . . on February 5th, 50 copies of the book were presented by Graunt to the Royal Society of Philosophers, and he was proposed as a candidate for membership. A Committee was appointed to examine the book and reported favourably; on February 26th, only a month after publication, Graunt was elected to the Royal Society. He continued as a Fellow for some years, and was a member of the Council of the Society from November 1664 until April 1666. [4, p. 539]

However, soon after this, Graunt suffered a series of misfortunes, causing a gradual descent into poverty. These started with the destruction of his "house and presumably much of its contents" [4, p. 538] in the Great Fire of London in 1666, though it was rebuilt with financial assistance from Petty. His businesses may have been adversely affected also, but the most devastating cause may have been his conversion to Roman Catholicism—at that time a very unpopular step. Ultimately, "the family house passed to Petty" [4, p. 539],—presumably as a way to raise money or pay off a loan.

In a summary, ref. [4, p. 537] provides this very laudatory assessment of *Observations*.

> John Graunt was a London draper who, three hundred years ago, published some "Natural and Political Observations on the Bills of Mortality." These observations represent the first, and an extremely competent, attempt to draw scientific conclusions from statistical data. The present study illustrates Graunt's careful scientific approach, his ability to extract the essence from what by modern standards are distinctly untrustworthy demographic data, and his intuitive appreciation of the amount of interpretation his findings would stand. Graunt's analysis was largely based upon ratios and proportions of vital events and consideration of the way in which these altered in different circumstances, and is remarkably free of major statistical errors. His statistical understanding was considerable; for example, we owe to him the first scientific estimates of population size, the concept of the life table, the idea of statistical association, the first studies of time series, and a pioneer attempt to draw a representative sample. Graunt's book is well worth reading today, not only for entertainment and instruction, but because it laid the foundations of the science of statistics."

There were several further editions of the book—a second edition in 1662, and a third and fourth in 1665. A fifth edition appeared in 1676, "two years after Graunt's death, probably having been seen through the press by Sir William Petty, it contains a very few 'further observations'" [4, p. 539].

It is very likely that Graunt and Petty discussed the material in the book during the period of its production, and there has grown up, in some, the opinion that it is really the work of Petty rather than Graunt. Reference [4, p. 553] remarks:

> This allegation was not made, as far as we know, until after Graunt's death, but it has persisted in some quarters until the present day.

However, a further quotation from Evelyn's diary for March 22, 1675 reads (referring to Sir William Petty):

> He is author of the ingenious deductions from the bills of mortality, which go under the name of Mr. Graunt. [1, p. 383]

This was, indeed, after the death of Graunt, but (as already noted) by less than twelve months, and Evelyn writes as if it were an already well-established fact. It seems that this opinion was widely held in the social circles wherein Evelyn (and Petty) moved. There may have been the feeling that such a highly esteemed work was less likely to have come from the impoverished Graunt than from the brilliant and popular Petty. However, opinion among scientists and statisticians (with the notable exception of E. Halley, of Halley's Comet fame) has favored Graunt.

Study of the interactions of these markedly different personalities and their contrasting careers—in which Petty started as the impecunious client of Graunt while later their roles were reversed—is a fascinating exercise in the broad range of character involved in developing scientific thinking. Graunt contributed the factor of careful analysis of data (highly necessary at the time, and still so at present). Although Petty is often represented as something of a charlatan, he contributed to the more widespread appreciation of possible applications of science in the community. He did not always seek to flatter the powerful in the medical hierarchy, as the following brief quotation indicates. He suggested that it be asked [3, p. 16]

> Whether of 1000 patients to the best physicians, aged of any decade, there do not die as many as out of the inhabitants of places where there dwell no physicians.
>
> Whether of 100 sick of acute diseases who use physicians, as many die and in misery, as where no art is used, or only chance.

Petty was buried in Romsey Abbey. A monument in his memory was erected there in the nineteenth century by a descendant, the third Marquis of Lansdowne [3, p. 8]. Graunt was buried in St. Dunstan-in-the-West, Fleet Street, London. The church was rebuilt and moved in 1830, and there is no record of any monument to Graunt [4, p. 539].

More detailed comments on the contents of the *Observations* are available in refs. [2] and [5] as well as refs. [3] and [4].

REFERENCES

[1] Evelyn, J. (1818). *The Diary of John Evelyn, Esq., F. R. S.,* W. Bray, ed. Frederick Warne and Co., London.

[2] Glass, D. V. (1950). Graunt's life table. *J. Inst. Actu. Lond.,* **76,** 60–64.

[3] Greenwood, M. (1948). *Medical Statistics from Graunt to Farr.* Cambridge University Press.

[4] Sutherland, I. (1963). John Graunt: a tercentenary tribute. *J. R. Statist. Soc. A.,* **126,** 537–556.

[5] Willcox, W. F. (1937). The founder of statistics. *Rev. Int. Inst. Statist.,* **5,** 321–328.

Helmert, Friedrich Robert

Born: July 31, 1843, in Freiberg (Saxony), Germany.
Died: June 15, 1917, in Potsdam (Prussia), Germany.
Contributed to: geodesy, sampling distribution, transformation.

Friedrich Robert Helmert was a mathematical physicist whose main research was in geodesy, although this led him to investigate several statistical problems. In his doctoral dissertation [2] he developed a theory of the "ellipse of error," and in 1872 [3] he used the method of least squares in an examination of measuring instruments. In 1872 he was appointed professor of geodesy at the technical school in Aachen, and following the favorable reception of his work *Die mathematischen und physikalischen Theorien der höheren Geodäsie* (Part I in 1880 and Part II in 1884). He became professor of advanced geodesy at the University of Berlin in 1887, and director of the Prussian Geodetic Institute [1].

Until the 1960s, the derivation of the chi-square (χ^2) distribution was frequently attributed in statistical literature to Helmert, although credit for this more properly belongs to Ernst Abbe* and, to a large extent to Irenée-Jules Bienaymé* (see Sheynin [7]). However, in 1876 Helmert [4] proved that $S = \sum_{i=1}^{n}(x_i - \bar{x})^2/\sigma^2$ has a χ^2_{n-1} distribution (with $n - 1$ degrees of freedom) if $x_1, \ldots, x_n$ is a random sample from a normal population having common variance σ^2 and an unknown common mean μ, and where $\bar{x} = \sum x_i/n$.

Helmert showed first that if $\lambda_i = x_i - \bar{x}$, $i = 1, \ldots, n$, then the joint density of $\lambda_1, \ldots, \lambda_{n-1}$ and $\bar{x}$ (with $\lambda_n = -\lambda_1 - \cdots - \lambda_{n-1}$) is proportional to

$$\exp\{-\tfrac{1}{2}\sigma^{-2}[\lambda_1^2 + \cdots + \lambda_n^2]\} \times \exp\{-\tfrac{1}{2}n\sigma^{-2}(\bar{x} - \mu)^2\}. \tag{1}$$

He thus established the independence of $\bar{x}$ and any function of $x_1 - \bar{x}, \ldots, x_n - \bar{x}$, including S.

In order to obtain the distribution of S, Helmert introduced the transformation (presented here exactly as he gave it)

$$\begin{aligned}
t_1 &= \sqrt{2}(\lambda_1 + \tfrac{1}{2}\lambda_2 + \tfrac{1}{2}\lambda_3 + \tfrac{1}{2}\lambda_4 \cdots + \tfrac{1}{2}\lambda_{n-1}),\\
t_2 &= \sqrt{\tfrac{3}{2}}(\lambda_2 + \tfrac{1}{3}\lambda_3 + \tfrac{1}{3}\lambda_4 \cdots + \tfrac{1}{3}\lambda_{n-1}),\\
t_3 &= \sqrt{\tfrac{4}{3}}(\lambda_3 + \tfrac{1}{4}\lambda_4 \cdots + \tfrac{1}{4}\lambda_{n-1}),\\
&\vdots\\
t_{n-1} &= \sqrt{n/(n-1)}\lambda_{n-1}.
\end{aligned} \tag{2}$$

This changes the joint density of $\lambda_1, \ldots, \lambda_{n-1}$, namely

$$\sqrt{n}\left(\frac{h}{\sqrt{\pi}}\right)^{n-1} \exp\{-h^2(\lambda_1^2 + \cdots + \lambda_{n-1}^2)\}, \tag{3}$$

into that of $t_1, \ldots, t_{n-1}$, i.e.,

$$\left(\frac{h}{\sqrt{\pi}}\right)^{n-1} \exp\{-h^2(t_1^2 + \cdots + t_{n-1}^2)\}, \tag{4}$$

where $h = 1/(\sqrt{2}\sigma)$ is the *modulus* of the distribution of each of $X_1, \ldots, X_n$, or for that matter of $t_1, \ldots, t_{n-1}$ (the term "modulus" and the notation being in general use during the nineteenth century). Since (4) is the joint distribution of $n - 1$ independent identical normal variates with zero means, and since

$$\sigma^{-2} \sum_{i=1}^{n-1} t_i^2 = \sigma^{-2} \sum_{j=1}^{n} \lambda_j^2 = S. \tag{5}$$

it follows that S has the χ_{n-1}^2 distribution.

Lancaster [6] points out that the term "Helmert transformation" refers either to the combined transformation

$$\mathbf{t} = \mathbf{A}\mathbf{x} \tag{6}$$

where $x' = (x_1 \ldots x_n)$ and $\mathbf{t}' = (t_1 \ldots t_{n-1})$, or to the inverse transformation. It could also describe the transformation

$$\begin{pmatrix} \sqrt{n}\,\bar{x} \\ \mathbf{t} \end{pmatrix} = \mathbf{B}\mathbf{x} \tag{7}$$

or its inverse, given by

$$\mathbf{x} = \mathbf{B}^{-1}\begin{pmatrix} \sqrt{n}\,\bar{x} \\ \mathbf{t} \end{pmatrix},$$

$$\mathbf{B}^{-1} = \begin{bmatrix} \frac{1}{\sqrt{n}} & \frac{1}{\sqrt{1 \times 2}} & \frac{1}{\sqrt{2 \times 3}} & \frac{1}{\sqrt{3 \times 4}} & \cdots & \frac{1}{\sqrt{(n-1)n}} \\ \frac{1}{\sqrt{n}} & -\frac{1}{\sqrt{1 \times 2}} & \frac{1}{\sqrt{2 \times 3}} & \frac{1}{\sqrt{3 \times 4}} & \cdots & \frac{1}{\sqrt{(n-1)n}} \\ \frac{1}{\sqrt{n}} & & -\frac{2}{\sqrt{2 \times 3}} & \frac{1}{\sqrt{3 \times 4}} & \cdots & \frac{1}{\sqrt{(n-1)n}} \\ \frac{1}{\sqrt{n}} & & & -\frac{3}{\sqrt{3 \times 4}} & \cdots & \frac{1}{\sqrt{(n-1)n}} \\ \vdots & & & & & \vdots \\ \frac{1}{\sqrt{n}} & & & & \cdots & \frac{1}{\sqrt{(n-1)n}} \end{bmatrix} = \mathbf{B}', \tag{8}$$

since **B** is orthogonal. When **x** is distributed as above, any orthogonal transformation $\mathbf{y} = \mathbf{C}\mathbf{x}$ preserves independence, and $\sum_{i=1}^{n} y_i^2 = \sum_{i=1}^{n} x_i^2$. The Helmert matrix **B**, however, is probably the simplest orthogonal matrix having $1/\sqrt{n}$ for each element in the first row; see Lancaster [5], who also defines and discusses a generalized Helmert transformation and applications.

REFERENCES

[1] Fischer, F. (1972). In *Dictionary of Scientific Biography,* Vol. 6. Scribner's, New York, pp. 239–241.

[2] Helmert, F. R. (1868). *Studien über rationelle Vermessungen der höheren Geodäsie*. University of Leipzig, Leipzig.

[3] Helmert, F. R. (1872). *Die Ausgleichsrechnung nach der Methode der kleinsten Quadrate mit Anwendungen auf die Geodäsie und die Theorie der Mesinstrumente*. Leipzig.

[4] Helmert, F. R. (1876). *Astron. Nachr.,* **88,** cols. 113–120.

[5] Lancaster, H. O. (1965). *Amer. Math. Monthly,* **72,** 4–12. (Discusses a general class of Helmert-type matrices.)

[6] Lancaster, H. O. (1966). *Aust. J. Statist.,* **8,** 117–126. (A historical account of precursors of the χ^2 distribution.)

[7] Sheynin, O. B. (1966). *Nature,* **211,** 1003–1004.

CAMPBELL B. READ

Huygens, Christiaan

Born: April 14, 1629, in The Hague, Netherlands.
Died: June 8, 1695, in The Hague, Netherlands.
Contributed to: probability theory, mathematics, physics, astronomy.

Christiaan Huygens

Huygens, who was descended from a politically and artistically prominent family, enjoyed a splendid education. He showed remarkable technical and scientific gifts at an early age. From 1645 to 1649 he studied mathematics and law in Leiden and Breda. In 1655 he stayed for a few months in Paris and took a doctor's degree at Angers. With this opportunity he learned about Pascal* and Fermat's achievements in probability. Back in Holland, he wrote a small treatise on probability—the first in history—*Van Rekeningh in Spelen van Geluck* (calculation in games of chance). He sent the treatise to Van Schooten, who was glad to incorporate it into a work he was just preparing to be published in Latin and Dutch (1657 and 1660, respectively), in the Latin version under the title *De ratociniis in ludo aleae.* An anonymous work of 1692 *Of the laws of chance . . .* , probably by John Arbuthnot*, contains a translation of Huygens' treatise. Pierre Rémond de Montmort* (1708) and Abraham De Moivre* (1711) were certainly acquainted with Huygens' treatise. James Bernoulli* inserted it, with numerous comments, as a first part in an uncompleted manuscript, *Ars Conjectandi,* which was published posthumously in 1713. [See Bernoulli (1975).]

Although for half a century the only work on probability, Huygens' treatise, compared with his achievements in mathematics, physics, and astronomy, is only a minor work. Nevertheless, it shows some remarkable features.

The first chapters constitute an axiomatic introduction (as it were) to probability. Huygens founded probability on what is now called *expectation.* The term "expectatio" was introduced by Van Schooten in his Latin version, albeit in the sense of payoff table of a game; our "expectation" is indicated in Van Schooten's text by terms that translated would read "value of expectation."

Indeed, it is Huygens' fundamental question to ask for the value of the prospect of receiving payments $a_1, \ldots, a_n$, which are equally likely. (This formulation is of course ours.) The answer is obtained by a remarkable transformation: the payoff table is replaced with an equitable n-person game with possible outputs $a_1, \ldots, a_n$; by definition, the value of the payoff table equals the stake required to participate in that n-person game.

To explain Huygens' procedure, consider the case $n = 2$. The payoff table consists of the payments a and b, which are equally likely to be won. Instead, let two persons play an equitable game with the stake u while agreeing that the winner will pay the amount b to the loser (which is again an equitable stipulation). This means that the winner will actually earn $2u - b$, while the loser earns b. In order for the winner to get the required a, we shall put $u = (a + b)/2$, which is the stake of the game, and hence by definition the value of the payoff table.

In a similar way Huygens deals with the case $n = 3$, and then the case of a payoff table that grants p times the payment a and q times the payment b, with of course $(pa + qb)/(p + q)$ as its value.

After this introduction Huygens continues with a number of cases of "le probléme des partis" (sharing the stakes if a sequence of games is interrupted prematurely) and an inductive rule how to solve the general case. Next in order, a large number of dice problems are proposed and solved. The treatise finishes with five problems for the reader, the last of which concerns a game of virtually infinite duration.

Huygen's probabilistic work does not contain mathematical-statistical elements. The first such arguments are found in a note (1712) of John Arbuthnot* (refuting the equiprobability of male and female births) and in James Bernoulli's *Ars Conjectandi* (estimating the ratio of black and white balls in a bag).

BIBLIOGRAPHY

Bernoulli, J. (1975). *Die Werke.* Vol. III. Basel: Birkhauser.

Freudenthal, H. (1980). Huygens' foundation of probability. *Historia Math.*, **7,** 113–117.

Huygens, C. (1914). *Oeuvres complétes,* Vol. XIV. The Hague.

HANS FREUDENTHAL

Lambert, Johann Heinrich

Born: 1728, in Mulhouse, Alsace.
Died: 1777, in Berlin, Germany.
Contributed to: philosophy, photometry, theoretical mathematics, theories of errors and of probability.

Johann Heinrich Lambert, a self-taught and remarkably broad scientist, mathematician, and philosopher, was born in Mulhouse, in the Alsace, in 1728. He grew up in impoverished circumstances. At the age of 12 he was forced to leave school in order to help his father, a tailor. But he persisted in his studies, and at the age of 20 he became a tutor to a wealthy Swiss family. During the 10 years he spent with the family he was able to study, travel, and even publish some of his work. In 1765, he finally obtained a permanent position, at the Royal Academy of Berlin. He died in Berlin in 1777, at the age of 49.

Lambert's maternal language was a dialect of German, but it is impossible to assign him a nationality in the modern sense of the term. Alsace is now part of France, but in Lambert's time Mulhouse was a free city associated with Switzerland. In his writing, Lambert used three languages: Latin, French, and German. He was interested in the development of German as a literary and scientific language, and much of his work, including his philosophical treatises, was written in German. But the approximately 50 papers he wrote for the academy at Berlin were in French. In French, his name is Jean-Henri Lambert.

Lambert left his mark on a wide variety of fields, ranging from philosophy and theoretical mathematics to very practical parts of science. He wrote extensively on logic and the philosophy of knowledge. He was the first to demonstrate the irrationality of π and e. He is remembered for his law of cometary orbits, his cosine law in photometry, him map projections, and his hygrometer. He was not the most outstanding scholar of his time in any single field, but he was almost unique among eighteenth-century scholars in the success with which he combined philosophical and broadly practical interests.

Lambert's most substantial contribution to statistics was the work on the theory of errors that is contained in his books *Photometria* (1760) and *Beyträge zum Gebrauche der Mathematik und deren Anwendung* (Vol. 1, 1765). This work was inspired by his own empirical investigations in photometry and geodesy. By the time Lambert began the work, Thomas Simpson had already given a probabilistic argument for the use of averages, and Boscovich* and Mayer had already developed algorithms for solving overdetermined sets of equations. Lambert pulled these threads together and created the idea of a general theory of errors. He discussed the problem of determining probability distribution for errors, and stressed the relevance of probability to the general problem of determining unknown constants. His specific proposals for error distributions and methods of estimation were not widely adopted. But his demonstration of the broad relevance of the theory of errors and his formulation of its problems helped set the stage for the more successful work of Laplace* and Gauss*.

In retrospect, the most striking part of Lambert's work on the theory of errors was his formulation, in *Photometria,* of what we now call the method of maximum likelihood for a location parameter. Lambert discussed this method only briefly, and he did not use or mention it in his later work. His friend Daniel Bernoulli* published his own account of the method in 1777. The method did not survive as an independent approach to statistical estimation; instead, it was incorporated into Laplace's Bayesian synthesis. (When a uniform prior distribution is adopted for a parameter, the maximum-likelihood estimate coincides with the mode of the posterior distribution, which Laplace called the "most probable value" of the parameter.) It was not until the twentieth-century work of R. A. Fisher* that the method of maximum likelihood was reextracted from the Bayesian framework.

Another remarkable anticipation of twentieth-century thought occurs in Lambert's treatment of nonadditive probabilities. In his philosophical treatise *Neues Organon* published in 1764, Lambert discussed several examples where evidence seems to justify only nonadditive probabilities. One example involves a syllogism in which the major premise is uncertain. If we have only a probability of $\frac{3}{4}$ for the statement that C is an A, then the further information that all A are B justifies a $\frac{3}{4}$ probability that C is B but does not justify any probability at all that C is not B. The numbers $\frac{3}{4}$ and 0 add to less than 1, so this is an example of nonadditive probability. Lambert also corrected and generalized James Bernoulli's rules for combining nonadditive probabilities, rules that could be used, for example, to combine the probabilities provided by the testimony of independent witnesses. As it turns out, Lambert's nonadditive probabilities have the structure of what we now call belief functions, and his rules for combining nonadditive probabilities are special cases of Dempster's rule of combination for belief functions.

Literature

Lambert's contributions to probability are described in detail in two recent articles in *Archive for History of Exact Sciences:* "J. H. Lambert's work on probability," by O. B. Sheynin, Vol. 7 (1971), pp. 244–256, and "Non-additive probabilities in the work on Bernoulli and Lambert," by Glenn Shafer, Vol. 19 (1978), pp. 309–370.

Christopher J. Scriba's article on Lambert in *The Dictionary of Scientific Biography,* Vol. 7 (Charles C. Gillispie, ed., Scribner's, New York, 1973), provides an excellent introduction to his life and work and includes a valuable bibliography. Publications too recent to be listed in that bibliography include "Johann Heinrich Lambert, mathematician and scientist, 1728–1777," by J. J. Gray and Laura Tilling, in *Historia Mathematica,* Vol. 5 (1978), pp. 13–41; *Le Savant et Philosophe Mulhousien Jean-Henri Lambert (1728–1777),* by Roger Jaquel, published in 1977 by Editions Ophrys, Paris; "Johann Heinrich Lambert (1728–1777)," by R. H. Daw, in the *Journal of the Institute of Actuaries,* Vol. 107 (1980), pp. 345–363; and two books published in 1980 by Editions Ophrys, Paris: *Colloque International et Interdisciplinaire J. H. Lambert, Mulhouse, 26–30 Septembre 1977, and Correspondance entre D. Bernoulli et J.-H. Lambert.*

GLENN SHAFER

Laplace, Pierre Simon

Born: March 23, 1749, in Beaumont-en-Auge, France.
Died: March 5, 1827, in Paris, France.
Contributed to: mathematics, theoretical astronomy, probability and statistics.

Pierre Simon Laplace

CAREER ADVANCE

Born into the French bourgeoisie, Laplace died a marquis, having lived one of the most influential and "successful" careers in science. His scientific life was dominated by three problems, and especially connections between them: methods of solution of differential equations, and related techniques; theoretical astronomy; and probability and statistics. All these interests were present in his earliest works, published in the early 1770s, and continued in a stream of papers for the rest of the century. During the 1790s and early 1800s he wrote his major works in astronomy: the *Exposition du Systéme du Monde* (1st ed., 1796; 6th posthumous ed., 1835), and especially the *Traité de Mécanique Céleste* (Vols. 1–4, 1799–1805).

The Revolution brought Laplace new chances of professional advance, which he grasped readily. A full member of the old Académie des Sciences in 1785, he was active in the scientific class of the Institut de France on its constitution in 1795, and in the restored Académie from 1816. He took the lead in the affairs of the Bureau des Longitudes, laying the emphasis strongly on theoretical astronomy. He exercised similar influence on the École Polytechnique, mainly through its powerful Conseil de Perfectionnement, formed in 1800; for he shifted the thrust of the curriculum toward its theoretical aspects, thus countering Monge's emphasis on practicalities.

When Bonaparte took power in 1799, he made Laplace Minister of the Interior—but he removed him six weeks later for trying to "carry the spirit of the infinitesimal into administration." However, he made Laplace Chancellor of the Senate in 1803, and a count in 1806.

CAREER EMINENCE

During the early 1800s Laplace came to his philosophy of physics, which has since been termed "Laplacian physics." His view was that *all* physical theories should be formulated in terms of binary forces acting between "molecules" of the matter involved. In this spirit he extended the range of his scientific interests, hitherto confined primarily to astronomy and planetary physics, and made useful contributions to capillary theory, sound, optics, and heat diffusion. He and his neighbor at Arcueil, the chemist Berthollet, sponsored a group of young scientists at this time which was called "The Society of Arcueil." Biot and Poisson were among its members.

Laplace was always adept at bending with the political wind. While L. Carnot and Monge lost their seats in the restored Académie des Sciences in 1816, Count Laplace not only kept his but was also elected to the Académie Française that year and even elevated to a marquisate in 1817. His three connected scientific interests continued to occupy much of his time. His treatise *Théorie Analytique des Probabilités* appeared in editions of 1812, 1814, and 1820 (with supplements in the 1820s), while an *Essai Philosophique sur les Probabilités,* serving as an introduction to the major work, was published in five editions between 1814 and 1825. In the 1820s he also put out a fifth volume of his *Mécanique Céleste,* consisting mostly of reprints or reworkings of earlier papers. He worked right to the end; two substantial papers were published in the year (1827) of his death.

STYLE

The problem of influence on Laplace is particularly hard to solve, partly because of the concurrency of others' work (during the 1770s and 1780s in particular, his work overlapped with that of Lagrange and Legendre in many respects) and especially because of his habit of rarely citing his sources, even his own previous writings. Todhunter has a delicious explanation: "We conclude that he supposed the erudition of his contemporaries would be sufficient to prevent them from ascribing to himself more than was justly due." Laplace often skipped details of derivations, about which his translator Bowditch spoke for all: "Whenever I meet in LaPlace with the words 'Thus it plainly appears,' I am sure that hours, and perhaps days, of hard study will alone enable me to discover *how* it plainly appears."

CALCULUS

Laplace introduced some new methods of solving differential equations: the methods of cascades and of successive approximations are the most important. By and large, however, he extended or varied techniques developed by others. For example, he was adept at using Lagrange's formal relation

$$u(x+h) - u(x) = \exp\left(\frac{hd}{dx} - 1\right) u(x) \tag{1}$$

between differences and differentials to devise means of solving not only differential equations but also difference and mixed equations, and to evolve methods of summing series [1].

ASTRONOMY

Laplace's principal motivation in astronomy was the mathematical analysis of all the known motions, especially the perturbations, of the heavenly bodies. He and Lagrange wished to *prove* the stability of the planetary system, and so refute the catastrophism embodied in the Newtonian tradition. His work also brought him to various aspects of planetary physics: tidal theory, the shape of the earth, planetary rotation, and so on. The connections with probability lay principally in his proposal, rather novel for the time, of using astronomical data to determine constants and numerical values for specific quantities; and in studying "population" problems such as the distribution of comets. With regard to error analysis, he developed minimax methods as his preferred linear regression techniques [2].

EARLY PROBABILITY

One of Laplace's discoveries in probability itself was Bayes' theorem, which he published in 1774 [3]. Although Bayes' paper had appeared (posthumously) ten years earlier, it is quite likely that the French were unaware of it. And there is no doubt that Laplace established Bayesian statistics as a body of knowledge in probability theory. His applications of it included standard combinatorial problems (drawing balls from from an urn, etc.), and demography (the ratio of male and female births, and later, population estimation for France). He sometimes used the beta distribution, and in the process contributed to the theory of the incomplete beta function [4]. Much of his later work switches between Bayesian and non-Bayesian methods, suggesting that for him the distinction between the two did not mark a dichotomy in the way that is often asserted today.

In 1782, Laplace introduced his "generating function"

$$f(t, x) \overset{\mathrm{DF}}{=} \sum_{r=0}^{\infty} y_r(x)t^r \tag{2}$$

as a technique for solving differential and related equations, and also for use in probability [5]. In the latter context it stood as a generating function for a discrete random variable, with an integral analogue for continuous variables. Transformation of (2) by $t = e^{iu}$ led him later to some consideration of characteristic functions [6], but he largely missed the bearing of harmonic analysis on such functions for continuous variables, and also the related topic of the Laplace transform.

LAPLACE TRANSFORM

What, then, did Laplace do with the Laplace transform? The term is something of a misnomer, since he only manipulated the *in*definite integrals

$$\int x^s \phi(x)\, dx \quad \text{and} \quad \int e^{-sx} \phi(x)\, dx \tag{3}$$

rather than the true Laplace transform. Some connections are there. For example, he began work with (3) in the 1780s in connection with his use of generating functions (2); and his purpose then was linked with asymptotic theory, and included various uses of the error function. In an extensive essay of 1785 on asymptotic theory he obtained expansions of $\int_0^T e^{-t^2}\,dt$ in powers of T or of T^{-1} according as T was small or large, and obtained definite integrals of related functions, such as [7]

$$\int_{-\infty}^{+\infty} z^{2r} e^{-z^2}\,dz = \frac{(2r)!\sqrt{\pi}}{4^r r!}. \tag{4}$$

In the 1800s he was aware of Fourier's work; indeed, he eventually encouraged Fourier's study of heat diffusion and trigonometric series and in 1809 used (4) to find a new integral solution to the diffusion equation [8]. This solution in turn led Fourier in 1811 to Fourier-transform theory. But the Laplace-transform did not come out as correlate theory; in particular, in his *Probabilités* Laplace came to the inverse transform, but, finding it to be an integral along a complex path, did not know how to proceed [9]. Even though Cauchy began to introduce complex-variable integrals within a few years, the potential of Laplace-transform theory remained hidden for many decades.

LATER STATISTICS

However, Laplace made other advances around 1810. One was his study of a certain urn problem of which the solution led him to the "Hermite polynomials," including the "Gram–Charlier expansion," both misnomers [10]. Inspiration from Fourier may again be evident, as his method imitates Fourier's treatment in 1807 of the (also misnamed) "Bessel function" $J_0(x)$. Another achievement was a form of a central limit theorem, where again the 1780s studies of the error function played a role (as also, the erudite reader suspects, did some work by Lagrange on determining the mean of a set of observations, Legendre's and Gauss's advocacy of least-squares regression earlier in the 1800s and De Moivre's previous work). After showing that the sum of two terms symmetrically placed in the binomial expression $(p+q)^{r+s}$ and $2l$ terms apart was given approximately by

$$\sqrt{\frac{2(r+s)}{\pi rs}}\exp\left[-\frac{(r+s)l^2}{2rs}\right], \qquad l \text{ large}, \tag{5}$$

he used the Euler–MacLaurin summation formula to prove that "the probability that the difference between the ratio of the number of times that the even a can happen to the total number $[n]$ of attempts and the facility p of its occurrence is less than l" was given by

$$\frac{2}{\sqrt{\pi}}\int_0^{\alpha} e^{-t^2}\,dt + \frac{\alpha}{l\sqrt{\pi}}e^{-\alpha^2} + O\left(\frac{1}{n}\right), \tag{6}$$

where

$$\alpha^2 = \frac{l^2 n}{2(np+z)(nq-z)}, p+q=1, \quad |z|<1. \tag{7}$$

Thus the probability was $O(1/\sqrt{n})$ [11].

During the last 15 years of his life, Laplace was mostly occupied with problems in Laplacian physics (and astronomy), but he also made further contributions to probability at times, especially in supplements to his treatise. For example, in 1818 he compared absolute deviation with least-squares deviation and in effect carried out an exercise in sufficient statistics (without, however, individuating the concept of sufficiency) [12]. Again, in his last year of life he handled a multiple regression analysis, in a paper which exemplified the continually interlocking interests of his whole career, for it was concerned with lunar tides in the atmosphere [13].

PHILOSOPHY OF PROBABILITY

Laplace discussed his philosophy of probability in various places, and took it as the theme of his *Essai*. He saw the parallels between the physical and social sciences, and thus the applicability of probability to each. Comets were distributed; so were births. The planetary system was stable; so were lottery receipts.

Laplace also took probability as the core of his philosophy of science; for he thought of scientific knowledge as only probable, though the degree of probability would be increased by confirmations. The world was fully determined, but *our* theories could only capture its course partially. "The curve described by a simple molecule of air or any gas is regulated in a manner as certain as the planetary orbits," he wrote, "the only difference between them lies in our ignorance. Probability relates partly to our ignorance, partly to our knowledge" [14].

INFLUENCE

The influence on others of Laplace's contributions to probability and statistics built up more slowly than his mathematical and astronomical achievements. He was working in a much less developed field, and so played the pioneer to a greater extent. His "thus-it-plainly-appears" style was especially unfortunate here, since the contexts in which the derivations occurred were much less familiar to the reader.

However, Laplace's contributions to probability and statistics were fundamental: not merely some new results but especially a much greater degree of mathematization. He also changed the emphasis of probability from its preoccupation with moral sciences and jurisprudence to include also applications in scientific contexts, whither it had hitherto infrequently strayed. His most important early successors were Quetelet* and Poisson; after them, both probability and statistics moved to adulthood in the family of sciences, and the heritage from Laplace began to be recognized.

Literature

A seven-volume edition of Laplace's *Oeuvres,* published 1843–1847, contains only the books mentioned in the text. The 14-volume *Oeuvres Complétes* (1878–1912) has another reprinting of these books, with the other seven volumes given over to his papers. The edition is poor even by the standards of works for French mathematicians; notations are sometimes modernized, editorial apparatus and commentary is almost nonexistent, and worst of all, it is *not* complete, lacking his first three papers and his first book [*Théorie du Mouvement et de la Figure Elliptique des Planètes* (1784)]. The only comprehensible edition of *Mécanique Céleste,* Vols. 1–4, is N. Bowditch's English translation *Celestial Mechanics* (4 vols., Boston, 1829–1839, reprinted in New York, 1966). Most of Laplace's manuscripts were destroyed long ago, although there are a few things here and there; in particular, C. C. Gillispie, "Mémoires inédits ou anonymes de Laplace . . . ," *Rev. Hist. Sci.,* **32** (1980), 223–280 includes a paper on determining the mean of a set of observations.

The most comprehensive single study of Laplace is C. C. Gillispie (in collaboration with others), "Laplace, Pierre-Simon, Marquis de," in *Dictionary of Scientific Biography,* Vol. 15 (1978, New York), 273–403, including extensive bibliographies of primary and secondary literature. Laplace's work on probability has been studied by various authors: a good start is provided by S. Stigler, "Napoleonic statistics: the work of Laplace," *Biometrika,* **62** (1975), 503–517; and O. B. Sheynin, "P. S. Laplace's work on probability," *Arch. Hist. Exact Sci.,* **16** (1976), 137–187 and "Laplace's theory of errors," *ibid.,* **17** (1977), 1–61.

REFERENCES

These are taken from *Oeuvres Complètes,* discussed above, in which the books on probability occupy Vol. 7. The volume numbers are to the edition; the dates are of the original publication of the work involved.

[1] **9,** 315–325 (1780).
[2] See, e.g., **2,** 246–314 (1799).
[3] **8,** 27–65 (1774).
[4] **9,** 422–429 (1781).
[5] **10,** 549 (1782).
[6] See, e.g., **12,** 309–319 (1810).
[7] **10,** 230, 269 (1785); cf. **7,** 104 (1812, 1820).
[8] **14,** 189–193 (1809).
[9] **7,** 136–137 (1812, 1820).
[10] **12,** 377–385 (1811).
[11] **7,** 281–287 (1812, 1820).
[12] **7,** 531–580 (1818, 1820).
[13] **13,** 342–353 (1827), and also in a supplement fo *Mécanique Celeste* [Vol. 5, pp. 481–505 (1827)] .
[14] **7,** viii (1814, 1825).

I. GRATTAN-GUINNESS

Newton, Isaac

Born: December 25, 1642, in Woolsthorpe, Lincolnshire, England.
Died: March 20, 1727, in London, England.
Contributed to: algebra, astronomy, infinitesimal calculus, numerical methods, mathematical and experimental physics.

INTRODUCTION

To probabilists and statisticians, Isaac Newton is known as an outstanding mathematician, the discoverer, together with G. W. Leibniz (1646–1716), of the infinitesimal calculus and the originator of the law of universal gravitation. While he does not appear to have taken an active part in the development of probability and statistics in the late seventeenth and early eighteenth centuries, it is clear that he was familiar with the probability calculus of the times and had encountered the problem of the variability of sample means.

A brief biography may prove useful (see e. g., Youschkevitch [13]). Newton was born at Woolsthorpe in Lincolnshire in 1642; after attending school at nearby Grantham, he went up to Trinity College, Cambridge, in 1661, and was granted his B.A. in 1665 and his M.A. in 1668. In 1669, at the age of 26, he was appointed Lucasian professor, succeeding Isaac Barrow, the first incumbent of that chair at the University of Cambridge. In 1672, he became a Fellow of the Royal Society of London. His lectures, deposited at the University Library, contained new work on optics that appeared in his *Opticks* (1704), on arithmetic and algebra, and on elements of the infinitesimal calculus, later published as Book I of the *Principia* in 1687. In this, his major work, he laid out some basic mathematical principles and rules for limits in Book I, considered the laws of motion of bodies in resisting media in Book II, and lastly in Book III outlined the laws of celestial mechanics and of universal gravitation. In 1696, he was appointed Warden of the Mint, and moved to London; he became Master of the Mint in 1699 and was knighted by Queen Anne in 1705. He was elected president of the Royal Society in 1703 and is reputed to have ruled it with an iron hand until his death in 1727. During his London period, despite his responsibilities at the Mint and the Royal Society, Newton maintained his scientific interests, publishing his *Arithmetica Universalis* in 1707, a second edition of the *Principia* in 1713, and an enlarged version of the *Opticks* in 1717.

Newton had worked on the binomial theorem in 1655 (see Whiteside [12]), and would have understood the uses of the binomial distribution. Such results were well known among mathematicians in Europe at the time, following the publication of Huygens' [7] work in 1657, and the results of James Bernoulli (1654–1705), which appeared in print posthumously [1] in 1713 (*see* HUYGENS, CHRISTIAAN and BERNOULLIS, THE). Although Newton made no original contributions to the theory of probability, a series of letters (see Turnbull [11]) exchanged with Samuel Pepys (1633–1703) in 1693 attest to his familiarity with contemporary probability calculations. Accounts of this correspondence, summarized in the following section, may be found in David [4, 5], Schell [9], Chaundy and Bullard [2], and Gani [6].

After his appointment to the Mint in 1696, Newton must have become familiar with the Trial of the Pyx, a sampling inspection scheme for coinage based on the aggregate weighing of a large number of coins selected at random (see Craig [3]). The concept of the Trial is similar to that of the modern sampling test procedure for means. Stigler [10] has presented some evidence that Newton, through his experience at the Mint and his studies of chronology, may well have had an understanding of the decrease in variability of means as the number of measurements averaged is increased. This is also briefly outlined in the section on statistics.

PROBABILITY: NEWTON'S SOLUTION OF A DICING PROBLEM

On November 22, 1693, Samuel Pepys addressed a letter to Isaac Newton at Cambridge, introducing its bearer Mr. John Smith, the Writing Master of Christ's Hospital School, as one who desired Newton's opinion on a question of dicing. The enquiry may well have resulted from the interest shown in lotteries at that time. Pepys formulated the dicing problem as follows:

The Question.

> *A*–has 6 dice in a Box, wth wch he is to fling a 6.
> *B*-has in another Box 12 Dice, wth wch he is to fling 2 Sixes.
> *C*-has in another Box 18 Dice, wth wch he is to fling 3 Sixes.
> Q. Whether *B* & *C* have not as easy a Taske as *A*, at even luck?

Newton in his reply to Pepys of November 26, 1693 wrote that the problem was "ill-stated," and took

> . . . the Question to be the same as if it had been put thus upon single throws.
> What is ye expectation or hope of *A* to throw every time one six at least wth six dyes?
> What is ye expectation or hope of *B* to throw every time two sixes at least wth 12 dyes?
> What is ye expectation or hope of *C* to throw every time three sixes or more than three wth 18 dyes?

He then stated "it appears by an easy computation that the expectation of *A* is greater that that of *B* or *C*," without giving any details. After further correspondence, Newton gave Pepys the details of his calculations on December 16, 1693. These were based on the following simple binomial results:

Pr{1 or more sixes in 1 throw of 6 dice}

$$= 1 - \left(\frac{5}{6}\right)^6 = 1 - a.$$

Pr{2 or more sixes in 1 throw of 12 dice}

$$= 1 - 12\left(\frac{5}{6}\right)^{11}\left(\frac{1}{6}\right) - \left(\frac{5}{6}\right)^{12}$$

$$= 1 - b.$$

Pr{3 or more sixes in 1 throw of 18 dice}

$$= 1 - \frac{18 \times 17}{1 \times 2}\left(\frac{5}{6}\right)^{16}\left(\frac{1}{6}\right)^{2} - 18\left(\frac{5}{6}\right)^{17}\left(\frac{1}{6}\right) - \left(\frac{5}{6}\right)^{18}$$

$$= 1 - c,$$

where

$$a = \left(\frac{5}{6}\right)^{6}, \qquad b = \left(\frac{5}{6}\right)^{12}\left(1 + \frac{12}{5}\right), \qquad c = \left(\frac{5}{6}\right)^{18}\left(1 + \frac{18}{5} + \frac{18 \times 17}{2 \times 5^{2}}\right).$$

The values of $1 - a$ and $1 - b$ were found to be

$$1 - a = \frac{31{,}031}{46{,}656}, \qquad 1 - b = \frac{1{,}346{,}704{,}211}{2{,}176{,}782{,}336},$$

but Newton did not give figures for the 18-dice case. His method of calculation would, however, have led to

$$1 - c = \frac{60{,}666{,}401{,}980{,}916}{101{,}559{,}956{,}668{,}416}.$$

In effect, A would have the most favorable throw, as $1 - a > 1 - b > 1 - c$.

STATISTICS: THE TRIAL OF THE PYX AND NEWTON'S CHRONOLOGY

Newton's position at the Mint clearly involved familiarity with the Trial of the Pyx. This sampling inspection scheme, which had been in operation since the thirteenth century, consisted of taking one gold coin out of roughly every 15 pounds of gold minted or one silver coin out of every 60 pounds of silver (one day's production) at random, over a period of time, and placing them in a box called the Pyx (after the Greek ı Øs for box). At irregular intervals of between one and several years, a Trial of the Pyx would be declared with an adjudicating jury selected from among established goldsmiths. At the Trial, the Pyx would be opened and its contents counted, weighed, and assayed in bulk to ensure that the gold and silver coins were within the allowed tolerances.

The aggregated weight of the sample of n coins was expected not to exceed n times the required tolerance for any single coin. This procedure was equivalent to carry out a rudimentary two-sided test, where the tolerances were in fact set so that only about 5% of a representative collection of coins would fail to satisfy them. Newton underwent one such Trial in 1710, when he successfully survived the charge that his gold coinage was below standard.

Statistical theory indicates that $\sqrt{n}$ times the tolerance for a single coin would have been a more appropriate measure of tolerance for the aggregate. Newton may possibly have had some understanding of this point. He is known to have emphasized a reduction in the variability of individual coins from the Mint, but the Trial of the Pyx must have raised in his mind the question of the variability of sample means.

Some circumstantial evidence for this is contained in Newton's last work "The Chronology of Ancient Kingdoms Amended," published posthumously in London in 1728. In this, Newton estimated the mean length of a king's reign "at about eighteen or twenty years a-piece." Stigler [10] points out that he repeated this phrase three times without ever quoting nineteen as the mean length; this mean was in fact 19.10 years, with a standard deviation of 1.01. Newton's interval of 18–20 corresponds to a band of one standard deviation about the mean or roughly a 65% confidence interval.

While we cannot be certain that Newton had in fact pondered the problem of significance tests, he was implicitly providing some form of interval estimate for the length of a king's reign. On this premise, Stigler argues that Newton "had at least an approximate intuitive understanding of the manner in which the variability of means decreased as the number of measurements averaged increased."

CONCLUDING REMARKS

The correspondence with Pepys, outlined earlier, provides convincing evidence that Newton was conversant with the calculus of probabilities of his time. In this, he was not alone; both Schell [9] and Chaundy and Bullard [2] refer to Pepys' simultaneous request to George Tollet, who obtained the same results as Newton.

The problem itself is of some intrinsic interest; it has been generalized by Chaundy and Bullard [2] in 1960 to take account of an s-faced die, $s \geq 2$. The authors study the asymptotic behavior of the probability

$$f(sn, n) = \Pr\{n \text{ or more of a selected face in 1 throw of } sn \text{ dice}\}$$

$$= \sum_{j=n}^{sn} \binom{sn}{j} \left(\frac{1}{s}\right)^{j} \left(1 - \frac{1}{s}\right)^{sn-j}.$$

Gani [6] has also recently considered de Méré's problem in this more general context. Here, one is concerned with the different question of the number n of repeated throws of $r = 1, 2, \ldots$ six-sided dice, which is necessary to achieve a successful throw of r sixes. In particular, de Méré was interested in the number n of throws required for

$$\Pr\{2 \text{ sixes in } n \text{ throws of } 2 \text{ dice}\} > \tfrac{1}{2};$$

this Pascal* showed to be $n = 25$.

The evidence for Newton's understanding of the statistical principles involved in the variability of sample means is less secure, but his familiarity with the Trial of the Pyx and his treatment of the mean length of a king's reign, indicate that he must at least have considered the problem.

It is interesting to speculate whether in other circumstances, Newton the mathematician might have become more active as a probabilist. Perhaps the simple answer is that in a choice between investigating a "System of the World" and problems which in his time were often related to gambling, Newton's fundamental seriousness would almost inevitably have caused him to select the first.

No account of Newton's contributions is complete without a mention of two approximation methods in mathematics that bear his name and that are also used in a statistical context. These are the Newton–Raphson method for approximating the roots of $f(x) = 0$, and the Gauss–Newton method for the replacement of a nonlinear function $g(x_1, \ldots, x_k)$ by its linear approximation

$$g(a_1, \ldots, a_k) + \sum_{i=1}^{k} (x_i - a_i) \frac{\partial g(a_1, \ldots, a_k)}{\partial a_i};$$

for details see Ortega and Rheinboldt [8].

REFERENCES

[1] Bernoulli, J. (1713). *Ars Conjectandi.*

[2] Chaundy, T. W. and Bullard, J. E. (1960). *Math. Gaz.*, **44,** 253–260.

[3] Craig, J. (1953). *The Mint.* Cambridge University Press, Cambridge.

[4] David, F. N. (1957). *Ann. Sci.*, **13,** 137–147.

[5] David, F. N. (1962). *Games, Gods and Gambling.* Griffin, London.

[6] Gani, J. (1982). *Math. Sci.*, **7,** 61–66.

[7] Huygens, C. (1657). *De Ratiociniis in Ludo Aleae.*

[8] Ortega, J. M. and Rheinboldt, W. C. (1970). *Iterative Solution of Nonlinear Equations in Serveral Variables.* Academic Press, New York.

[9] Schell, E. D. (1960). *Amer. Statist. Ass.*, **14**(4), 27–30.

[10] Stigler, S. M. (1977). *J. Amer. Statist. Ass.*, **72,** 493–500.

[11] Turnbull, H. W., ed. (1961). *The Correspondence of Isaac Newton 1668–1694,* Vol. III. Cambridge University Press, London.

[12] Whiteside, D. T. (1961). *Math. Gaz.,* **45,** 175–180.

[13] Youschkevitch, A. P. (1974). Isaac Newton, *Dictionary of Scientific Biography,* Vol. X. Scribner's New York.

J. GANI

Pascal, Blaise

Born: June 19, 1623, in Clermont-Ferrand, France.
Died: August 19, 1662 in Paris, France.
Contributed to: mathematics, computing machines, physics, philosophy.

The history of probability is traditionally taken to begin in 1654, the year of the correspondence between Pascal and Pierre Fermat. Pascal's central notion was that of mathematical expectation, and he developed decision-theoretic ideas based thereon. He did not use the term *probability* [12]. Nevertheless, there are narrowly probabilistic ideas in Pascal's writings: the notions of equally likely events, of the random walk, and the frequency interpretation of probability. Through this last Pascal seems to have had influence on the first substantial landmark of probability, Bernoulli's theorem.

The de Méré–Pascal double-six situation [7] is encountered at the beginning of courses in probability: this concerns the minimum number of throws of a pair of fair dice to ensure at least one double six with a better than even chance. The number of tosses until a double-six occurs has a geometric distribution, which may explain why the geometric distribution and, by extension, the negative-binomial distribution, are sometimes named in Pascal's honor. In regard to probability theory, however, Pascal's direct concern was with the equitable division of total stake when a game is interrupted before its agreed completion (*problème de partis,* or the problem of points). One problem of this kind solved by Pascal and Fermat (by correspondence and by different methods), although for some particular cases, marks an epoch in probability theory because of difficult structural features. At each of a number of trials, each of two players has probability $\frac{1}{2}$ of winning the trial. It is agreed that the first player with n wins gains the total stake. The game is interrupted when player A needs a trial wins to gain the stake and player B needs b. How should the stake be divided? The obvious sample space, of sequences leading to game termination, consists of sample sequences of unequal length and therefore unequal probabilities. Pascal expands the sample space to sequences each of length $a + b - 1$, which are equiprobable; and the probability of a win by A, and hence the proportion of stake presently due to him,

$$\sum_{r=a}^{a+b-1} \binom{a+b-1}{r} \left(\frac{1}{2}\right)^{a+b-1},$$

is then easily evident, and can easily be obtained from "Pascal's triangle." We may identify the complicating factor in this situation as the stopping time defined on an infinite random sequence. Pascal's study of the rule of formation of "Pascal's triangle" (which he did not discover) helped him to discover and understand the principle of mathematical induction ("reasoning by recurrences").

Pascal discovered the principle of a calculating machine when he was about 20, and in physics his name occurs in Pascal's principle (or the law of pressure).

However, he is most widely known for his philosophical and theological later writings. He was influenced by Jansenism, a religious movement associated with Catholicism and distinguished by its piety; the night of November 23–24, 1654 [his famous (second) "conversion"] resulted in his joining the lay community associated with the movement at Port-Royal des Champs. Becoming interested in pedagogic research, he contributed to a "Logic," which appeared anonymously in 1662 entitled *La Logique ou L'Art de Penser,* often referred to as the *Logique de Port-Royal.* This is now attributed, in a recent reprinting, entirely to Arnauld and Nicole [1], and Pascal's part in it remains relatively obscure. In this 1970 reprint, the fourth part, entitled *De la méthode,* likely due to Pascal, contains elements of probabilistic thinking—for the first time in contexts other than games of chance—that illustrate most clearly his decision-theoretic ideas, and, according to Jacob Bernoulli, partly motivated Bernoulli's weak law of large numbers. In a passage of his *Ars Conjectandi,* reproduced in Uspensky [13], Bernoulli [2] calls the unknown author "*magni acuminis et ingenii vir*" (a man of great acumen and ingenuity). In particular, as regards decision theory, this section of the *Logique* contains allusions to Pascal's famous passage from his *Pensées: De la nécéssité du pari* or, as it is generally called in English, *the Wager*. Essentially a philosophical argument, this can be expressed in terms of a 2×2 loss table with a prior distribution over the states of nature and puts forward that decision as solution which minimizes expected loss [5, 10].

During the Port-Royal period also, Pascal, in correspondence to Fermat, proposed the "gambler's ruin" problem in disguised form; in a random walk on the integers between two absorbing barriers, starting in the middle, with unequal probabilities of movement a unit left or right (and a positive probability of no movement at all from any intermediate position), the probability of ruin is sought [7].

The usual assessments of Pascal's probabilistic contributions have been based on the Pascal–Fermat correspondence, and on his *Traité du Triangle Arithmétique,* discovered after his death. This treats extensively the formulation and solution of the equitable division of stakes problem in association with Pascal's triangle. One of the first substantial mathematical treatments of such problems arising out of Pascal's work, particularly the problem of points, is due to Montmort,* an extensive summary of whose book is given as Chapter 8 of ref. [12]. Realization of the probabilistic significance of the philosophical later writings (not referred to in historical treatments of probability such as David [4] and Todhunter [12]) is of recent origin [3, 5, 7].

Pascal's mother died when he was three; his father took charge of his education. In 1631 the family moved to Paris, where one sister, Jacqueline, entered the convent of Port-Royal in 1652, an event that seems to have had substantial significance on Pascal's intense religious development, of which some mention has already been made. At this distance in time it is as difficult to assess the validity of Pascal's scientific standing as it is to assess his personality from the extant fragments of his writings. This writer is inclined to agree with the conclusions of F. N. David* [4, p. 97]; French writers incline to a more sympathetic view (e. g., [11, p. 337]). General references on his life may be found in [6, §5.8].

REFERENCES

[1] Arnauld, A. and Nicole, P. (1970). *La Logique ou L'Art de Penser.* Georg Olms, Hildesheim, W. Germany. (Reproduction of first anonymous Paris edition of 1662. An edition was also published in 1970 by Flammarion, Paris. There is an English translation in its fifth edition by P. and J. Dickoff, New York, 1964.)

[2] Bernoulli, J. (1713). *Ars Conjectandi.* Basileae Impensis Thurnisiorum Fratrum, Basel.

[3] Coumet, E. (1970). *Ann. Econ. Soc. Civilis.,* **5,** 574–598.

[4] David, F. N. (1962). *Games, Gods and Gambling: The Origins and History of Proability and Statistical Ideas from the Earliest Times to the Newtonian Era.* Charles Griffin, London. (Contains an entertaining assessment of the Pascal–Fermat correspondence and Pascal's scientific standing and personality.)

[5] Hacking, I. (1975). *The Emergence of Probability.* Cambridge University Press, London. (Focuses on philosophical aspects. Chapter 8 deals with "the Wager" as a decision-theoretic problem.)

[6] Heyde, C. C. and Seneta, E. (1977). *I. J. Bienaymé: Statistical Theory Anticipated.* Springer, New York. (Section 5.8 gives a biographical account in the context of probability and statistics.)

[7] Ore, O. (1960). *Amer. Math. Monthly,* **67,** 409–419. (The first more-or-less complete account of Pascal's contributions to probability and decision theory.)

[8] Pascal, B. (1904–1925). *Oeuvres,* Vols. 1–14, L. Brunschvieg, P. Boutroux, and F. Gazier, eds., Les Grands Ecrivains de France, Hachette, Paris. (Collected works with editorial comments: a standard version.)

[9] Pascal, B. (1970). *Oeuvres Complétes,* Vol. 2, J. Mesnard, ed. Desclée de Brouwer, Paris. (From a more recent collection than ref. [8]. Possibly the best edition to date of complete writings with extensive editorial comments. Volume 1 was published in 1964. Volume 2 contains Pascal's work on the equitable division of stakes.)

[10] Seneta, E. (1979). In *Interactive Statistics,* D. McNeil, ed. North-Holland, Amsterdam. (Pascal's mathematical and philosophical writings pertaining to aspects of probability and decision theory are sketched and considered from a modern viewpoint.)

[11] Taton, R. (1974). In *Dictionary of Scientific Biography,* Vol. 10, C. C. Gillispie, ed. pp. 330–342. (Complete survey of Blaise Pascal by an eminent French historian of science.)

[12] Todhunter, I. (1865). *A History of the Mathematical Theory of Probability from the Time of Pascal to that of Laplace.* Cambridge University Press, London and Cambridge. (Reprinted in 1949 and 1961 by Chelsea, New York. Standard reference on the early history.)

[13] Uspensky, J. V. (1937). *Introduction to Mathematical Probability.* McGraw-Hill, New York, pp. 105–106.

E. SENETA

Quetelet, Lambert Adolphe Jacques

Born: February 22, 1796, in Ghent, Belgium.
Died: February 17, 1874, in Brussels, Belgium.
Contributed to: descriptive statistics, demography, vital statistics, statistics in the social sciences.

Adolphe Quetelet was one of the nineteenth century's most influential social statisticians. He was born Lambert Adolphe Jacques Quetelet in Ghent, Belgium on February 22, 1796. He received a doctorate of science in 1819 from the University of Ghent, with a dissertation on conic sections. From 1819 on he taught mathematics in Brussels, founded and directed the Royal Observatory, and he dominated Belgian science for a half century, from the mid 1820s to his death in 1874.

Early in 1824 Quetelet spent three months in Paris, where he studied astronomy and probability and learned what he could about the running of an observatory. Upon his return to Brussels he convinced the government to found an observatory, and for most of his life he operated from this base, giving particular attention to its meteorological functions. But one science was not enough to contain his energy and interests, and from about 1830 he became heavily involved in statistics and sociology.

Quetelet's international reputation was made in 1835 with the publication of a treatise where he coined the term "social physics" [7, 8, 12]. This work is today best known for the introduction of that now famous character, the "average man" ("l'homme moyen"). The average man began as a simple way of summarizing some characteristic of a population (usually a national population), but he took on a life of his own, and in some of Quetelet's later work he is presented as an ideal type, as if nature were shooting at the average man as a target and deviations from this target were errors. The concept was criticized by Cournot and others, for example on the grounds that an individual average in all dimensions might not even be biologically feasible (the average of a set of right triangles may not be a right triangle).

In 1846 he published a book [9, 10] on probability and social science in the form of a series of letters to two German princes he had tutored (one of them, Albert, had married Queen Victoria of England in 1840). That book exerted a significant influence on social statistics by demonstrating that as diverse a collection of human measurements as the heights of French conscripts and the chest circumferences of Scottish soldiers could be taken as approximately normally distributed. This gave a further dimension to the idea of an average man—deviations from the average were normal, just as errors of observation deviated normally from their mean. Quetelet devised a scheme for fitting normal distributions (actually, symmetric binomial distributions with $n = 999$) to grouped data that was essentially equivalent to the use of normal probability paper [14]. The appearance of the normal curve in such areas so far from astronomy and geodesy had a powerful influence on Francis Galton's* thinking, and reading of this work of Quetelet may have inspired James Clerk Maxwell* in formulating his kinetic theory of gases.

Quetelet made few technical contributions to statistics, although in 1852 he did anticipate some later work on the use of runs for testing independence. Quetelet derived the expected numbers of runs of different lengths both for independent and for simple Markov sequences, and compared them with rainfall records, concluding that there was strong evidence of persistence in rainy or dry weather [11, 14]. In his earlier 1835 treatise, Quetelet was led to a forerunner of a measure of association in 2×2 tables, although the measure was neither developed nor expressed algebraically [2]. In other work (e. g., ref. [9]) Quetelet gave much attention to classifying sources of variation as due to accidental, periodic, or constant causes, in a sort of informal precursor to the analysis of variance or the decomposition of time series.

Quetelet was a prolific writer and editor. He wrote a dozen books, founded and wrote much of the material for several journals, and still found time to fill the pages of the official publications of the Belgian Académie Royale des Sciences. In addition, he carried on an immense correspondence with scientists and others all over Europe [1, 15]. He was a highly successful entrepreneur of science who was instrumental in the founding of the Statistical Society of London, the International Statistical Congresses, and the Statistical Section of the British Association for the Advancement of Science, not to mention several Belgian bureaus and commissions, and similar activities in meteorology. He was the first foreign member of the American Statistical Association. The historian of science George Sarton has called him the "patriarch of statistics" [13].

REFERENCES

[1] Diamond, M. and Stone, M. (1981). *J. R. Statist. Soc. A,* **144,** 66–79, 176–214, 332–351. (Presents manuscript material on the relationship between Quetelet and Florence Nightingale*.)

[2] Goodman, L. and Kruskal, W. H. (1959). *J. Amer. Statist. Ass.,* **54,** 123–163.

[3] Hankins, F. H. (1908). *Adolphe Quetelet as Statistician.* Longman, New York.

[4] Landau, D. and Lazarsfeld, P. F. (1978). In *International Encyclopedia of Statistics,* Vol. 2, pp. 824–834. (Reprinted from The International Encyclopedia of the Social Sciences, with minor additions.)

[5] Lazarsfeld, P. F. (1961). *Isis,* **52,** 277–333. (An excellent discussion of Quetelet's role in the quantification of sociology.)

[6] Lottin, J. (1912). *Quetelet, Statisticien et Sociologue.* Alcan, Paris/Institut Supérieur de Philosophie, Louvain.

[7] Quetelet, A. (1835). *Sur l'homme et le développement de ses facultés, ou Essai de physique sociale,* 2, vols. Bachelier, Paris.

[8] Quetelet, A. (1842). *A Treatise on Man and the Development of his Faculties.* W. & R. Chambers, Edinburgh. (Reprinted by Burt Franklin, 1968.)

[9] Quetelet, A. (1846). *Lettres à S. A. R. le Duc Régnant de Saxe-Cobourg et Gotha, sur la Théorie des Probabilité, appliquée aux Sciences Morales et Politiques.* Hayez, Brussels.

[10] Quetelet, A. (1849). *Letters Addressed to H. R. H. the Grand Duke of Saxe Coburg and*

Gotha, on the Theory of Probabilities as Applied to the Moral and Political Sciences. Layton, London. (A mediocre translation of ref. [9].)

[11] Quetelet, A. (1852). *Bulletins de l'Académie Royale de Belgique,* **19,** 303–317.

[12] Quetelet, A. (1869). *Physique Sociale, ou Essai sur le Développement des Facultés de l'Homme,* 2 vols. Muquardt, Brussels. (An expanded second edition of ref. [7].)

[13] Sarton, G. (1935). *Isis,* **23,** 6–24.

[14] Stigler, S. M. (1975). *Bull. Int. Statist. Inst.,* **46,** 332–340.

[15] Wellens-De Donder, L. (1966). *Mémoires de l'Académie royale de Belgique,* **37**(2), 1–299. (A catalog of an extensive archive of Quetelet's correspondence.)

STEPHEN M. STIGLER

'sGravesande, Willem Jacob

Born: September 26, 1688 in 'sHertogenbosch, Netherlands
Died: February 28, 1742 in Leiden, Netherlands
Contributed to: demography, tests of significance

Willem Jacob 'sGravesande

William (Willem, Guillaume) 'sGravesande (1688–1742) was a Dutch scientist influential in introducing Isaac Newton's "experimental philosophy" into the Netherlands. In 1718 he became professor of mathematics, astronomy, and philosophy at the University of Leiden. 'sGravesande's *Oeuvres* [2], published more than 30 years after his death, contained an elementary exposition of probability. His place in the history of probability and statistics is merited through his involvement in a debate provoked by John Arbuthnot's* paper in the *Philosophical Transactions of the Royal Society of London,* dated 1710 (publication was actually 1711) [1]. Arbuthnot's paper contained what is now generally recognized as the first statistical significance test, and purported to show that the consistent excess of male over female births in London, and elsewhere, was due to divine providence rather than to chance. Arbuthnot's statistical "proof" was based on the records of christenings in London over the 82–year period 1629–1710, which showed an excess of male over female christenings in every year.

Arbuthnot held that, if "chance" determined sex at birth, the probabilities of male and female births would be equal. Consequently, the probability would be no greater than one-half that any one year would have an excess of male rather than female births. He proceeded to calculate an upper limit for the probability of 82 consecutive "male" years, $(\frac{1}{2})^{82}$, and concluded that this probability was so small it cast doubt on the hypothesis that sex was determined by "chance."

Arbuthnot's paper aroused the interest of 'sGravesande and his colleague, the Dutch physician and mathematician Bernard Nieuwentijt (1654–1718). Nieuwentijt had been sent a copy of Arbuthnot's paper by William Burnet, a Fellow of the Royal Society. 'sGravesande and Nieuwentijt were evidently in sympathy with Arbuthnot's conclusions, and with his general approach. They felt, though, that Arbuthnot could have put the case for divine providence even more strongly, had he troubled to engage in a more detailed calculation. Arbuthnot had been impressed not only by the unbroken run of "male" years, but also by the narrow limits within which the ratio of male to female births appeared to fall, year after year. 'sGravesande set about incorporating the latter aspect into the argument.

'sGravesande adopted the same interpretation of "chance" as Arbuthnot. He based his calculations on a notional year in which the total number of christenings was equal to the 82-year average of 11,429 total births. Using the binomial distribution with parameters 11,429 and one-half, he calculated the probability that in any one year, the number of male births would lie within the two observed extremes. His figure (0.2917) was quite an accurate approximation. Raising this to the power 82 gave a probability of the order of 10^{-44} or 10^{-45}, much smaller even than the probability Arbuthnot had calculated.

'sGravesande's probability calculation was a painstaking affair, which involved summing all the relevant terms of the binomial, from 5,745 male births to 6,128 male births (in a total of 11,429 births). He adopted several mathematical sophistications to cut down the calculation involved. Throughout most of the calculation, he worked with *relative* values for the binomial coefficients, observing that all terms in the binomial expansion, for his model of "chance," involved the factor $(\frac{1}{2})^{82}$. Moreover, he noted that adjacent binomial coefficients bear a simple numerical relationship to each other. He gave the highest value coefficient in the distribution the arbitrary value 100,000, and proceeded in a recurrent fashion from there.

His calculations and results were set out in a paper (Demonstration mathématique du soin que Dieu prend de diriger ce qui se passe dans ce monde, tirée du nombre des garçons et des filles qui naissent journellement) which he circulated privately to colleagues, including Nicholas Bernoulli.* The latter spent some time in England in 1712, and en route had met 'sGravesande, whom he knew well. Bernoulli was stimulated to do some work himself, and he corresponded on the subject with 'sGravesande, Burnet and Pierre Rémond de Montmort. His view was that Arbuthnot and 'sGravesande had taken a too restrictive view of "chance," and that the observed pattern of christenings could be adequately explained by a representation of chance based on a multifaceted die with 18 "male" sides and 17 "female" sides. Apart from this important difference of binomial probability parameter, the framework of Bernoulli's calculations was similar to 'sGravesande's. Bernoulli, however, made his calculations more elegant by deriving an expression that gave him an approximation to the sum of terms in a binomial expansion [3].

The full text of 'sGravesande's calculations was not published until his *Oeuvres* appeared. However, an account appeared in 1715 in a book by his colleague Nieuwentijt, *Het regt gebruik der wereldbeschouwingen* (The right use of the contemplation of nature). This was translated into English by John Chamberlayne in 1718 under the title *The religious philosopher: or, the right use of contemplating the works of the Creator* [4]. Nieuwentijt included an account of Arbuthnot's "proof", and a table showing the data on which it was based. He followed this with a summary of 'sGravesande's calculations and conclusions.

It is tempting, in retrospect, to see 'sGravesande's reasoning as bringing out more clearly than Arbuthnot's the notion and relevance of a tail-area probability. 'sGravesande's summation was *not* over a complete tail area, because its upper limit was 6,128 male births, rather than 11,429. For all practical purposes, though, it can be regarded as such, because the probability of observing more than 6,128 male births is so small. So far as we can tell, 'sGravesande did not think of it this way, though it is evident that he knew the probabilities beyond 6000 male births to be virtually negligible.

REFERENCES

[1] Arbuthnot, J. (1710). An argument for divine providence, taken from the constant regularity observ'd in the births of both sexes. *Phil. Roy. Soc. London,* **27,** 186–190. Reprinted in M. G. Kendall and R. L. Plackett, eds. *Studies in the History of Statistics and Probability,* vol. 2, pp. 30–34. Griffin, London (1977).

[2] 'sGravesande, W. J. (1774). *Oeuvres philosophiques et mathématiques de Mr G. J. 'sGravesande,* J. N. S. Allemand, ed., 2 vols. Amsterdam.

[3] Hald, A. (1984). Nicholas Bernoulli's theorem. *Int. Statist. Rev.,* **52,** 93–99.

[4] Nieuwentijt, B. (1715). *Het regt gebruik der wereldbeschouwingen.* Amsterdam. *The religious philosopher: or, the right use of contemplating the works of the Creator,* John Chamberlayne, trans. London (1718).

BIBLIOGRAPHY

Hald, A. (1990). *A History of Probability and Statistics and their Applications before 1750.* Wiley, New York.

Shoesmith, E. (1985). Nicholas Bernoulli and the argument for divine providence. *Int. Statist. Rev.,* **53,** 255–259.

Shoesmith, E. (1987). The continental controversy over Arbuthnot's argument for divine providence. *Historia Math.,* **14,** 133–146.

E. SHOESMITH

Sinclair, John

Born: May 10, 1754, in Thurso Castle, Caithness, Scotland.
Died: December 21, 1835, in Edinburgh, Scotland.
Contributed to: government statistics, census taking, agricultural statistics.

Sir John Sinclair was an active worker in public life, both as a politician and as an administrator, in the late eighteenth and early nineteenth centuries. He was, *inter alia,* the first President of the Board of Agriculture.

Quoting from ref. [4]:

> John was educated at the high school of Edinburgh, and at the universities of Edinburgh, Glasgow, and Oxford, where he matriculated as a gentleman commoner at Trinity College on 26 Jan. 1775. He read for the law, though with no intention of practising, and in the same year became a member of the faculty of advocates at Edinburgh. In November 1774 he entered Lincoln's Inn, and in 1782 he was called to the English bar. . . .
>
> On 26 March 1776 Sinclair married Sarah, daughter of Alexander Maitland; and in 1780 he became member of parliament for Caithness. Almost his first political action was to volunteer to second the address at the opening of the session of 1781, an offer politely refused by Lord North. Sinclair then made an abortive attempt to form a clique of his own. He devoted considerable attention to naval affairs, which formed the subject of his maiden speech and of one of his earliest pamphlets. The even balance of parties towards the close of North's administration gave considerable influence to independent members, and in 1782 Sinclair obtained a grant of 15,000£, towards the relief of a serious famine in the north of Scotland. Although his attitude as a party politician was never very decisive, he was through life an ardent advocate of parliamentary reform (*Lucubrations during a Short Recess,* 1782; *Thoughts on Parliamentary Reform,* 1831), and he was so strongly in favour of peace with America and France as to suggest the expediency of surrendering Gibraltar (*Propriety of retaining Gibraltar considered,* 1783). Caithness having only alternate representation with Bute, Sinclair contested Kirkwall unsuccessfully against Fox at the election of 1784; but he secured the seat for Lostwithiel in Cornwall.
>
> In 1785 Sinclair lost his first wife, and, abandoning public life for a time, started on a foreign tour, in the course of which he met Necker and Buffon. Next year he made a seven months' journey through the north of Europe. He visited the courts of most of the northern states, and had audiences with Gustavus III of Sweden, the Empress Catherine of Russia, Stanislaus, king of Poland, and the Emperor Joseph. Shortly after his return Sinclair married (6 March 1788) Diana, the daughter of Lord Macdonald, by whom he had a numerous family.

Quoting, now, from Plackett [2]:

> Sinclair's statistical work arose naturally from his environment, personal circumstances and abundant energy. . . . His lifelong enthusiasm for collecting "useful information" led him in particular to note the use of "Statistik" by the Göttingen school of political economists. Early in 1790 he began work on forming a complete account of the Scottish nation.

The Statistical Account of Scotland, known today as the Old Statistical Account, was published between 1791 and 1799 in 21 octavo volumes, each of 600–700 pages. Sinclair's procedure is explained in Volume Twentieth.

> . . . by statistical is meant in Germany, an inquiry for the purpose of ascertaining the political strength of a country, or questions respecting *matters of state*; whereas, the idea I annex to the term, is an inquiry into the state of a country, *for the purpose of ascertaining the quantum of happiness enjoyed by its inhabitants, and the means of its future improvement.*

His purpose is thus related to the utilitarian ideas of the time, and expresses the view that statistics is an instrument of social progress. The inquiry consisted of 166 questions: 40 respecting geography and natural history, 60 respecting population, 16 respecting productions, together with 44 miscellaneous and 6 additional questions. Answers to the schedule generally took the form of comments on a variety of topics chosen individually. For example, the account of East Kilbride by the Rev. David Ure A. M. (Volume third, 1792) has sections headed: name, situation, and extent; heritors and rent; population; agriculture and roads; trees; commerce; mechanics and manufactures; minerals; wages, provisions, and education; general character; eminent men; diseases; church; poor; miscellaneous observations. The section on *Disease* is a mixture of old attitudes and new ideas. . . .

. . . Replies were published when sufficient material had accrued, and so the work is a sequence of reports on individual parishes without any organization. The quality of the replies depended on the ability of the ministers and their degree of cooperation, both of which varied considerably. Non-respondents were visited by *Statistical Missionaries,* and over eight years some twenty letters were addressed to them. . . .

An Analysis of the Statistical Account of Scotland appeared thirty years later, and the general layout is dependent on the Germany tradition. Sinclair declares his intention to supply a "condensed view . . . containing the substance of the *information he had procured, respecting the state of the several parishes of Scotland.*" He asserts that "the foundation of all human knowledge . . . must be laid in the examination of *particular facts*". The view that statistics equals facts was expressed by Achenwall* in the middle of the eighteenth century, and reappeared yet again some eight years after the *Analysis* was published, among the reasons given for establishing the Statistical Society of London, namely to procure, arrange and publish "Facts calculated to illustrate the Condition and Prospect of Society." Sinclair's *Analysis* is a factual description of Scottish society at the end of the eighteenth century, with conclusions of practical value, topics which merit further examination, and judgments which now seen bizarre. For example, the following observation anticipates the idea of regression to the mean (Vol. 1, p. 99).

It is a singular circumstance, that Sir John Gordon of Park, about a century ago, introduced a breed of tall men into his estate in the parish of Ordiquhill collected from different parts of Scotland; but their descendants of the third generation have generally come down to the usual size of the other inhabitants of the country."

Pritchard [3] remarks that "Sir John Sinclair was one of the first data collectors and it is he who introduced the words statistics and statistical, as we now understand them, into the English language." He adds that

> Sinclair explained his adoption of the words "statistical" and "statistics" thus: Many people were at first surprised at my using the words "statistics" and "statistical", as it was supposed that some term in our own language might have expressed the same meaning. But in the course of a very extensive tour through the northern parts of Europe, which I happened to take in 1786, I found that in Germany they were engaged in a species of political enquiry to which they had given the name "statistics," and though I apply a different meaning to that word—for by "statistical" is meant in Germany an inquiry for the purpose of ascertaining the political strength of a country or questions respecting matters of state—whereas the idea I annex to the term is an inquiry into the state of a country, for the purpose of ascertaining the quantum of happiness enjoyed by its inhabitants, and the means of its fiture improvement; but as I thought that a new word might attract more public attention, I resolved on adopting it, and I hope it is now completely naturalised and incorporated with our language.

Plackett [2] says, "There had been earlier uses noted for example by Hilts (1978) but the books he cites have few readers today." However,

> When the Statistical Society of London was founded in 1834, Sinclair—at the age of 80—was the oldest original member. In the same year, he presented a paper on agriculture at the Edinburgh meeting of the British Association for the Advancement of Science, but this was found to lack "facts which can be stated numerically." Statistics had begun to move away from statistical accounts.

This note has been concerned primarily with the statistical work of John Sinclair—just one part of a life with many facets. Many details of his political and public service contributions are described in ref. [4].

REFERENCES

[1] Hilts, V. (1978). *Aliis Exterendum,* or the origins of the Statistical Society of London, *Isis,* **69,** 21–43.

[2] Plackett, R. L. (1986). The Old Statistical Account, *J. R. Statist. Soc. A,* **149,** 247–251.

[3] Pritchard, C. (1992). The contributions of four Scots to the early development of statistics, *Math. Gazette,* **76,** 61–68.

[4] (1917). [Entry on] Sir John Sinclair. In *The Dictionary of National Biography,* vol. 18, Oxford University Press, Oxford, pp. 301–305.

Süssmilch, Johann Peter

Born: September 3, 1707, in Berlin, Germany.
Died: March 22, 1767, in Berlin, Germany.
Contributed to: population statistics.

Born into a wealthy family, Süssmilch received a classical education and, after giving up legal studies, took to theology at Halle (1727). In 1728 he moved to Jena; there, he continued his previous studies, gained knowledge in philosophy, mathematics, and physics, and even tutored mathematics. In 1732, he defended his "Dissertatio physica de cohaesione et attractione corporum" and returned to Berlin, becoming tutor to the eldest son of Field Marshal von Kalkstein. In 1736, against his best wishes, he was ordained chaplain. As such he participated in the First Silesian War (of Prussia against Austria). In 1742 he became minister and, in 1750, a member of the Oberconsistorium, the directorate of church affairs. He married in 1737 and fathered ten children. One of his sons was Oberbürgermeister of Breslau, and Baumann, mentioned in ref. [16], was his son-in-law. His health deteriorated in 1763 when he suffered from a stroke. A second one followed in 1766, and he died soon afterwards.

Süssmilch naturally believed that multiplication of mankind was a divine commandment (Gen. i.28), and that rulers, therefore, must foster marriages and take care of their subjects. He condemned wars and excessive luxury and declared that the welfare of the poor is to the advantage of the state and in the self-interest of the rich. His pertinent appeals brought him into continual strife with municipal (Berlin) authorities and ministers of the state (Prussia).

Süssmilch published many contributions on population statistics, political arithmetic in general, and linguistics. In 1745, chiefly for his researches in the last-mentioned discipline, he was elected to the Berlin Academy of Sciences (Class of Philology). His main work, which largely embodied most of his writings in population statistics, was the "Göttliche Ordnung" [16]. Euler (whom Süssmilch repeatedly mentioned) actively participated in preparing its second edition, was coauthor of at least one of the chapters of this ("Von der Geschwindigkeit der Vermehrung und der Zeit der Verdoppelung [of the population]", partly reprinted in ref. [8, pp. 507–532]), and elaborated it in his own memoir [7]. One of the conclusions made by Süssmilch and Euler, viz., that the population increases, roughly, in a geometric proportion, was picked up by Malthus.

Süssmilch's driving idea, which is revealed in the title of his main contribution [16], was to prove divine providence as manifested in the laws of vital statistics. He thus followed a well-established tradition and appropriately referred to Graunt* and to Derham, an author of two influential books [3, 4]. These statistical "laws," though, were nothing but the stability of certain ratios (marriages to population; births to marriages; or deaths to births). They look thin compared with the laws which Kepler and Newton* considered of divine origin. The theological approach, which soon went out of fashion, was, if anything, a disadvantage. (Even the pious Empress Maria Theresia banned the "Göttliche Ordnung" in Austria and Hungary—for being too Protestant.)

The influence of Süssmilch is mainly due to the mass of data collected rather than to their professional treatment (combining data from towns and villages, say, without taking weighted means, etc.) See Westergaard* [20] for various critical remarks. Even so, the book marks the beginning of demography. It paved the way for Quetelet,* and its life table was in use well into the nineteenth century.

Without information on the age structure of the population, a comprehensive life table had to be based on statistics of births, and of deaths, classified according to age and sex. Such computations were first carried out by Halley [10], and were familiar to contemporaries of Süssmilch, like Kersseboom and Struyck. Although Süssmilch had studied some mathematics, he made no contribution to a more sophisticated computation of life tables.

REFERENCES

[1] Birg, H., ed. (1986). *Ursprünge der Demographic in Deutschland. Leben und Werk J. P. Süssmilch's.* Campus, Frankfurt/Main–New York.

[2] Crum, F. S. (1901). The statistical work of Süssmilch. *Publ. Amer. Statist. Assoc. N. S.* **7,** 335–380.

[3] Derham, W. (1713). Physico-theology, or a demonstration of the being and attributes of God from his work of creation. Innys, London. (About 15 editions appeared until 1798.)

[4] Derham, W. (1714). Astrotheology, or a demonstration of the being and attributes of God from a survey of the heavens. Innys, London. (About 12 editions appeared until 1777.)

[5] Döring, H. (1835). Süssmilch. In *Die gelehrten Theologen Deutschlands im 18. und 19. Jahrhundert.* Saur, Neustadt an der Orla vol. 4, pp. 451–456. Reprinted in Deutsches biogr. Archiv, a microfilm edition, B. Fabian, ed., München, ca 1990.

[6] Esenwein-Rothe, I. (1967). J. P. Süssmilch als Statistiker. In *Die Statistik in der Wirtschaftsforschung. Festgabe für Rolf Wagenführ,* H. Strecker and W. R. Bihn, eds. Duncker and Humblot, Berlin, pp. 177–201.

[7] Euler, L. (1760). Recherches générales sur la mortalité et la multiplication du genre humain. Reprinted in [8, pp. 79–100] . English transl. (1977) in *Mathematical Demography,* D. Smith, and N. Keyfitz, eds. Springer, Berlin, pp. 83–92.

[8] Euler, L. (1923). *Opera Omnia, Ser. 1*, vol. 7, L. -G. Du Pasquier, ed. Teubner, Leipzig–Berlin.

[9] Förster, J. C. (1768). *Nachricht von dem Leben und Verdiensten Süssmilchs.* Reprinted: Cromm, Göttingen, (1988).

[10] Halley, E. (1694). An estimate of the degrees of mortality of mankind drawn from curious tables of the births and funerals of the city of Breslaw. *Phil. Trans. R. Soc. London.* **17,** 596–610. Reprinted, Johns Hopkins, Baltimore (1942).

[11] Hecht, J. (1987). J. P. Süssmilch: A German prophet in foreign countries. *Population Stud.* **41,** 31–58.

[12] John, V. (1884). *Geschichte der Statistik.* Encke, Stuttgart.

[13] John, V. (1894). Süssmilch. *Allg. Deutsche Biogr.* **37,** 188–195.

[14] Paevsky, V. V. (1935). Les travaux démographiques de L. Euler. In *Recueil des articles et matériaux en commémoration du 150–anniversaire du jour de sa mort.* Akad. Nauk SSSR, Moscow–Leningrad, pp. 103–110. (In Russian; title of volume and contributions also in French.)

[15] Pearson, K. (1978). *The History of Statistics in the 17th and 18th Centuries,* Lectures read in 1921–1933, E. S. Pearson, ed. Griffin, London–High Wycombe.

[16] Süssmilch, J. P. (1741). Die Göttliche Ordnung in den Veränderungen des menschlichen Geschlechts, aus der Geburt, dem Tode und der Fortpflanzung desselben. Buchladen d. Realschule, Berlin. (Later editions: 1765 and 1775 (almost identical). Posthumous edition by C. J. Baumann: 1775–1776, in 3 vols. The third volume contains "Ammerkungen und Zusätze" to the first two of them, largely written by Baumann. He reprinted this edition twice or thrice (1787–1788; 1790–1792 (?) and 1798 (?)). Reprint of the first edition: Berlin, 1977; reprint of the third (1765) edition together with Baumann's vol. 3 (1776): Göttingen–Augsburg, 1988.]

[17] Süssmilch, J. P. (1758). Gedancken von den epidemischen Kranckheiten und dem grösseren Sterben des 1757ten Jahres. (Reprinted in [19, pp. 69–116].)

[18] Süssmilch, J. P. (1979–1984). "L'ordre divin" aux origines de la démographie, J. Hecht, ed. Institut Nationale d'Études Démographiques, Paris. [Volume 1 (1979)) includes French translations of several contributions on Süssmilch; his biography; a bibliography of his writings; and an extensive bibliography of works devoted to him. Volume 2 (1979), with consecutive paging, is a French translation of selected sections from the "Göttliche Ordnung" (1765), and vol. 3 (1984) contains indices.]

[19] Süssmilch, J. P. (1994). *Die königliche Residenz Berlin und die Mark Brandenburg im 18. Jahrhundert. Schriften und Briefe,* J. Wilke, ed. Akademie Verlag, Berlin. (Includes a reprint of Süssmilch's "Gedancken" [17], publications of a few of his manuscripts and of some of his correspondence, his biography based on archival sources, and a list of selected references.)

[20] Westergaard, H. (1932). *Contributions to the History of Statistics.* King, London. (Reprinted: New York, 1968, and The Hague, 1969. The chapter devoted to Süssmilch is reprinted in *Studies in the History of Statistics and Probability,* Sir Maurice Kendall & R. L. Plackett, eds. Griffin, London–High Wycombe, 1977, vol. 2, pp. 150–158.

J. PFANZAGL
O. SHEYNIN

Wargentin, Pehr Wilhelm

Born: September 11, 1717 [Julian calendar], in Sunne (Jämtland), Sweden.
Died: December 13, 1783, in Stockholm, Sweden.
Contributed to: astronomy, demography, population statistics, promotion of science.

Pehr Wilhelm Wargentin

Pehr Wargentin was born in 1717 in the beautiful countryside near Lake Storsjön in the country of Jämtland in northern Sweden. His father, Wilhelm Wargentin, was of German descent and had studied science and theology in Åbo, Uppsala, and Lund. Wilhelm Wargentin was appointed vicar in Sunne on the condition that he married his deceased predecessor's widow, who after a year gave birth to Pehr Wargentin. Wargentin was taught first by his father, then at a well-regarded primary school near his home (*Frösö Trivialskola*) and at a secondary school in Härnösand.

At the age of eleven Wargentin saw a lunar eclipse, which fascinated him. After finishing school in 1735 he went to Uppsala University, where he studied astronomy and other sciences. One of his teachers was Anders Celsius, known for his geodesic measurements and for the centigrade thermometer scale. As suggested by Celsius, Wargentin wrote a master thesis (*magisterarbete*) in 1741 on the motion of the moons of Jupiter, which was a topic of great international scientific interest at that time, useful for the determination of longitudes of geographical places. In his master thesis Wargentin succeeded in constructing a table of unprecedented accuracy. Later he further improved on these contributions, and he continued his work in astronomy to the end of his life.

In 1749 Wargentin was appointed Secretary of the Royal Swedish Academy of Sciences in Stockholm, which had been set up in 1739, with the Royal Society in London as a model, when Linnaeus returned from England. Wargentin retained this position until his death more than three decades later, and during this period he made a great and successful effort to develop the young Academy into a powerful platform for promotion of the new sciences and their utilization in the country. He had an important role in several different projects such as the erection of a new observatory in Stockholm (where he became superintendent), the establishment of regular meteorological observations in Stockholm, the editing of an annual almanac, the late-coming transition to the Gregorian calendar in the country, a canal construction project, and population statistics.

Wargentin's contributions to statistics dealt with the development of a national system for population statistics. At the death of King Charles XII in 1718, the Parliament (consisting of the Four Estates) seized power from the Monarch, while the country suffered the consequences from a long, devastating, and finally lost war against its neighbors, notably Russia. There was concern about the sparsity of the

Swedish population, and it was thought that a greatly increased population and a better distribution of people over regions and occupations would bring wealth and power to the country. Partly inspired by the ideas of political arithmetic in England and other countries, there was a growing urge for data on the population, today known as population statistics.

By the Church Act of 1686, the Swedish State Church had already conducted a nationwide continuous population registration, which turned out to be an excellent basis for population statistics. The Parliament initiated a Royal Decree of 1748 on the annual compilation of statistical data, which became effective the following year. The compiled data comprised a tabular summary of the following items:

1. The number of infants baptized per month, with distribution by sex and by marital status of the mother.
2. The number of persons buried per month, with distribution by sex and by subdivision into three categories, viz., children under 10 years old, adolescents and other single persons, and married persons.
3. The number of marriages per month, and that of marriages dissolved by death.
4. The number of deaths per month, with distribution by sex and by age in five-year groups, children under one year of age stated separately. Furthermore, causes of death were given in 33 categories.
5. The size of the population at the end of the year, with distribution by sex and by age in five-year groups, children under one year of age stated separately.
6. The size of the population at the end of the year, with distribution by sex and by estate (*stånd*) or occupation. The latter were given in 61 categories.

A unique feature of this tabular system, known as *Tabellverket,* was that it contained both vital statistics and census data, i.e., data on both the changes and the size of the population, thus providing the basis for a complete demographic description of the population. The data on the size of the population were kept secret during the first years, since it was feared that enemies of the country might take advantage of knowing how small its population was (little more than two million in Sweden, including Finland).

Forms for the tables were sent to the clergy all over the country, to be filled in by them for each parish. From these tables, summary tables were compiled for each deanery and then for each diocese. The latter summaries were sent to the County Governors, who forwarded their county summaries to the Royal Chancellery in Stockholm. Though the data collection was annual from the start, in view of the burden on the clergy the census data (items 5 and 6) were from 1752 collected only every three years, and later only every five years.

From 1753 Wargentin played a key role in organizing and developing the system. The plan for it had been worked out by Pehr Elvius, Wargentin's predecessor as Secretary of the Academy. Wargentin became the key figure in the Royal Commission on Tabulation, which was set up for the work in 1756, and which was the predecessor of today's Statistics Sweden. He wrote long reports to the Parliament in 1755–1765, and several essays (in Swedish) on demography in the Proceedings of the Academy in 1754–1782.

Especially notable are a series of six essays from 1754–1755 entitled "The importance of annual summaries on births and deaths," and one essay from 1766 entitled "Mortality in Sweden." The 1754–1755 essays, still easy to read and free from theological speculations, treat many different topics, such as the importance of an increase in the population, the regularity of different phenomena in the demographic field, life insurance and mortality, and discussion of Halley's method to calculate the size of a population by using only data about deaths and births. Other topics were causes of death and steps to reduce mortality. The population pyramid was described. The papers take up ideas of well-known writers on political arithmetic and related fields, such as Graunt*, Petty*, Halley, Süssmilch*, Déparcieux,* and Kersseboom, who are extensively cited.

Wargentin's 1766 paper "Mortality in Sweden" is his most important one. He made use of data on deaths for the years 1755–1763, which he related to the size of the population data in 1757, 1760, and 1763. His tables show the relative number of deaths for each age group and sex group. This was the first time an accurate national mortality table was given. He demonstrated how the mortality numbers could be used for several interesting mortality comparisons between regions, such as Stockholm vs. the countryside. He also gave long discussions on the reliability of the data. By the efforts of an English actuary, Richard Price, Wargentin's tables became transformed to life tables and known abroad. Furthermore Süssmilch, who laid the foundation for demographic methodology in his work on the "Divine Order" (*Göttliche Ordnung*), made essential use of Wargentin's mortality tables in his developments in later editions of that work.

Later papers by Wargentin deal with seasonal variations in births, showing that conceptions peaked in December, and with the extent of the emigration, which was shown to be substantially smaller than had been previously thought. Like contemporary writers in other countries, Wargentin had advocated a prohibition of emigration, but his new findings made him change his opinion.

The political interest in population statistics faded with time, especially after King Gustav III restored much of the Monarch's former power in 1772, although the latter action actually had the sympathy of Wargentin. After Wargentin's death and toward the end of the century the tabular system decayed somewhat. But it survived, so that there are complete time series of Swedish population statistics from 1749 until today. An edited presentation of historical statistics, with some reconstruction to fill some gaps in the set of variables, was later given by Gustav Sundbärg in the journal *Statistisk Tidskrift* around 1900; cf. ref. [4].

Wargentin was a highly respected man of an honest and faithful character, with modest and cheerful manners, and with an extraordinary capacity for work. He took an important role in many pioneering endeavors of his time. He had a most intense correspondence on scientific matters with scholars in may countries, and although his published statistical papers are in Swedish, some of them were translated into German and French and became internationally known. During the last months of his life he suffered from illness but still continued his hard work and daily observations and planned new projects. In 1783 he died quietly in his home in the observatory, at the age of 66.

On June 1, 1756 Pehr Wargentin married Christina Magdalina Raab (1724–1769). They had three daughters.

Wargentin's patron was Andres Celsius (1702–1744), who unfortunately died at a young age, and then, later on, the famous botanist Carl von Linné (1707–1778). These three were, at one time, the only Swedish members of the [scientific] French Academy. Wargentin was also a member of the Gottingen and St. Petersburg Academies and of the Royal Society of London. He was instrumental in introducing the Gregorian calendar into Sweden, in 1753.

W. de Sutter used Wargentin's astronomical tables, many years after their publication, and found them to be very accurate when establishing oscillations in the Earth's rotation. Wargentin carried out an extensive correspondence with Richard Price—well known in statistical circles for his work in connection with Thomas Bayes'* paper in the *Philosophical Transactions of the Royal Society*. [7]

REFERENCES

[1] Hofsten, E. and Lundström, H. (1976). *Swedish Population History: Main Trends from 1750 to 1790.* Urval No. 8, SCB [Statistics Sweden], Stockholm.

[2] Nordenmark, N. V. E. (1939). *Pehr Wilhelm Wargentin, Kungl. Vetenskapsakademins sekreterare och astronom.* Almqvist & Wiksell, Uppsala (In Swedish, with a summary in French.)

[3] Pearson, K. (1978). *The History of Statistics in the 17th and 18th Centuries.* Charles Griffin, London.

[4] Sundbärg, G., (1907). *Bevölkerungsstatistik Schwedens 1750–1900.* SCB [Statistics Sweden], Stockholm. (In German.) Reprinted (with preface and vocabulary in Engligh), Urval No. 3, SCB [Statistics Sweden], Stockholm (1970).

[5] Statistiska centralbyrån (1983). *Pehr Wargentin, den Svenska Statistikens Fader.* SCB [Statistics Sweden], Stockholm. (In Swedish, with a summary in English. Contains reprints of seven papers by Wargentin.)

[6] Westergaard, H., (1932). *Contributions to the History of Statistics.* King & Son, London. Reprinted, New York 1970.

[7] Grünlund, O. (1946). *Pehr Wargentin och Befolkningsstatistiker,* Stockholm, Sweden: P. A. Norstedt and Sons.

LENNART BONDESSON
MARTIN G. RIBE

SECTION **2**

Statistical Inference

Birnbaum, Allan

Born: May 27, 1923, in San Francisco, California, USA.
Died: July 1, 1976, in London, England.
Contributed to: statistical inference, foundations of statistics, statistical genetics, statistical psychology, history of science.

Allan Birnbaum was one of the most profound thinkers in the field of foundations of statistics. In spite of his short life span, he made a significant impact on the development of statistical ideas in the twentieth century. His tragic, and apparently self-inflicted, death in 1976 was a severe blow to the progress of research into foundations for statistical science and the philosophy of science in general.

Born in San Francisco in 1923 of Russian-Jewish orthodox parents, he studied mathematics as an undergraduate in the University of California at Berkeley and at Los Angeles, completing simultaneously a premedical program in 1942, and a first degree in 1945. He remained at Berkeley for the next two years, taking graduate courses in science, philosophy, and mathematics. He was influenced by the work of Ernest Nagel and Hans Reichenbach. Following the latter's advice, he moved to Columbia University, New York, in 1947. Here he obtained a Ph. D. degree in mathematical statistics in 1954 (partly under the guidance of E. L. Lehmann, who was spending a semester at Columbia University in 1952). He had already been a faculty member for three years and continued at Columbia until 1959. During this time he was active in research on classification and discrimination, and development of mathematical models and statistical techniques for the social sciences. In the latter, he interacted with such luminaries as Duncan, Luce, Nagel, and Lazarsfeld. In 1959, Birnbaum moved to the Courant Institute of Mathematical Sciences, where he remained (officially) until 1974. During this period he spent considerable periods as a visiting professor in several places—notably Stanford University, Imperial College (London), for most of 1971–1974 as a visiting Fellow at Peterhouse, Cambridge, and also (in 1974) as a visiting professor at Cornell University; and he was also associated with the Sloan-Kettering Cancer Research Center. He finally left the Courant Institute in 1974, reputedly, in part, due to an apparent lack of appreciation for statistics by mathematicians at the Institute. Birnbaum was appointed Professor of Statistics at City University, London, in 1975, less than a year before his death in 1976.

Birnbaum combined a forceful personality and great intellectual honesty and power with a quiet and unassuming nature, and was a sensitive and kind person. He had a great variety of interests, and his subtle sense of humor made him especially effective on informal occasions and in private conversations. The author of this note remembers well how Birnbaum arranged for him to visit the Courant Institute in 1965 to learn about the status of statistics in the Soviet Union at that time. Allan was very much attached to his only son, Michael, born late in his life.

He was elected a Fellow of the Institute of Mathematical Statistics, the American Statistical Association, and the American Association for the Advancement of Science. His scientific output includes 41 papers on statistical theory and applications listed in a memorial article by G. A. Barnard and V. P. Godambe in the *Annals of Statistics* (1982), **10,** 1033–1039. His early work was in the spirit of E. L. Lehmann's famous textbook *Testing of Statistical Hypotheses,* and is, indeed, referred to in that book. His most notable contribution is generally considered to be a paper originally published in the *Journal of the American Statistical Association* (1962), **57,** 269–306 (discussion, pp. 307–326), which was reprinted in vol. I of *Breakthroughs in Statistics,* Samuel Kotz and Norman L. Johnson, eds., Springer-Verlag, New York (1992), pp. 478–518, with a perceptive introduction by J. F. Bjornstad (pp. 461–477). This paper studies the likelihood principle (LP) and its application in measuring the evidence about unknown parameter values contained in data. Birnbaum showed that the LP is a consequence of principles of sufficiency and conditionality and claimed that this implies, inter alia, the irrelevance, at the inference stage, of stopping rules. (Birnbaum interpreted the LP as asserting that if two experiments produce proportional likelihood functions, the same inference should be made in each case.) Birnbaum's result is regarded by many as one of the deeper theorems of theoretical statistics (with a remarkably simple proof). The result has given rise to continuing controversy, and claims have been made (notably by the prominent statistician S. W. Joshi) that it may be false. L. J. Savage* remarked, on the other hand, that appearance of the paper was an event "really momentous in the history of statistics. It would be hard to point to even a handful of comparable events" (obituary notice in *The Times* of London). Indeed, the paper has had substantial effects on estimation in the presence of nuisance parameters, on prediction in missing-data problems, and in metaanalysis (combination of observations).

Another groundbreaking initiative by Birnbaum, in the field of foundations of statistics, in his paper on Concepts of Statistical Evidence, published in *Philosophy, Science and Method: Essays in Honor of Ernest Nagel* (St. Martin's, New York). His ideas on statistical scholarship were expressed in an article published in 1971 in the *American Statistician,* **25,** pp. 14–17. His final views on the Neyman–Pearson theory are discussed in a paper which appeared posthumously (with a discussion) in *Synthèse,* **36,** 19–49 (1978), having previously been rejected by the Royal Statistical Society.

Further important contributions include an article on conceptual issues in experimental genetics [*Genetics,* **72,** 734–758 (1972)] and historical articles on John Arbuthnot* [*American Statistician,* **21,** 23–25 and 27–29 (1961)] and on statistical biographies [*Journal of the Royal Statistical Society, Series A,* **133,** 1265–1282 (1970)].

BIBLIOGRAPHY

Barnard, G. A. and Godambe, V. P. (1982). Allan Birnbaum 1923–1976, *Ann. Statist.,* **10,** 1033–1039

Bjørnstad, J. F. (1992). Introduction to Birnbaum. In *Breakthroughs in Statistics, 1,* S. Kotz and N. L. Johnson, eds. Springer-Verlag, New York, pp. 461–467.

Lindley, D. V. (1978). Birnbaum, Allan. In *International Encyclopedia of Statistics,* W. H. Kruskal and J. Tanur, eds. Free Press, New York, vol. 1, pp. 22–23.

Norton, B. (1977). Obituaries: *J. R. Statist. Soc. A,* **140,** 564–565; *The Times.*

Cox, Gertrude Mary

Gertrude M. Cox

Born: January 13, 1900, in Dayton, Iowa, USA.
Died: October 17, 1978, in Duke University Hospital, Durham, North Carolina, USA.
Contributed to: psychological statistics; development of test scores; experimental designs (incomplete block designs); factor and discriminant analysis; statistics training programs; statistical computing; administration of statistical programs; international programs; statistical consulting.

Many of the remarks furnished here have been excerpted from a 1979 obituary prepared with Larry Nelson and Robert Monroe in *Biometrics,* **35:** 3–7 and a biography by me in *Biographical Memoirs,* **59,** 117–132 (National Academy Press, 1990).

I became acquainted with Gertrude Cox in 1936 when I started graduate work at Iowa State College, from which she had secured a Master's degree in statistics in 1931. After that she began work for a Ph.D. in psychological statistics at the University of California in Berkeley; she gave that up in 1933 to return to Iowa State to direct the Computing Laboratory of the newly created Statistical Laboratory under George Snedecor.* She became interested in the design of experiments, for which she developed and taught graduate courses. Her courses were built around a collection of real-life examples in a variety of experimental areas. She taught from mimeographed materials, which formed part of the famous Cochran–Cox *Experimental Designs* (Wiley, New York, 1950). She had three major principles in setting up an experiment:

1. The experimenter should clearly set forth his or her objectives before proceeding with the experiment.
2. The experiment should be described in detail.
3. An outline of the analysis should be drawn up before the experiment is started.

She emphasized the role of randomization and stressed the need to ascertain if the size of the experiment was sufficient to demonstrate treatment differences if they existed.

In 1940, Snedecor responded to a request for suggestions on possible candidates to head the new Department of Experimental Statistics in the School of Agriculture at North Carolina State College (in Raleigh); upon seeing his list of all males, Gertrude asked why he had not included her name. He then inserted a footnote which stated that if a woman could be considered, he recommended her. This footnote has become a statistical landmark, because Gertrude was selected. She started staffing her department with statisticians who had majors or strong minors in applied fields. In 1942, I became the mathematical statistician and the one who consulted with the economists. In 1944 the President of the Consolidated University of North Carolina established an all-University Institute of Statistics with Gertrude as head, and in 1945 she obtained funds from the General Education Board to establish graduate programs at N.C. State and, in a newly created Mathematical Statistics Department, at Chapel Hill.

In 1949, Gertrude gave up the Headship at N.C. State to devote full time to the Institute, including development of strong statistics programs throughout the South. This latter development was augmented by an arrangement with the Southern Regional Education Board to establish a Committee on Statistics. From 1954–1973, the Committee sponsored a continuing series of six-week summer sessions and is now cosponsoring (with the American Statistical Association [ASA]) a Summer Research Conference.

One of Gertrude Cox's major achievements was the development of strong statistical computing programs. N.C. State was a leader in the use of high-speed computers; it had one of the first IBM 650s on a college campus and developed the initial SAS programs. One of her strongest points was her ability to obtain outside financial support. She persuaded the Rockefeller Foundation to support a strong program in statistical genetics, and the Ford Foundation one in dynamic economics. In 1958, Iowa State University conferred upon her an honorary Doctorate of Science as a "stimulating leader in experimental statistics . . . outstanding teacher, researcher, leader and administrator. . . . Her influence is worldwide, contributing to the development of national and international organizations, publications and councils of her field."

Starting in 1958, Dr. Cox and other members of the N.C. State statistics faculty developed procedures to establish a Statistical Division in the not-for-profit Research Triangle Institute (RTI) in the Research Triangle Park (RTP) between Raleigh, Chapel Hill, and Durham; Gertrude retired from the University in 1960 to direct this division. She retired from RTI in 1965, but continued to teach at N.C. State and consult on research projects. RTP has developed into a world-recognized research park.

Gertrude Cox was a consultant before and after retirement to many organizations, including the World Health Organization, the U.S. Public Health Service, and the government of Thailand, and on a number of U.S. Government committees for the Bureau of the Budget, National Institute of Health, National Science Foundation, Census Bureau, and Agricultural Department. She was a founding member of the International Biometric Society in 1947, which she served as president in 1968–1969, on its council three times, and as the first editor of its journal, *Biometrics.* She was an active member of the International Statistical Institute and was President of the ASA in 1956.

In 1970, North Carolina State University designated the building in which statistics was housed as Cox Hall, and in 1977 a Gertrude M. Cox Fellowship Fund was established for outstanding graduate students in statistics. Her election to the National Academy of Sciences in 1975 was a treasured recognition of her many contributions.

In this review I have included only one reference to her substantial published work. A complete bibliography is included in the *Biographical Memoirs.*

I will conclude with the closing remarks in those memoirs. This excerpt is reproduced here with permission of the National Academy of Sciences.

> Gertrude Cox loved people, expecially children. She always brought back gifts from her travels and was especially generous at Christmas time. She considered the faculty members and their families to be her family and entertained them frequently. She was an excellent cook and had two hobbies that she indulged during her travels: collecting dolls and silver spoons. She learned chip carving and block printing at an early age and spent many hours training others in these arts. She loved gardening, and, when she had had a particularly hard day with administrators, would work off her exasperation in the garden. She had a fine appreciation for balance, design and symmetry.
>
> In 1976, Gertrude learned that she had leukemia but remained sure that she would conquer it up to the end. She even continued construction of a new house, unfortunately, not completed until a week after her death. While under treatment at Duke University Hospital she kept detailed records of her progress, and her doctor often referred to them. With characteristic testy humor she called herself "the experimental unit," and died as she had lived, fighting to the end. To those of us who were fortunate to be with her through so many years, Raleigh will never be the same.

RICHARD L. ANDERSON

Cramér, Harald

Harald Cramér.

Born: September 25, 1893, in Stockholm, Sweden.
Died: October 5, 1985, in Stockholm, Sweden.
Contributed to: actuarial mathematics, analytic number theory, central-limit theory, characteristic functions, collective-risk theory, mathematical statistics, stationary processes.

(Carl) Harald Cramér spent almost the whole of his professional life in Stockholm. He entered Stockholm University in 1912, studying chemistry and mathematics. Although he worked as a research assistant in biochemistry, his primary interest turned to mathematics, in which he obtained a Ph.D. degree in 1917, with a thesis on Dirichlet series.

He became an assistant professor in Stockholm University in 1919, and in the next seven years he published some 20 papers on analytic number theory. During this time, also, Cramér took up a position as actuary with the Svenska Life Insurance Company. This work led to a growing interest in probability and statistics, as a consequence of which Cramér produced work of great importance in statistical theory and methods over the next 60 years.

In 1929, he was appointed to a newly created professorship in "actuarial mathematics and mathematical statistics" (sponsored by Swedish Life Insurance Companies). At this time, also, he was appointed actuary to the Sverige Reinsurance Company. His work there included new developments in premium loadings for life insurance policies and ultimately, after many years, to his book, *Collective Risk Theory* [5].

The 20 years following 1929 were his most intensely productive period of research. *Random Variables and Probability Distributions* [3], published in 1937, provided a fresh, clearly expressed foundation for basic probability theory as used in the development of statistical methods. The seminal book *Mathematical Methods of Statistics* [4], written during enforced restrictions of international contacts during World War II, presented a consolidation of his studies, and has been of lasting influence in the development of statistical theory and practice. During these 20 years, also, Cramér built up a flourishing institute, providing conditions wherein workers in many fields of statistics could find encouragement to develop their ideas. To this period, also, belongs the "Cramér–Wold device" [7] for establishment of asymptotic multidimensional normality.

In 1950, Cramér was appointed President of Stockholm University, and in the period until his retirement from this office in 1961 a substantial proportion of his time was occupied with administrative duties, with consequent diminution in research activities.

However, after 1961, he returned to a variety of research endeavors including participation in work for the National Aeronautics and Space Administration at the

Research Triangle Institute in North Carolina, during the summers of 1962, 1963, and 1965. During this time also, in collaboration with Leadbetter, he produced the book *Stationary and Related Stochastic Processes* [6]. Collective risk theory (see ref. [5], referred to earlier) is concerned with the progress through time of funds subject to inputs (premiums and interest) and outputs (claims), constituting a special type of stochastic process, and Cramér's attention to this field may be regarded as a natural long-term development.

For fuller accounts of Cramér's life and work, see the obituaries by Blom [1] and Leadbetter [8]. Blom and Matérn [2] provide a bibliography of Cramér's publications.

REFERENCES

[1] Blom, G. (1987). *Ann. Statist.,* **15,** 1335–1350.

[2] Blom, G. and Matérn, B. (1984). *Scand. Actu. J.,* 1–10.

[3] Cramér, H. (1937). *Random Variables and Probability Distributions.* Cambridge Tracts, **36.** Cambridge University Press, London.

[4] Cramér, H. (1945). *Mathematical Methods of Statistics.* Almqvist and Wiksell, Uppsala, Sweden; Princeton University Press, Princeton, N.J.

[5] Cramér, H. (1955). *Collective Risk Theory.* Skandia Insurance Company, Stockholm, Sweden.

[6] Cramér, H. and Leadbetter, M. R. (1967). *Stationary and Related Stochastic Processes.* Wiley, New York.

[7] Cramér, H. and Wold, H. O. A. (1936). *J. Lond. Math. Soc.,* **11,** 290–294.

[8] Leadbetter, M. R. (1988). *Int. Statist. Rev.,* **56,** 89–97.

David, Florence Nightingale

Born: August 23, 1909, in Ivington, England.
Died: July 18, 1993, in Kensington, California, USA.
Contributed to: distribution theory, combinatorics, statistics and genetics, robustness, history of statistics.

F. N. David

No account of the history of statistics in the early twentieth century would be complete without mention of Florence Nightingale David. In a career spanning six decades and two continents, she authored or edited 9 books, 2 monographs, and over 100 journal articles. (Her books are listed at the end of this article.) Her coauthors included Karl Pearson,* Jerzy Neyman,* Maurice Kendall,* Norman Johnson, Evelyn Fix, David Barton, and Colin Mallows.

Her book *Probability Theory for Statistical Methods* is a blend of the theory of probability and statistics as of 1949, with practical applications such as "Elimination of races by selective breeding (p. 92)." (The reader will be pleased to know that she was referring to cattle breeding.) Her book *Games, Gods and Gambling* provides a delightful account of the early history of probability. *Combinatorial Chance,* coauthored with David Barton, is an excellent reference for the full spectrum of results in combinatorics as of 1960.

F. N. David graduated from Bedford College for Women in 1931. Shortly afterwards she was hired as research assistant to Karl Pearson at University College, London (UCL). She remained there after Pearson retired, becoming Lecturer in 1935 and receiving her Ph.D. in 1938. Much of her time as Pearson's assistant was spent on calculations for her first book, *Tables of the Correlation Coefficient,* published in 1938.

Dr. David also published six research papers during her early years at UCL, four in *Biometrika* and two in its short-lived companion (started by Karl Pearson*'s son Egon* and Jerzy Neyman*), *Statistical Research Memoirs*. Among those six papers were her only joint publications with two of her most influential mentors, Karl Pearson and Jerzy Neyman, although she later edited a *Festschrift for J. Neyman,* published in 1966 by John Wiley and Sons.

While Dr. David's work in the 1930s was of a theoretical nature, she turned her attention to more practical problems in the service of her country in World War II. She worked as a Senior Statistician, first for the Ordnance Board, Ministry of Supply, then with Sir Austin Bradford Hill* at the Ministry of Home Security.

Her work during the war focused on problems related to bombs. She published three technical reports for the Civil Defence Research Committee, with F. Garwood, all under the title "Applications of mathematical statistics to the assessment of the efficacy of bombing."

After the war Dr. David returned to University College. She continued to consult with both the British and American governments, and authored 37 papers that are still classified.

During the 1950s and 1960s Dr. David was extremely productive, publishing a wide variety of papers ranging from "A note on the evaluation of the multivariate normal integral" (*Biometrika,* 1953, **40,** 458–459) to "The analysis of abnormal chromosome patterns" (with D. E. Barton, *Annals of Human Genetics,* 1962). Her favorite outlet was *Biometrika,* in which almost half of her papers appear.

In 1962 Dr. David was named Professor at University College, only the second woman to be so named. In 1967 she moved to the University of California (UC) at Riverside to establish a department of statistics. Following the groundwork laid by Professor Morris Garber, F. N. David built the department from two professors and two lecturers to a full-fledged Department of Statistics, with ten professors and five part-time lecturers.

Dr. David received numerous awards during her career. She was a Fellow of the Institute of Mathematical Statistics and the American Statistical Association. Although she claimed to dislike teaching, her classes were very popular and in 1971 she was awarded the UC Riverside Academic Senate Distinguished Teaching Award. In 1992 she was named the first recipient of the Elizabeth L. Scott Award, an award given by the Committee of Presidents of Statistical Societies to someone who has helped foster opportunities for women in statistics.

While at UC Riverside Dr. David supervised six Ph.D. students, and worked behind the scenes for the advancement of women. In an interview with Nan Laird [1] in 1988, she said,

> Female assistant professors would come to me and weep because I was on the President's University Committee for Affirmative Action. I used to be very nasty to the complaining person . . . and then I would ring up the chap who was doing it and say, "do you want me to attack you on the Senate floor?" (p. 243)

In 1977 Dr. David retired and moved to Berkeley, being named both Professor Emeritus at UC Riverside and Research Associate at UC Berkeley. She continued her long-time collaboration with the United States Forestry Service at Berkeley and was active in both teaching and research after her retirement.

For more information about the life of F. N. David, read the delightful interview with Nan Laird [1] in *Statistical Science* or the obituary by J. Utts [2] in *Biometrics.* For a taste of the breadth and depth of her contributions, pick up any volume of *Biometrika* from 1950 to 1964, and scattered volumes from 1932 to 1972 and you will find an interesting paper by Florence N. David.

REFERENCES

[1] Laird, N. M. (1989). A conversation with F. N. David. *Statist. Sci.* **4,** 235–246.

[2] Utts, J. (1993). Florence Nightingale David 1909–1993: Obituary. *Biometrics,* **49,** 1289–1291.

BIBLIOGRAPHY

David, F. N. (1938). *Tables of the Correlation Coefficient.* Biometrika Trust.

David, F. N. (1949). *Probability Theory for Statistical Methods.*Cambridge University Press. (2nd ed., 1951.)

David, F. N. (1953). *Elementary Statistical Exercises.* University College London. (Printed in revised form by Cambridge University Press, 1960.)

David, F. N. (1953). *A Statistical Primer.* Charles Griffin, London.

David, F. N. (1962) and Barton, D. E. *Combinatorial Chance.* Charles Griffin, London.

David, F. N. (1962). *Games, Gods and Gambling.* Charles Griffin, London.

Barton, D. E., and Kendall, M. G. (1966). *Symmetric Functions and Allied Tables.* Cambridge University Press, New York.

JESSICA UTTS

de Finetti, Bruno

Born: June 13, 1906, in Innsbruck, Austria.
Died: July 20, 1985, in Rome, Italy.
Contributed to: Bayesian inference, pedagogy, probability theory, social justice, economics.

Bruno de Finetti.

Bruno de Finetti was born in Innsbruck of Italian parents in 1906, graduated in mathematics from the University of Milan in 1927, and then, having attracted the attention of Corrado Gini, went to the National Institute of Statistics in Rome, where he stayed until 1931. From there he went to Trieste to work for an insurance company, staying until 1946. During this time he also held chairs in Trieste and Padova. He then moved to the University of Rome, first in Economics and later in the School of Probability in the Faculty of Science.

For statisticians, his major contributions to knowledge were his view of probability as subjective, and his demonstration, through the concept of exchangeability, that this view could embrace standard frequency statistics as a special case. One of the aphorisms of which he was especially fond says "Probability does not exist"; by which he meant that probability does not, like the length of a table, have an existence irrespective of the observer. Nor is probability a property purely of the observer: Rather it expresses a relationship between the observer and the external world. Indeed, probability is the way in which we understand that world. It is the natural language of knowledge. This view had been held by others, but de Finetti showed that probability is the only possible language: that our statements of uncertainty must combine according to the rules of probability. He also showed that special cases, called *exchangeable,* lead an observer to appreciate chance in natural phenomena, so that frequency views are merely the special, exchangeable case of probability judgments. The unity given to disparate approaches to probability was an enormous advance in our understanding of the notions and enabled a totally coherent approach to statistics, a subject which is usually presented as a series of ad hoc statements, to be given. This is Bayesian statistics, though de Finetti was careful to distinguish between Bayesian ideas and Bayesian techniques, the latter being merely the calculations necessary to solve a problem, the former being the notions expressing the observer's view of the situation. The latter does not distinguish between sex and drawing pins—both are technically binomial (n, p)—whereas the former does.

De Finetti's view of mathematics was quite different from that dominant today, especially in the United States. His emphasis was on ideas and applications, not on technicalities and calculations. He was fond of quoting the saying of Chisini, a teacher of his, that "mathematics is the art that teaches you how *not* to make calculations." An unfortunate consequence of this is that, for those used to the modern mathematical style, he is difficult to read. Theorem, proof, and application are blended into a unity that does not allow one aspect to dominate another. He was severely critical, as are many others, of the very technical papers that one is supposed to accept because no slip has been found in an insipid and incomprehensible chain of syllogisms. Nomenclature was important to him. He insisted on "random quantity," not "random variable": for what varies?

His main contribution to applied probability was the introduction of a scoring rule. In its simplest form, if a person gives a probability p for an event A, he is scored $(p - 1)^2$ or p^2, according to A is subsequently found to be true or false. This simple idea has led to extensions and to the appreciation of what is meant by a good probability appraiser.

Another major interest of his was teaching. He was emphatic that children should be taught to think probabilistically at an early age, to understand that the world is full of uncertainty and that probability was the way to handle and feel comfortable with it. A child should not be taught to believe that every question has a right answer, but only probabilities for different possibilities. For example, in a multiple-choice test, an examinee should give probabilities for each of the suggested solutions, not merely select one. He should not predict but forecast through a probability distribution.

He did a substantial amount of actuarial work. He was vitally interested in economics and social justice. He believed in an economic system founded on the twin ideas of Pareto optimality and equity. This, he felt, would produce a better social system than the greed that he felt to be the basis of capitalism. His views have been described as naive, but this is often a derogatory term applied by self-interested people to those who dare to disturb that self-interest. He stood as a candidate in an election (and was defeated) and was arrested when supporting what he saw to be a worthy cause.

De Finetti was kind and gentle, yet emphatic in his views. He was interested in the unity of life. Although he will best be remembered for theoretical advances, his outlook embraced the whole spectrum of knowledge.

BIBLIOGRAPHY

Probability, Induction and Statistics: The Art of Guessing. Wiley, London, 1972.

Theory of Probability: A Critical Introductory Treatment. Wiley, London, Vol. 1, 1974; Vol. 2, 1975.

"Contro disfunzioni e storture: Urgenza di riforme radicali del sistema." In *Lo Sviluppo della Società Italiana nei Prossimi Anni.* Accademia Nazionale dei Lincei, Roma, 1978, pp. 105–145.

Il Saper Vedere in Matematica. Loescher, Torino, 1967.

Matematica Logico-Intuitiva. Cremonese, Roma, 1959.

DENNIS V. LINDLEY

Elfving, Gustav

Born: June 28, 1908, in Helsinki, Finland.
Died: March 25, 1984, in Helsinki, Finland.
Contributed to: complex analysis, Markov chains, counting processes, order statistics, optimum design of experiments, nonparametric statistics, sufficiency and completeness, expansion of distributions, quality control, Bayes statistics, optimal stopping.

Gustav Elfving.

Erik Gustav Elfving was a son of Fredrik Elfving and Thyra Ingman. The father was a long-time professor of Botany at the University of Helsinki, a demanding teacher who in his twenties had raised money in Finland for Darwin's memorial in London. Gustav's childhood home was the professor's residence in the new Institute of Botanics, which had largely been brought about by the energetic father's efforts. Gustav Elfving married Irene (Ira) Aminoff in 1936.

Like his father he chose an academic career, but majored in mathematics; his other subjects were physics and astronomy. Elfving wrote his doctoral thesis [1] on a subject now known as Nevanlinna theory, under the guidance of Rolf Nevanlinna, an eminent representative of the school of complex analysis founded in Helsinki by Ernst Lindelöf. But his interest soon turned to probability theory, to which subject he had been introduced by Jarl Waldemar Lindeberg (1876–1930). An outline of his career is quickly drawn.

He held a lecturership in Turku (Åbo Akademi) in 1935–1938 and in the Technical University of Helsinki in 1938–1948. In 1948 he was appointed professor of mathematics at the University of Helsinki, and was retired in 1975 from this position. Elfving worked several times abroad. In 1946–1947 he was a stand-in for Harald Cramér in Stockholm University. He also made visits to the United States (Cornell University 1949–1951, Columbia University 1955, and Stanford University 1960 and 1966).

The stimulus for Elfving's move into probability and statistics occurred under anecdotal circumstances, in 1935. He had joined an expedition of the Danish Geodetic Institute to Western Greenland. The team were forced to stay in their tents for three solid days due to rain. Then, to pass time, Elfving started to think about least-squares problems.

In his first paper in probability [2] he introduced the difficult problem of the *embeddability* of a discrete-time Markov chain into a continuous-time one in the finite-state time-homogeneous setting. The problem has been of concern e.g. in the modeling of social and population processes [10], but a complete solution is to date still lacking. In another contribution [6] he introduced a stopping problem of the seller of a valuable item. The model involves discounting and bids coming in according to a point process.

Pioneering work was done by Elfving in several papers on the methods of computing optimum regression designs. The "Optimum allocation in linear regression theory" of 1952 [3] is without doubt Elfving's single most influential paper, inaugurating that line of research. The researcher wants to estimate two unknown parameters. He has at his disposal a finite number of different experiments, each repeatable any number of times and yielding observations having expectations linear in the two parameters and a common variance. The problem is to allocate a given number of observations among these experiments. Given a criterion of optimality defining the goal of the investigation, a well-defined design problem arises. In this paper Elfving studied what are now known as C-optimal and A-optimal designs. He characterized the optimum designs using an elegant geometric criterion, a result often referred to as "Elfving's theorem." Elfving's approach differed from the other early efforts in the optimum design of experiments in that it dealt with not an isolated design problem but a whole class of them, providing flexible criteria and characterizations of solutions. It also readily admitted generalizations beyond the simple setup of ref. [3], as in ref. [4]. The survey [5] sums up Elfving's work in this field. Subsequent developments, including generalizations of the design problem and systematic examination of criteria of optimality, were due to or initiated by Jack Kiefer, and many other authors followed.

In his science, and outside, Elfving was first and foremost a man of learning and ideas. The topics of his original papers were varied. As noted by Johan Fellman [9], Elfving usually focused on the new and basic ideas, keeping his presentation simple and unobscured by mathematical technicalities; but a few papers display his substantial analytical technique. Often he was content with a sketch of the generalizations, or he left them to others. In class the beauty of a completed proof would not be enough; he regularly supplemented proofs with an intuition about the basic ideas. He was successful as a lecturer and in addressing a general audience. He gave insightful and witty general talks and wrote surveys on topics such as probability, theory of games, decision theory, and information theory. He pondered the relationship of entropy and esthetic evaluation and wrote an anthropological essay on modern man's longing for a primitive way of life.

After retirement Elfving undertook the writing of a monograph [7] on the history of mathematics in Finland, 1828–1918, a period of Finland's autonomy under Russia, on behalf of the Finnish Society of Sciences. Among the some fifty mathematicians covered are Lindelöf, Hjalmar Mellin (of the Mellin transformation), and Lindeberg; early contributions to statistics are discovered and described; see also ref. [8]. Reflecting Elfving's depth of interests, ref. [7] is more than just an account of the mathematical achievements of a few successful scientists. Elfving provides necessary background of past and contemporary research, at home and abroad, but also of Finnish culture, society, and politics. His writing is unpretentious, factual, and entertaining, and the treatment of individual scientists, seen as phenomena of their time, characteristically unbiased.

REFERENCES

[1] Elfving, G. (1934). Über eine Klasse von Riemannschen Flächen und ihre Uniformisierung. *Acta Soc. Sci. Fennicae N. S. A,* **II**(3), 1–60. (Doctoral Dissertation.)

[2] Elfving, G. (1937). Zur Markoffschen Ketten. *Acta Soc. Sci. Fennicae. N. S. A,* **II**(8), 1–17.

[3] Elfving, G. (1952). Optimum allocation in linear regression theory. *Ann. Math. Statist.,* **23,** 255–262.

[4] Elfving, G. (1954). Geometric allocation theory. *Skand. Aktuarietidskrift,* 170–190.

[5] Elfving, G. (1959). Design of linear experiments. In *Probability & Statistics. The Harald Cramér Volume,* Ulf Grenander, ed. pp. 58-74.

[6] Elfving, G. (1967). A persistency problem connected with a point process. *J. Appl. Probab.* **4,** 77–89.

[7] Elfving, G. (1981). *The History of Mathematics in Finland 1828–1918.* Soc. Sci. Fennicae, Helsinki.

[8] Elfving, G. (1985). Finnish mathematical statistics in the past. *Proc. First Tampere Seminar on Linear Statistical Models and Their Applications 1983.* Department of Mathematical Sciences/Statistics, University of Tampere, Tampere, pp. 3–8.

[9] Fellman, J. (1991). Gustav Elfving and the emergence of optimal design theory. *Working paper 218,* Swedish School of Economics and Business Administration, Helsinki, 7 pp.

[10] Singer, B. and Spilerman, S. (1976). The representation of social processes by Markov models. *Amer. J. Sociol.* **82**(1), 1–54. (Embeddability.)

BIBLIOGRAPHY

Chow, Y. S., Robbins, H., and Siegmund D., (1971). *Great Expectations: The Theory of Optimal Stopping.* Houghton-Mifflin, Boston. (Elfving's problem)

Elfving, G. (1970). Research—experiences and views. In *Festschrift in Honour of Herman Wold,* T. Dalenius, G. Karlsson, and S. Malmqvist, eds. Almqvist & Wiksell, Uppsala.

Johansen, S., Ramsey, F. L. (1979). A bang-bang representation for 3 × 3 embeddable stochastic matrices. *Z. Warsch. Verw. Geb.* **47**(1), 107–118.

Pukelsheim, F. (1993). *Optimal Design of Experiments.* Wiley, New York.

TIMO MÄKELÄINEN

Fisher, Ronald Aylmer

R. A. Fisher

Born: February 17, 1890, in East Finchley, London, England.
Died: July 29, 1962, in Adelaide, Australia.
Contributed to: mathematical statistics, probability theory, genetics, design of experiments

Ronald Aylmer Fisher achieved original scientific research of such diversity that the integrity of his approach is masked. Born into the era of Darwin's evolutionary theory and Maxwell's theory of gases, he sought to recognize the logical consequences of a world containing indeterminism, whose certainties were essentially statistical. His interests were those of Karl Pearson,* who dominated the fields of evolution, biometry, and statistics during his youth, but his perspective was very different. His ability to perceive remote logical connections of observation and argument gave his conceptions at once universal scope and coherent unity; consequently, he was little influenced by current scientific vogue at any period of his life.

Fisher was (omitting his stillborn twin) the seventh and youngest child of George Fisher, fine-arts auctioneer in the West End, and Katie, daughter of Thomas Heath, solicitor of the City of London. His ancestors showed no strong scientific bent, but his uncle, Arthur Fisher, was a Cambridge Wrangler.

In childhood, Fisher met the misfortune first of his poor eyesight, and the eyestrain that was always to limit his private reading, and he learned to listen while others read aloud to him. In 1904, his beloved mother died suddenly of peritonitis. In 1906, his father's business failure required him to become largely self-supporting. His general intelligence and mathematical precocity were apparent early. From Mr. Greville's school in Hampstead, he went on to Stanmore in 1900, and entered Harrow in 1904 with a scholarship in mathematics. In his second year he won the Neeld Medal in mathematical competition with all the school. To avoid eyestrain he received tuition in mathematics under G. H. P. Mayo without pencil, paper, or other visual aids. Choosing spherical trigonometry for the subject of these tutorials, he developed a strong geometrical sense that was greatly to influence his later work. In 1909, he won a scholarship in mathematics to Cambridge University. In 1912, he graduated as a Wrangler and, being awarded a studentship for one year, studied the theory of errors under F. J. M. Stratton and statistical mechanics and quantum theory under J. Jeans.

In April 1912, Fisher's paper [3] was published, in which the *method of maximum likelihood* was introduced (though not yet by that name). As a result, that summer Fisher wrote to W. S. Gosset ("Student") questioning his divisor $n - 1$ in the formula for the standard deviation. He then reformulated the problem in an entirely different and equally original way, in terms of the configuration of the sample in n-dimensional space, and showed that the use of the sample mean instead of the population mean was equivalent to reducing the dimensionality of the sample space by one; thus he recognized the concept of what he later called *degrees of freedom*.

Moreover, the geometrical formulation immediately yielded Student's distribution, which Gosset had derived empirically, and in September Fisher sent Gosset the mathematical proof. This was included in Fisher's paper when, two years later, using the geometrical representation, he derived the general sampling distribution of the correlation coefficient [4].

Fisher's mathematical abilities were directed into statistical research by his interest in evolutionary theory, especially as it affected man. This interest, developing at Harrow, resulted in the spring of 1911 in the formation of the Cambridge University Eugenics Society at Fisher's instigation. He served on the Council even while he was chairman of the undergraduate committee; he was the main speaker at the second annual meeting of the society. While famous scientists wrangled about the validity either of evolutionary or genetical theory, Fisher accepted both as mutually supportive; for he saw that if natural variation was produced by genetic mechanism, every evolutionary theory except natural selection was logically excluded. While the applicability of genetic principles to the continuous variables in man was disputed on biometrical grounds, Fisher assumed that the observed variations were produced genetically, and in 1916 [5] justified this view by biometrical argument.

In its application to man, selection theory raised not only scientific but practical problems. The birthrate showed a steep and regular decline relative to increased social status. This implied the existence throughout society of selection against every quality likely to bring social success. Fisher believed, therefore, that it must result in a constant attrition of the good qualities of the population, such as no civilization could long withstand. It was important to establish the scientific theory on a firm quantitative basis through statistical and genetic research, and, more urgently, to publicize the scientific evidence so that measures should be taken to annul the self-destructive fertility trend.

Fisher accepted at once J. A. Cobb's suggestion [2] in 1913 that the cause of the dysgenic selection lay in the economic advantage enjoyed by the children of small families over those from larger families at every level of society. Later he proposed and urged adoption of various schemes to spread the financial burden of parenthood—so that those who performed similar work should enjoy a similar standard of living, irrespective of the number of their children—but without success; and the family allowance scheme adopted in Great Britain after World War II disappointed his hopes.

To further these aims, on leaving college he began work with the Eugenics Education Society of London, which was to continue for 20 years. From 1914 he was a regular book reviewer for the *Eugenics Review;* in 1920 he became business secretary and in 1930 vice-president of the society; and he pursued related research throughout. Major Leonard Darwin, the president, became a dear and revered friend, a constant encouragement and support while Fisher was struggling for recognition, and a stimulus to him in the quantitative research that resulted in *The Genetical Theory of Natural Selection* [14].

In 1913, Fisher took a statistical job with the Mercantile and General Investment Company in the City of London. He trained with the Territorial Army and, on the outbreak of war in August 1914, volunteered for military service. Deeply disappointed by his rejection due to his poor eyesight, he served his country for the next five years by teaching high school physics and mathematics. While he found teaching unattractive, farming appealed to him both as service to the nation and as the one life in which a numerous family might have advantages. When, in 1917, he married Ruth Eileen, daughter of Dr. Henry Grattan Guinness (head of the Regions Beyond Missionary Union at the time of his death in 1915), Fisher rented a cottage and small holding from which he could bicycle to school, and with Eileen and her sister, began subsistence farming, selling the excess of dairy and pork products to supply needs for which the family could not be self-sufficient. Their evening hours were reserved for reading aloud, principally in the history of earlier civilizations.

In these years Fisher's statistical work brought him to the notice of Karl Pearson.* In 1915, Pearson published Fisher's article on the general sampling distribution of the correlation coefficient in *Biometrika* [4], and went on to have the ordinates of the error of estimated correlations calculated in his department. The cooperative study [21] was published in 1917, together with a criticism of Fisher's paper not previously communicated to its author. Pearson had not understood the method of maximum likelihood Fisher had used, and condemned it as being inverse inference, which Fisher had deliberately avoided. Fisher, then unknown, was hurt by Pearson's highhandedness and lack of understanding, which eventually led to their violent confrontation. Meanwhile, Pearson ignored Fisher's proposal to assess the signficance of correlations by considering not the correlation r itself but

$$z = \tfrac{1}{2}\ln\frac{1+r}{1-r},$$

a remarkable transformation that reduces highly skewed distributions with unequal variances to distributions to a close approximation normal with constant variance. Fisher's paper on the correlation between relatives on the supposition of Mendelian inheritance [5], submitted to the Royal Society in 1916, had to be withdrawn in view of the referee's comments. (Knowing that Pearson disagreed with his conclusions, Fisher had hoped that his new method, using the analysis-of-variance components, might be persuasive.) In this paper the subject and methodology of biometrical genetics was created. These facts influenced Fisher's decision in 1919 not to accept Pearson's guarded invitation to apply for a post in his department.

In September 1919, Fisher started work in a new, at first temporary post as statistician at Rothamsted Experimental Station, where agricultural research had been in progress since 1843. He quickly became established in this work. He began with a study of historical data from one of the long-term experiments, with wheat on Broadbalk, but soon moved on to consider data obtained in current field trials, for which he developed the *analysis of variance*. These studies brought out the inadequacies of the arrangement of the experiments themselves and so led to the evolution of the science of experimental design. As Fisher worked with experimenters using successively improved designs, there emerged the principles of *randomization,* adequate *replication, blocking* and *confounding,* and randomized blocks, Latin squares, factorial arrangements, and other designs of unprecedented efficiency. The statistical methods were incorporated in successive editions of *Statistical Methods for Research Workers* [10]. The 11-page paper on the arrangement of field experiments [11] expanded to the book *The Design of Experiments* [16]. These volumes were supplemented by *Statistical Tables for Biological, Agricultural and Medical Research,* coauthored by Frank Yates [19].

Following up work on the distribution of the correlation coefficient, Fisher derived the sampling distributions of other statistics in common use, including the *F*-distribution and the multiple correlation coefficient. Using geometrical representations, he solved (for normally distributed errors) all the distribution problems for the general linear model, both when the null hypothesis is true and when an alernative hypothesis is true [9, 12].

Concurrently, the theory of estimation was developed in two fundamental papers in 1922 [7] and in 1925 [8]. Fisher was primarily concerned with the small samples of observations available from scientific experiments, and was careful to draw a sharp distinction between sample statistics (estimates) and population values (parameters to be estimated). In the method of maximum likelihood he had found a general method of estimation that could be arrived at from the mathematical and probabilistic structure of the problem. It not only provided a method to calculate unique numerical estimates for any problem that could be precisely stated, but also indicated what mathematical function of the observations ought to be used to estimate the parameter. It thus provided a criterion for the precise assessment of estimates, a revolutionary idea in 1920.

Using this method to compare two estimates of the spread σ of a normal distribution, Fisher [6] went on to show that the sample standard deviation s was not only better but uniquely best; because the distribution of any other measure of spread conditional on s does not contain the parameter σ of interest, once s is known, no other estimate gives any further information about σ. Fisher called this quality of s *sufficiency.* This finding led to his introduction of the concept of the amount of *information* in the sample, and the criteria of *consistency, efficiency,* and sufficiency of estimates, measured against the yardstick of available information. He exploited the asymptotic efficiency of the method of maximum likelihood in 1922, and, extending consideration to small samples in 1925, observed that small-sample sufficiency, when not directly available, was obtainable via *ancillary statistics* derived from the likelihood function.

Thus, seven years after moving to Rothamsted, Fisher had elucidated the underlying theory and provided the statistical methods that research workers urgently needed to deal with the ubiquitous variation encountered in biological experimentation. Thereafter, he continued to produce a succession of original research on a wide variety of statistical topics. For example, he initiated nonlinear design, invented *k*-statistics, and explored extreme-value distributions, harmonic analysis, multivariate analysis and the discriminant function, the analysis of covariance, and new elaborations of experimental design and of sampling survey.

So diverse and fundamental were Fisher's contributions to mathematical statistics that G. A. Barnard [1] wrote that "to attempt, in a short article, to assess the contributions to the subject by one largely responsible for its creation would be futile."

Fisher's central contribution was surely, as Barnard wrote, "his deepening of our understanding of uncertainty" and of "the many types of measurable uncertainty." "He always ascribed to Student the idea that, while there must necessarily be uncertainty involved in statistical procedures, this need not imply any lack of precision—the uncertainty may be capable of precise quantitative assessment. In continuing work begun by Student, of deriving the exact sampling distributions of the quantities involved in the most common tests of significance, Fisher did much to give this idea form and reality." In developing the theory of estimation, he explored the type of uncertainty expressible precisely in terms of the likelihood; and his ideas on the subject never ceased to evolve. From the beginning, however, he distinguished likelihood from mathematical probability, the highest form of scientific inference, which he considered appropriate only for a restricted type of uncertainty. He accepted classical probability theory, of course, and used Bayes' theorem in cases in which there was an observational basis for making probability statements in advance about the population in question; further, he proposed the fiducial argument as leading to true probability statements, at least in one common class of cases.

Fisher introduced the fiducial argument in 1930 [13]. In preparing a paper on the general sampling distribution of the multiple correlation coefficient in 1928, he noticed that in the test of significance the relationship between the estimate and the parameter was of a type he later characterized as "pivotal." He argued that if one quantity was fixed, the distribution of the other was determined; consequently, once the observations fixed the value of the observed statistics, the whole distribution of the unknown parameter was determined. Thus, in cases in which the *pivotal relationship* existed, true probability statements concerning continuous parameters could be inferred from the data. Exhaustive estimates were required [15].

Controversy arose immediately. The fiducial argument was proposed as an alternative to the argument of the title "inverse probability," which Fisher condemned in all cases in which no objective prior probability could be stated. H. Jeffreys* led the debate on behalf of less restrictive use of inverse probability, while J. Neyman* developed an approach to the theory of estimation through sampling theory, which in some instances led to numerical results different from Fisher's. For many years the debate focused on the case of estimating the difference between two normally distributed populations with unknown variances not assumed to be equal (Behrens' test). This led Fisher to introduce the concept of the *relevant reference set.* He argued that the sampling-theory approach ignored information in the sample, and that of all possible samples, only the subset yielding the observed value s_1^2/s_2^2 was relevant. Later problems with the fiducial argument arose in cases of multivariate estimation, because of nonuniqueness of the pivotals. Different pairs of pivotals could be taken that resulted in different and nonequivalent probability statements. Fisher did not achieve clarification of criteria for selection among such alternative pivotals; he was working on the problem at the end of his life.

In proposing the fiducial argument in 1930, Fisher highlighted the issues of scientific inference and compelled a more critical appreciation of the assumptions made, and of their consequences, in various approaches to the problem. In reviewing the subject in *Statistical Methods and Scientific Inference* [18], he distinguished the conditions in which he believed significance tests, likelihood estimates, and probability statements each still had an appropriate and useful role to play in scientific inference.

In his genetical studies, having demonstrated the consonance of continuous variation in man with Mendelian principles, and having thereby achieved the fusion of biometry and genetics [5], in the 1920s Fisher tackled the problems of natural selection expressed in terms of population genetics. Of this work, K. Mather [20] wrote:

> Fisher's contribution was basic, characteristic and unique. It is set out in *The Genetical Theory of Natural Selection.* He pointed out that natural selection is not evolution, that in fact evolution is but one manifestation of the operation of natural selection, and that natural selection can and should be studied in its own right. Having delimited his field, he proceeded to cultivate it as only he could, with all his resources of mathematics allied to experimentation carried out in circumstances which most of us would have regarded as prohibitively difficult. Again he went beyond merely harmonizing to fusing the principles of genetics and natural selection. His well-known theory of the evolution of dominance (so sharply criticized, yet so widely accepted) is but one facet of his work: he formulated his fundamental theorem of natural selection, equating the rate of increase in fitness to the genetic variance in fitness; he considered sexual selection and mimicry; and he extended his discussion to man, society and the fate of civilizations. It was a truly great work.

The book was written down by his wife at his dictation during evenings at home; for a while it took the place of the reading and conversation that ranged from all the classics of English literature to the newest archeological research and that centered on human evolution. At home, too, was Fisher's growing family: his oldest son, George, was born in 1919; then a daughter who died in infancy, a second son, and in the end six younger daughters. True to his eugenic ideal, Fisher invested in the future of the race, living simply under conditions of great financial stringency while the children were reared. He was an affectionate father, and especially fond of George, who was soon old enough to join him in such activities as looking after genetic mouse stocks. Wherever possible, he brought the children into his activities, and he answered their questions seriously, with sometimes brilliant simplicity; he promoted family activities and family traditions. Domestic government was definitely patriarchal, and he punished larger offences against household rules, though with distaste. As long as possible the children were taught at home, for he trusted to their innate curiosity and initiative in exploring their world rather than to any imposed instruction. He had no sympathy with lack of interest, or fear of participation; and if his motive force was itself frightening, learning to deal with it also was a part of the child's education. In fact, he treated his children like his students, as autonomous individuals from the beginning, and encouraged them to act and think on their own responsibility, even when doing so involved danger or adult disapproval.

In 1929, Fisher was elected a Fellow of the Royal Society as a mathematician. The influence of his statistical work was spreading, and he was already concerned that statistics should be taught as a practical art employing mathematical theory. In 1933, Karl Pearson retired and his department at University College London was split; E. S.

Pearson* succeeded as head of the statistics department, and Fisher as Galton Professor of Eugenics, housed on a different floor of the same building. For both men, it was an awkward situation. While others gave their interpretation of Fisher's ideas in the statistics department, he offered a course on the philosophy of experimentation in his own. After J. Neyman* joined the statistics department in 1934, relations between the new departments deteriorated, and fierce controversy followed.

Fisher continued both statistical and genetical research. In 1931 and again in 1936, he was visiting professor for the summer sessions at Iowa State University at Ames, Iowa, at the invitation of G. W. Snedecor, director of the Statistical Computing Center. In 1937–1938, he spent six weeks as the guest of P. C. Mahalanobis,* director of the Indian Statistical Institute, Calcutta. In his department, where Karl Pearson had used only biometrical and genealogical methods, Fisher quickly introduced genetics. Work with mouse stocks, moved from the attic at his home, was expanded, and experimental programs were initiated on a variety of animal and plant species, for example, to study the problematical tristyly in *Lythrum salicaria.* Fisher was very eager also to initiate research in human genetics.

Sponsored by the Rockefeller Foundation, in 1935 he was able to set up a small unit for human serological research under G. L. Taylor, joined by R. R. Race in 1937. In 1943, Fisher interpreted the bewildering results obtained with the new Rh blood groups in terms of three closely linked genes, each with alleles, and predicted the discovery of two new antibodies and an eighth allele—all of which were identified soon after. Fisher's enthusiasm for blood-group polymorphisms continued to the end of his life, and he did much to encourage studies of associations of blood groups and disease.

In 1927, Fisher proposed a way of measuring selective intensities on genetic polymorphisms occurring in wild populations, by a combination of laboratory breeding and field observation, and by this method later demonstrated very high rates of selective predation on grouse locusts. E. B. Ford was one of the few biologists who believed in natural selection at the time; in 1928 he planned a long-term investigation of selection in the field, based on Fisher's method. To the end of his life, Fisher was closely associated with Ford in this work, which involved development of capture–recapture techniques and of sophisticated new methods of statistical analysis. The results were full of interest and wholly justified their faith in the evolutionary efficacy of natural selection alone.

Forcibly evacuated from London on the outbreak of war in 1939, Fisher's department moved to Rothamsted, and finding no work as a unit, gradually dispersed; Fisher himself could find no work of national utility. In 1943, he was elected to the Balfour Chair of Genetics at Cambridge, which carried with it a professorial residence. Lacking other accommodation, he moved his genetic stocks and staff into the residence, leaving his family in Harpenden. Estranged from his wife, separated from home, and deeply grieved by the death in December 1943 of his son George on active service with the Royal Air Force, Fisher found companionship with his fellows at Caius College, and with the serological unit (evacuated to Cambridge for war work with the Blood Transfusion Service), which planned to rejoin his department after the war.

There was little support after the war for earlier plans to build up an adequate genetics department. No bid was made to keep the serological unit. No departmental building was erected. Support for the research in bacterial genetics initiated in 1948 under L. L. Cavalli (Cavalli-Sforza) was withdrawn in 1950 when Cavalli's discovery of the first Hfr strain of *Escherichia coli* heralded the remarkable discoveries soon to follow in bacterial and viral genetics. Fisher cultivated his garden, continued his research, published *The Theory of Inbreeding* [17] following his lectures on this topic, and built a group of good quantitative geneticists. He attempted to increase the usefulness of the university diploma of mathematical statistics by requiring all diploma candidates to gain experience of statistical applications in research, in a scientific department. Speaking as founding president of the Biometric Society, as president of the Royal Statistical Society, and as a member or as president of the International Statistical Institute, he pointed out how mathematical statistics itself owes its origin and continuing growth to the consideration of scientific data rather than of theoretical problems.

His own interests extended to the work of scientists in many fields. He was a fascinating conversationalist at any time, original, thoughtful, erudite, witty, and irreverent; with the younger men, his genuine interest and ability to listen, combined with his quickness to perceive the implications of their research, were irresistible. He encouraged, and contributed to, the new study of geomagnetism under S. K. Runcorn, a fellow of his college. He was president of Gonville and Caius College during the period 1957–1960.

He received many honors and awards: the Weldon Memorial Medal (1928), the Guy Medal of the Royal Statistical Society in gold (1947), three medals of the Royal Society, the Royal Medal (1938), the Darwin Medal (1948), and the Copley Medal (1956); honorary doctorates from Ames, Harvard, Glasgow, London, Calcutta, Chicago, the Indian Statistical Institute, Adelaide, and Leeds. He was Foreign Associate, United States National Academy of Sciences; Foreign Honorary Member, American Academy of Arts and Sciences; Foreign Member, American Philosophical Society; Honorary Member, American Statistical Association; Honorary President; International Statistical Institute; Foreign Member, Royal Swedish Academy of Sciences; Member, Royal Danish Academy of Sciences; Member, Pontifical Academy; Member, Imperial German Academy of Natural Science. He was created Knight Bachelor by Queen Elizabeth in 1952.

After retirement in 1957, Sir Ronald Fisher traveled widely before joining E. A. Cornish in 1959 as honorary research fellow of the C. S. I. R. O. Division of Mathematical Statistics in Adelaide, Australia. He died in Adelaide July 29, 1962.

REFERENCES

The collections of Fisher's papers referred to throughout this list are described in full in the subsequent Bibliography.

[1] Barnard, G. A. (1963). *J. R. Statist. Soc. A,* **216,** 162–166.

[2] Cobb, J. A. (1913). *Eugen. Rev.,* **4,** 379-382.

[3] Fisher, R. A. (1912). *Messeng. Math.,* **41,** 155–160. [Also in Fisher (1971–1974), No. 1.]

[4] Fisher, R. A. (1915). *Biometrika,* **10,** 507–521. [Also in Fisher (1971–1974), No. 4 and No. 1.]

[5] Fisher, R. A. (1918). *Trans. R. Soc. Edinb.,* **52,** 399-433. [Also in Fisher (1971–1974), No. 9.]

[6] Fisher, R. A. (1920). *Monthly Not. R. Astron. Soc.,* **80,** 758-770. [Also in Fisher (1971–1974), No. 2.]

[7] Fisher, R. A. (1922). *Philos. Trans. A,* **222,** 309-368. [Also in Fisher (1971–1974), No. 18 and No. 10.]

[8] Fisher, R. A. (1925). *Proc. Camb. Philos. Soc.,* **22,** 700-725. [Also in Fisher (1971–1974), No. 42 and No. 11.]

[9] Fisher, R. A. (1925). *Metron,* **5,** 90-104. [Also in Fisher (1971–1974), No. 43.]

[10] Fisher, R. A. (1925). *Statistical Methods for Research Workers.* Oliver & Boyd, Edinburgh. (Subsequent editions: 1928, 1930, 1932, 1934, 1936, 1938, 1941, 1944, 1946, 1950, 1954, 1958, 1970.)

[11] Fisher, R. A. (1926). *J. Minist. Agric. Gr. Brit.,* **33,** 503–513. [Also in Fisher (1971–1974), No. 48 and No. 17.]

[12] Fisher, R. A. (1928). *Proc. R: Soc. A,* **121,** 654-673. [Also in Fisher (1971–1974), No. 61 and No. 14.]

[13] Fisher, R. A. (1930). *Proc. Camb. Philos. Soc.,* **26,** 528-535. [Also in Fisher (1971–1974), No. 84 and No. 22.]

[14] Fisher, R. A. (1930). *The Genetical Theory of Natural Selection.* Oxford University Press, London. (Reprinted by Dover, New York, 1958.)

[15] Fisher, R. A. (1934). *Proc. R. Soc. A,* **144,** 285-307. [Also in Fisher (1971–1974), No. 108 and No. 24.]

[16] Fisher, R. A. (1935). *The Design of Experiments.* Oliver & Boyd, Edinburgh.

[17] Fisher, R. A. (1949). *The Theory of Inbreeding.* Oliver & Boyd, Edinburgh.

[18] Fisher, R. A. (1956). *Statistical Methods and Scientific Inference.* Oliver & Boyd, Edinburgh.

[19] Fisher, R. A. and Yates, F. (1938). *Statistical Tables for Biological, Agricultural and Medical Research.* Oliver & Boyd, Edinburgh.

[20] Mather, K. (1963). *J. R. Statist. Soc. A,* **216** 166-168.

[21] Soper, H. E., Young, A. W., Cave, B. M., Lee, A., and Pearson, K. (1916). *Biometrika,* 11, 328–413.

BIBLIOGRAPHY

Box, J. (1978). *R. A. Fisher: The Life of a Scientist.* Wiley, New York. [Complementary to Fisher (1971–1974) in showing the context of his statistical innovations.]

Cochran, W. G. (1976). *Science,* **156,** 1460–1462. (Cochran's encounters with Fisher make an amusing, and telling, character sketch.)

Finney, D. J. (1964). *Biometrics,* **20,** 322–329. [This volume, "In Memoriam R. A. Fisher,

1890–1962," includes articles on his contributions to various branches of statistical science: e.g., Rao (1964) and Yates (1964).]

Fisher, R. A. (1971–1974). *Collected Papers of R. A. Fisher,* 5 vols. J. H. Bennett, ed. University of Australia, Adelaide, Australia. [This compilation includes a complete bibliography of Fisher's published work and the biographical memoir by Yates and Mather (1963), with 291 of Fisher's papers, presented chronologically. A primary source for all students of Fisher or of the historical development of statistical theory through diverse scientific applications, 1912 to 1962.]

Fisher, R. A. (1950). *Contributions to Mathematical Statistics.* Wiley, New York. (Out of print, this volume contains 43 of Fisher's major articles on mathematical statistics, with his introductory comments.)

Rao, C. R. (1964). *Biometrics,* **20,** 186–300.

Savage, L. J. (1976). On rereading R. A. Fisher [J. W. Pratt, ed.], *Ann. Math. Statist.,* **4,** 441–500. (The talk was given in 1970; the article grew ever more comprehensive thereafter, and was published posthumously. A stimulating commentary with references to many others.)

Yates, F. (1964). *Biometrics,* **20,** 307–321.

Yates, F. and Mather, K. (1963). *Biogr. Mem. Fellows R. Soc. Lond.,* **9,** 91–120. (An excellent summary.)

JOAN FISHER BOX

Galton, Francis

Born: February 16, 1822, near Birmingham, England.
Died: January 17, 1911, in Haslemere, England.
Contributed to: statistical methods, genetics.

Sir Francis Galton came from an intellectual family; his mother was Violetta Darwin (aunt of Charles Darwin), and his grandfather Samuel Galton was a Fellow of the Royal Society (F. R. S.). He was the youngest of a family of nine brought up in a large house near Birmingham, where his father, Samuel Tertius Galton, ran a bank. He showed early intellectual powers; at 5 he could "add and multiply by all numbers up to 8" and "say all the Latin substantives and adjectives and active verbs." But he failed to qualify in medicine, as his father had hoped, or to complete an honors degree in mathematics at Cambridge. This mattered little, as he had independent means. In 1858 he married Louisa, sister of Montague Butler, a future Master of Trinity College, Cambridge.

He was a late developer, with published work beginning only at 28. His first interests were in exploration of South West Africa (later, Egypt and elsewhere), geography, and meteorology, subjects that were major interests throughout his life; he was the inventor of the word "anticyclone." He was on the management committee of Kew Observatory, and for much of his life served on the Council of the Royal Geographic Society. He was elected a fellow of the Royal Society for his geographical studies at the age of 38. From the age of 43 genetics and the use of statistical methods for the study of all kinds of questions became major occupations. These statistics were sometimes of measurable quantities, such as the heights of parents and children. But quite often he used subjective evaluation (at one time he carried a device to record whether the women he met were pretty, plain, or ugly, reaching the conclusion that London women were some of the prettiest in Britain). He also used questionnaires, although the wording of the questions might hardly meet modern criteria. In his book *Natural Inheritance* he wrote:

> Some people hate the very name of statistics but I find them full of beauty and interest. Whenever they are not brutalized, but delicately handled by the higher methods, and are warily interpreted, their power of dealing with complicated phenomena is extraordinary. [2, p. 62]

His major contributions to statistics were in connection with genetics and psychology. He used the normal (Gaussian) distribution a great deal. Essentially, he supposed that where a character could be ranked but not directly measured, he could usually assign it a value corresponding to a normal deviate with the same expected rank. He first used the word *correlation* in his book *Hereditary Genius* [1] in a general statistical sense, saying that the characteristic of strong morality and moral instability are in no way correlated. In 1877 he introduced a numerical measure of *regression* or *reversion,* effectively the modern regression coefficient, as the average deviation from the mean of children of a given parent as a fraction of that parent's deviation. With help from a Cambridge mathematician, Hamilton Dixon, he explained this in terms of a bivariate (normal) distribution with elliptical contours. He obtained the axis of the ellipse, thus anticipating principal components. The fact that the regression of y on x differed from that of x on y worried him persistently until in 1877 by a flash of inspiration (at Haworth Castle) he saw that the solution was to normalize x and y in terms of their own variability. The regression coefficient he then called the "co-relation" (or correlation). Since it was a regression, he called it r; whence the modern symbol r for correlation. (Galton's coefficient was not quite identical with the modern one, since he measured variability by interquartile distance rather than by standard deviation.)

Galton's mathematical and statistical ideas were often simple, though profound in their implications. In his later years he was a close friend of Karl Pearson,* then Professor of Applied Mathematics at University College, London. In 1901 he gave a sum of £200 to found the journal *Biometrika.* His Eugenics Record Office was later combined with Karl Pearson's Biometric Laboratory, to form what came to be the Galton Laboratory. Francis Galton's contributions to the advance of genetics and statistics did not end with his death, for he left £45,000 to found the Galton Chair of National Eugenics. Its first holder was Karl Pearson, who combined the studies of statistics and genetics in a single department at University College (still existing but as two separate departments).

Galton had in later life a very wide range of other interests, including psychology (memory, intelligence, perception, association, etc.), human faculty (visualization, hearing, etc.), education, fingerprints, and others, to most of which he tried to adapt statistical methods and mechanical inventions. He unsuccessfully tried to do arithmetic by smell. Many of his ideas were well ahead of their time; he suggested a coin called a "groat" (= a present-day penny), referred to the three dimensions of space and one of time, and speculated on interplanetary communication by coded signals related to simple arithmetic.

He was the author of over 300 publications, including 17 books.

REFERENCES

[1] Galton, F. (1869). *Hereditary Genius.* Macmillan, London. (Uses the word "correlation," but not with precise definition.)

[2] Galton, F. (1889). *Natural Inheritance.* Macmillan, London.

BIBLIOGRAPHY

The first four papers by Galton give quantitative values to qualitative characters, and the sixth and seventh introduce regression as a statistical concept.

Forrest, D. W. (1974). *Francis Galton, the Life and Work of a Victorian Genius.* Elek, London/Taplinger, New York. (Excellent, semipopular biography.)

Galton, F. (1872). *Fortnightly Rev.,* **12.** (Controversial application of statistical analysis to prayer.)

Galton, F. (1873). *Educational Times*. (First thoughts on the probability of survival of families.)

Galton, F. (1874). *Nature (Lond.),* **9,** 342–343.

Galton, F. (1875). *Phil. Mag.,* **49,** 33–46.

Galton, F. (1885). *J. Anthropol. Inst.,* **15,** 246–263.

Galton, F. (1886). *Proc. R. Soc.,* **40,** 42–63.

Galton, F. (1889). *Proc. R. Soc.,* **45,** 135–145. (The origin of a precise definition of correlation.)

Galton, F. (1895). *Nature (Lond.),* **51,** 319.

Galton, F. (1901). *Biometrika,* **1,** 7–10. (Recognition of biometry as a new, useful discipline.)

Galton, F. (1907). *Biometrika,* **5,** 400-404.

Galton, F. (1907). *Probability, the Foundation of Eugenics.* Henry Froude, London.

Pearson, K. (1914, 1924, 1930). *The Life, Letters and Labours of Francis Galton.* 3 vols. Cambridge University Press, Cambridge.

Watson, H. W. and Galton, F. (1874). *J. Anthropol. Inst. Gt. Brit. Ireland,* **4,** 138–144. (Introduces the Galton–Watson problem of the chance of survival of families when the distribution of the number of children is known.)

CEDRIC A. B. SMITH

GUTTMAN, LOUIS

Louis Guttman

Born: February 10, 1916, in Brooklyn, New York, USA.
Died: October 25, 1987, in Minneapolis, Minnesota, USA.
Contributed to: Scaling theory (including the Guttman scale), factor analysis, reliability theory, methodology and theory construction in social and psychological research (facet theory).

Louis Guttman was one of the most prominent psychometricians of this century and a promoter of formalization in the social and psychological sciences. He grew up in Minneapolis, Minnesota. He received his Ph.D. in sociology in 1942 from the University of Minnesota with a thesis on the algebra of factor analysis. With his academic base at Cornell University at Ithaca, New York, Guttman served during World War II as an expert consultant at the Research Branch of the Information and Education Division of the War Department, where he developed the scale bearing his name [8]. In 1947 he immigrated to Israel, where he founded and directed the Israel Institute of Applied Social Research (later the Guttman Institute of Applied Social Research). From 1955 he served also as Professor of Social and Psychological Assessment at the Hebrew University of Jerusalem, until his death on October 25, 1987. Guttman published in numerous journals and books of sociology, psychology, and statistics covering half a century from 1938, both as a sole author and in collaboration with others. Many of this earlier papers are still quoted as relevant to current statistical and mathematical advances. The development of scaling theory by Louis Guttman and Clyde Coombs is one of 62 "major advances in social science" for the period 1900–1965, identified and analyzed in *Science* [3].

While still a graduate student, Guttman undertook the clarification and formalization of intuitive—and often erroneous—techniques of data analysis that were beginning to pervade sociology and psychology [6]. Here lay the foundations of much of his later work on scale analysis, reliability theory, factor analysis, and nonmetric data analysis.

Guttman's early work focused on the linear algebra of factor analysis, frequently in relation to multiple regression (e.g. [5, 7, 9–13, 15, 16]. It ranged from developing computational formulas (for example, for the inverse of a correlation matrix [5] and for lower bounds for the number of factors [11]) to investigating fundamental issues concerning the logic of factor analysis (especially the indeterminancy of factor scores [12]; see also [30]). His works were highly valued by psychometricians, both practitioners and theoreticians.

Later on, in the early 1950s, Guttman proposed a new way of looking at factor analysis, initiating a radical shift in the way multivariate data can be analyzed and interpreted. Revising Spearman's notion of a hierarchy among intelligence tests,

Guttman conceived of tests that increase in complexity, as if they successively activated a sequence of "bonds" in the human mind: as tests become more complex, additional centers, along a given path, are employed. This consideration resulted in the number of factors being equal to the number of tests, in seeming contradiction to the very purpose of factor analysis, which seeks to identify a small number of underlying factors. However, a simple (unidimensional) ordering among tests is in itself a parsimonious representation of a new kind, which holds regardless of the number of factors [10, 25, 28]. This representation, termed the (parametrized) *simplex,* implies certain relationships among correlation coefficients: tests may all be mapped (as points) into a straight line, so that the larger the correlation coefficient between any two tests, the closer they are on the line [10, 29]. Similarly, the circumplex configuration (tests circularly ordered in the plane) has been formulated and discovered in psychological data (e.g. of color perception [24]). The *radex*—a two-

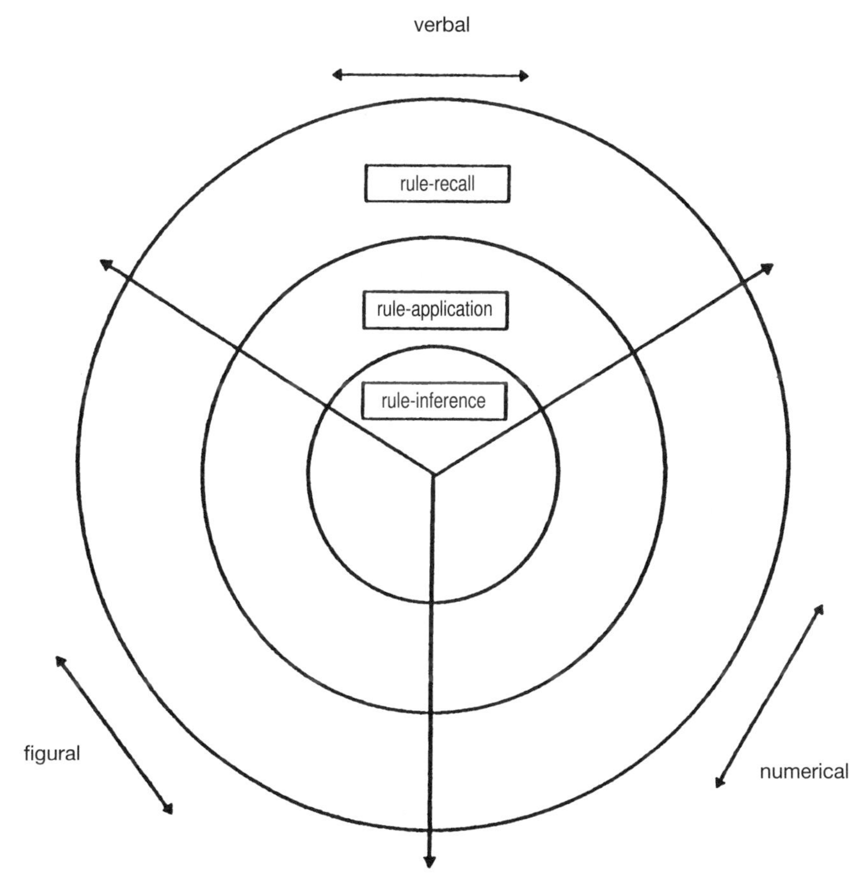

Figure 1. The radex theory of intelligence. The material facet is angular, the task facet is radial. One implication: rule recall differentiates among abilities better than rule inference.

dimensional concatenation of simplexes and circumplexes, wherein each test is a member of a simplex and of a circumplex—was introduced in its parametrized form by Guttman [16] and then corrected and developed further by others (see [13] for an excellent review of the parametrized formulation of some basic test configurations). In 1968 Guttman published in *Psychometrika* his much quoted paper "A general nonmetric technique for finding the smallest coordinate space for a configuration of points," an algorithm for mapping tests in the space of the smallest dimensionality capable of reflecting pairwise similarity (e.g. correlations) between them [18]. This data-analytic procedure has become known as *smallest-space analysis* (SSA) and has been computer-programmed and included in statistical packages as a multidimensional scaling (MDS) technique.

Guttman's true concern, however, was not with statistical or data-analytic procedures as such. Rather, he sought to combine such procedures, and the aspect of empirical data they invoke, with conceptual–definitional framework of a substantive domain of research, in order to discover lawfulness, and contribute to theory construction in social and psychological domains of research. Indeed, in order to clarify his research strategy, Guttman felt he should propose a definition for the very term *theory* (in the context of the empirical sciences):

> An hypothesis of a correspondence between a definitional system of observations and an aspect of the empirical structure of those observations, together with a rationale for such an hypothesis (e.g. ref. [25]).

The set of all items—observable variables—that pertain to an investigated concept (such as marital adjustment of couples, or intelligence of individuals) concern all acts of that concept (i.e. acts of adjustment, or acts of intelligence). Guttman called that set the "universe of content." Indeed, as a research strategy, Guttman's *facet theory* requires, first, the definition of the concept studied in terms of its content universe, i.e., its items. For example, he defined attitude items as those whose range (set of response categories, or values) is ordered from very positive to very negative behavior towards an object; and intelligence items, as those whose range is ordered from very correct to very incorrect performance with respect to an objective rule. The specification of such a common meaning to item ranges provides a rationale for "laws of monotonicity." For example:

> *The law of intelligence monotonicity:* If any two items are selected from the universe of intelligence items, and if the population observed is not selected artificially, then the population regression between these items will be monotone with positive or zero sign (e.g. [25, 29]).

This law, on the one hand, summarizes in a formal fashion findings that have been noted for some time by intelligence researchers, and , on the other hand (as an incumbent hypothesis), tells us what to expect in the future observations. Since the law specifes a correspondence between a definitional framework (i.e. the common range of intelligence items, as defined) and an aspect of the empirical data (i.e. the correlation sign), it qualifies as a "theory" according to the above-stated definition.

Guttman suggested that finer laws can be proposed by classifying items according to aspects of their *content*. One such classification for intelligence items may be the *material facet*—i.e. whether the items deal with verbal, numerical, figural, or some other kind of material. Another, independent classification, is that of the *task facet*—i.e. whether items require rule recall, rule application, or rule inference.

The unique feature of Guttman's facet theory as a methodological philosophy and a research strategy stems from the recognition that behavioral scientific concepts are typically manifested by an infinite number of items (observational variables) but only a finite sample of items can be actually observed. Facet theory copes with this challenge in both its aspects: the research design and the data analytic. For the research design Guttman proposes the *mapping sentence* technique [14, 25, 29, 1], which focuses the researcher's attention on a finite number of relevant conceptual facets (content classifications) rather than on the infinite number of items. Multivariate data analysis has coped with the question of item sampling by the gradual shift from traditional factor analysis, through parametrized test configuration and the contiguity principle [4], to regional partition patterns in continuous SSA spaces [26, 29].

There remained, of course, the problem of scoring individuals with respect to the studied concept. Guttman insisted that single scores can be assigned to subjects only if observed profiles (rows in data matrix) turn out in fact to form a unidimensional (Guttman) scale (also known as cumulative scale). Most often they do not, and the methods of multiple scaling by partial-order scalogram analysis must be used [27].

Guttman was strongly individualistic and creative, yet committed to the scientific tradition; kindly in his daily contacts, yet argumentative, sometimes to the point of arrogance, in insisting on the principles in which he so deeply believed. His conviction in the formalized integration of conceptual framework with data analysis has placed him in the seemingly dual position of "a statistican among psychologists" opposed to untestable "theories," and a "psychologist among statisticians" opposed to routine "number crunching." Colleagues and students who were not alienated by his style benefited from his seminal insights and provocative presentations.

Guttman's work has had an influence on intelligence research, attitude research, environmental psychology, the study of general behavioral systems, and other fields. In recognition of his scientific contributions Guttman was awarded the Rothschild Prize for Social Sciences, and the Israel Prize. He was elected to memberships in the Israel Academy of Science and Humanities, and to foreign honorary membership of the American Academy of Arts and Sciences. He held the Andrew White Professorship-at-Large at Cornell University, received the Outstanding Achievement Award of the University of Minnesota and the 1984 Educational Testing Service Award for distinguished service to measurement. The latter's citation recognized that "a central theme in Guttman's work [is] that measurement is not merely the assignment of numbers but the construction of structural theory." [19]

REFERENCES

[1] Borg, I. and Shye, S. (1995). *Facet Theory: Form and Content.* Sage Thousand Oaks, Calif. [A mathematical formulation of facet theory, and algorithms for computing optimal regions in SSA (MDS).]

[2] Canter, D., ed. (1985). *Facet Theory: Approaches to Social Research.* Springer-Verlag, New York. (A selection of applications of facet theory to psychology.)

[3] Deutsch, K. W., Platt, J., and Senghan D. (1971). Conditions favoring major advances in social sciences. *Science,* **171,** 450–459.

[4] Foa, U. G. (1958). The contiguity principle in the structure of interpersonal behavior. *Human Relations,* **11,** 229–238.

[5] Guttman, L. (1940). Multiple rectilinear prediction and the resolution into components. *Psychometrika,* **5,** 75–99.

[6] Guttman, L. (1941). The quantification of a class of attributes: A theory and method of scale construction. In *The Prediction of Personal Adjustment,* P. Horst, L. Guttman, et al., eds. Social Science Research Council, New York.

[7] Guttman, L. (1944). General theory and methods for matric factoring. *Psychometrika,* 9, 1–16.

[8] Guttman, L., Stouffer, S. A., et al. (1950). *Measurement and Prediction* Studies in Social Psychology in World War II, vol. 4. Princeton University Press, Princeton, N. J.

[9] Guttman, L. (1952). Multiple-group methods for common-factor analysis. *Psychometrika,* **17,** 209–222.

[10] Guttman, L. (1954). A new approach to factor analysis: The radex. In *Mathematical Thinking in the Social Sciences,* P. F. Lazarsfeld, ed. Free Press, New York.

[11] Guttman, L. (1954). Some necessary conditions for common-factor analysis. *Psychometrika,* **19,** 149–161.

[12] Guttman, L. (1955). The determinacy of factor score matrices with implications for five other basic problems of common-factor theory. *Brit. J. Statist. Psychol.,* **8,** 65–81.

[13] Guttman, L. (1956). "Best possible" systematic estimates of communalities. *Psychometrika,* **21,** 272–228.

[14] Guttman, L. (1957). Introduction to facet design and analysis. *Proc. Fifteenth Int. Congress of Psychology, Brussels.* North-Holland, Amsterdam.

[15] Guttman, L. (1957). Simple proofs of relations between the communality problem and multiple correlation. *Psychometrika,* **22,** 147–157.

[16] Guttman, L. (1958). To what extent can communalities reduce rank? *Psychometrika,* **23,** 297–308.

[17] Guttman, L. (1965). A faceted definition of intelligence. In *Studies in Psychology, Scripta Hierosolymitana,* vol. 14. Hebrew University, Jerusalem, pp. 166–181.

[18] Guttman, L. (1968). A general nonmetric technique for finding the smallest coordinate space for a configuration of points. Psychometrika, **33,** 469–506.

[19] Guttman, L. (1971). Measurement as structural theory. Psychometrika, 36, 329–347.

[20] Guttman, L. and Levy, S. (1991). Two structural laws for intelligence tests. *Intelligence,* **15,** 79–103.

[21] Levy, S. (1985). Lawful roles of facets in social theories. In *Facet Theory: Approach to Social Research,* D. Canter, ed. Springer-Verlag, New York, pp. 59–96.

[22] Levy, S., ed. (1994). *Louis Guttman on Theory and Methodology: Selected Writings.* Dartmouth, Aldershot, England. (A good selection of Guttman's original papers.)

[23] Schlesinger, I. M. and Guttman, L. (1969). Smallest space analysis of intelligence and achievement tests. *Psychol. Bull.,* **71,** 95–100.

[24] Shepard, R. N. (1978). The circumplex and related topological manifolds in the study of perception. In *Theory Construction and Data Analysis in the Behavioral Sciences,* S. Shye, ed. Jossey-Bass, San Francisco.

[25] Shye, S. (1978). On the search for laws in the behavioral sciences. In *Theory Construction and Data Analysis in the Behavioral Sciences,* S. Shye, ed. Jossey-Bass, San Francisco.

[26] Shye, S. (1978). Facet analysis and regional hypothesis. In *Theory Construction and Data Analysis in the Behavioral Sciences,* S. Shye, ed. Jossey-Bass, San Francisco.

[27] Shye, S. (1985). *Multiple scaling: The Theory and Application of Partial Order Scalogram Analysis.* North Holland, Amsterdam. (Scalogram algebra extending the Guttman scale to spaces of higher dimensionalities, with examples.)

[28] Shye, S. (1988). Inductive and deductive reasoning: A structural reanalysis of ability tests. *J. Appl. Psycho.,* **73,** 308–311. (With appendix comparing SSA and factor analysis.)

[29] Shye, S. and Elizur, D. (1994). *Introduction to Facet Theory: Content Design and Intrinsic Data Analysis in Behavioral Research.* Sage, Calif. Thousand Oaks, (A basic textbook.)

[30] Steiger, J. and Schonemann, P. (1978). A history of factor indeterminacy. In S. Shye, ed. *Theory Construction and Data Analysis in the Behavioral Sciences,* Jossey-Bass, San Francisco.

[31] Van den Wollenberg, A. L. (1978). Nonmetric representation of the radex in its factor pattern parametrization. In *Theory Construction and Data Analysis in the Behavioral Sciences,* S. Shye, ed. Jossey-Bass, San Francisco.

SAMUEL SHYE

Hoeffding, Wassily

Born: June 12, 1914, in Mustamaki, Finland, near St. Petersburg, Russia.
Died: February 28, 1991, in Chapel Hill, North Carolina, USA.
Contributed to: mathematical statistics, probability theory, mathematics, numerical analysis.

Wassily Hoeffding

Wassily Hoeffding spent the early part of his life in the St. Petersburg area. His father, of Danish origin, was an economist, and his mother specialized in medicine. The Hoeffding family moved to Germany, via Denmark, when Wassily was only about 6 years old, and, after finishing high school in 1933, he proceeded to higher education in economics. However, a year later he switched to mathematics, and he earned his Ph.D. degree, from Berlin University in 1940, with a dissertation on nonparametric measures of association and correlation. In fact, the identity he established, for bivariate measures of association [14], is popularly known as the Hoeffding lemma. During World War II, Hoeffding continued to live in Berlin and worked as an editorial assistant for the *Jahrbuch über die Fortschritte der Mathematik* at the Prussian Academy of Sciences (1940–1945), and also for the Berliner Hochschulinstitut für Versicherungswissenschaft (1941–1945). He migrated to the USA in 1946, and after a sojourn at Columbia University, New York, he settled in Chapel Hill, North Carolina. The Hoeffding family was divided: Wassily's two brothers, one a physician and the other an economist, were in other places, but his mother lived with him in Chapel Hill until her death, some 20 years later. He was research associate, 1946–1979, assistant professor, 1948–1952, associate professor, 1952–1956, and professor, 1956–1979, and he retired from active service as Kenan Professor of Statistics, University of North Carolina, Chapel Hill, in 1979. He was a professor emeritus there for the next 12 years. His occasional visits to other campuses during this longtime residence in Chapel Hill include Columbia and Cornell Universities in New York, the Steklov Institute in St. Petersburg, Russia, and the Indian Statistical Institute, Calcutta. A person of Danish ancestry Russian by birth and educated in Germany, Hoeffding had a remarkable appreciation for European literature (Russian was his favorite) and a very gentle sense of humor. In spite of working out a novel and highly original doctoral dissertation in Berlin, he felt that [2] ". . . probability and statistics were very poorly represented in Berlin at that time (1936–1945) . . . ," and it was only after moving to the USA that he started to appreciate the full depth of probability theory and statistics. In Chapel Hill, his creative work, though never very prolific (especially in the last twenty years of his life), culminated at an extraordinary level of perfection and novelty, covering a broad spectrum of mathematical statistics and probability theory. He was actively involved with reviewing published papers for *Mathematical Reviews* as well as manuscripts submitted to various journals in statistics and probability theory; translating Russian publications into English in this greater domain; and, in collaboration with Dr. S. Kotz, compiling the much needed *Russian–English Statistical Directory and Reader,* published by the University of North Carolina Press in 1964. His meticulous lectures on statistical inference, sequential inference and decision theory, offered year after year in Chapel Hill, earned the recognition of excellence from his numerous doctoral students, and from colleagues as well. In the 1970s, he started offering a highly motivated course on asymptotic statistical methods, and although some students were keen enough to prepare a mimeographed lecture-note version, Hoeffding did not develop his courses into monographs. During his tenure at Chapel Hill, he received numerous honors, including prestigious memberships in the National Academy of Sciences and American Academy of Arts and Sciences. He was a Fellow of the Institute of Mathematical Statistics, served as its president (1969), and delivered the Wald Memorial Lectures in 1967. He was also a Fellow of the American Statistical Association and an Honorary Fellow, Royal Society.

Among the diverse areas enriched by Hoeffding's outstanding research contributions, we may specially mention the following: nonparametrics, sequential analysis, statistical decision theory, probability inequalities, and central limit theorems. Hoeffding started his postdoctoral career with some original work on nonparametric measures of association and stochastic dependence. Some statistics arising in this context are not expressible as sums or averages of independent random variables, and for them distribution theory (except under suitable hypotheses of invariance) was not precisely known at that time. Later on, in 1948, Hoeffding laid down the foundation of the general theory of U-statistics, which deals with the basic formulation of statistical parameters in a nonparametric setup, construction of suitable unbiased and symmetric estimators of such functionals, and a thorough and well-unified treatment of their (sampling) distributional properties with direct access to the related asymptotic theory. In fact, the novel projection result Hoeffding considered for such functionals led to the general decomposition of possibly nonlinear statistics into linear ones and higher-order terms, and at the present time, this is referred to as the Hoeffding (or H) decomposition (see van Zwet [35]). Judged from a broader perspective, Hoeffding's U-statistics paper has been a landmark, and we may refer to an introduction [34] emphasizing its significance in *Breakthroughs in Statistics,* vol. 1, in which it is included. Hoeffding had a lot of affection for this fundamental work. He stated [4], "I like to think of this paper as my real Ph.D. dissertation." Had it not been for World War II, he might have achieved this accomplishment even earlier.

During 1947–1966, Hoeffding's creativity emerged in several directions. In addition to his extensive and highly original research work in the general area of nonparametrics and order statistics, his contributions to distribution theory, probability inequalities, large-deviation probabilities, statistical decision theory, and sequential analysis were most noteworthy. The projection technique he developed for *U*-statistics led him immediately to study nonnormal distribution theory for some tests for independence in bivariate distributions [9], and to collaborate with Herbert Robbins in a novel result on a central limit theorem for *m*-dependent random variables [29]. This was the time when permutational central limit theorems for linear statistics were being explored by A. Wald,* J. Wolfowitz,* W. Madow, and G. E. Noether in increasing generality. Hoeffding [11] not only brought in new developments in this sector, but also extended the theory to cover more general bilinear statistics under a Lindeberg-type regularity assumption; the ultimate answer to this query was provided more than a decade later by Jájek [3]. Although the permutational central limit theorems were useful in the study of the asymptotic distribution theory of various test statistics under suitable hypotheses of invariance, they were not of much use for the study of their power properties even in an asymptotic setup. Hoeffding attacked the problem from two related directions. He developed general asymptotics for the power of permutation tests against suitable parametric alternatives [12], and he reinforced the local optimality of the Fisher–Yates rank test for normal shift alternatives, extended this characterization to a general regression setup, and covered a broad class of underlying densities [10]. In this context [13] he developed some nice asymptotic properties of "expected order statistics" and illustrated their basic role in nonparametrics. His novel work in collaboration with his advisee J. R. Rosenblatt [30] on asymptotic relative efficiency also deserves special mention. Among other nonparametric works of Hoeffding, mention may be made of the centering of a simple linear rank statistic [24], wherein he made use of some elegant results on the L_1-norm of certain approximations for Bernstein-type polynomials and splines with equidistant knots [23, 25].

Hoeffding's work on statistical decision theory and sequential analysis was select, but fundamental. He developed some lower bounds for the expected sample size and the average risk of a sequential procedure [14, 18], stressed the role of assumptions in statistical decisions [17], and with J. Wolfowitz [33] studied distinguishability of sets of distribution in the case of i.i.d. r.v.'s. A decade later, he came up, with G. Simons, with an interesting result on unbiased coin tossing with a biased coin. He had also worked on incomplete and boundedly complete families of distributions [26, 27] as well as range-preserving estimators [28].

Hoeffding had a genuine interest in distribution theory for parametric as well as nonparametric statistics. With S. S. Shrikhande [31] he studied some bounds for the distribution of a sum of i.i.d. r.v.'s, and later on provided a more in-depth study of the Bernoulli case [16] and also the case of random vectors [19]. His classical paper [20] on probability inequalities for sums of bounded random variables has been indeed a milestone in large-sample theory and nonparametrics. They are affectionately known as the Hoeffding inequalities and have also been adopted in many nonstandard situations.

In 1965 Hoeffding came up with another outstanding piece of research work on asymptotically optimal tests for multinomial distributions [21], extended further to cover the case of other distributions [22], and in both cases, he laid special emphasis on the use of "large-deviation probabilities" in the characterization of such asymptotic optimality properties. His research opened up a broad avenue of fruitful research work in asymptotic statistical inference in the next two decades.

Hoeffding had five entries [on (1) asymptotic normality, (2) Hajek's projection lemma, (3) his 1948 test of independence, (4) probability inequalities for sums of bounded random variables, and (5) range-preserving estimators] in the *Encyclope-*

dia of Statistical Sciences, edited by S. Kotz, L. Johnson, and C. B. Read. These articles are reproduced in his collected works [1], and they also reveal his own contributions in these areas. Good reviews of his research work appeared in the form of three expository articles in [1], and the following (selected) bibliography is mainly adapted from there.

REFERENCES

[1] Fisher, N. I. and Sen, P. K., eds. (1994). *The Collected Works of Wassily Hoeffding.* Springer-Verlag, New York. [In addition to the collected works of Wassily Hoeffding, this volume contains a set of three expository papers written by (1) K. Oosterhoff and W. van Zwet, (2) G. Simons, and (3) P. K. Sen, wherein the significance of Hoeffding's work in different areas stressed.]

[2] Gani, J., ed. (1982). *The Making of Statisticians.* Springer-Verlag, New York.

[3] Hájek, J. (1961). *Ann. Math. Statist.,* **32,** 506–523.

[4] Hoeffding, W. (1940). *Schrift. Math. Inst. Und Inst. Angew. Math. Univ. Berlin,* **5**(3), 180–233.

[5] Hoeffding, W. (1941). *Arch. Math. Wirtsch. Sozial.,* **7,** 49–70.

[6] Hoeffding, W. (1942). *Skand. Aktuar.,* **25,** 200–227.

[7] Hoeffding, W. (1947). *Biometrika,* **34,** 183–196.

[8] Hoeffding, W. (1948). *Ann. Math. Statist.,* **19,** 293–325.

[9] Hoeffding, W. (1948). *Ann. Math. Statist.,* **19,** 546–557.

[10] Hoeffding, W. (1951). *Proc. 2nd Berkeley Symp. Math. Statist. Probab.,* pp. 83–92.

[11] Hoeffding, W. (1951). *Ann. Math. Statist.,* **22,** 558–566.

[12] Hoeffding, W. (1952). *Ann. Math. Statist.,* **23,** 169–192.

[13] Hoeffding, W. (1953). *Ann. Math. Statist.,* **24,** 93–100.

[14] Hoeffding, W. (1953). *Ann. Math. Statist.,* **24,** 127–130.

[15] Hoeffding, W. (1955). *Ann. Math. Statist.,* **26,** 268–275.

[16] Hoeffding, W. (1956). *Ann. Math. Statist.,* **27,** 713–721.

[17] Hoeffding, W. (1956). *Proc. 3rd Berkeley Symp. Math. Statist. Probab.,* vol. 1, pp. 105–114.

[18] Hoeffding, W. (1960). *Ann. Math. Statist.,* **31,** 352–368.

[19] Hoeffding, W. (1961). *Proc. 4th Berkeley Symp. Math. Statist. Probab.,* vol. 2, pp. 213–236.

[20] Hoeffding, W. (1963). *J. Amer. Statist. Assoc.,* **58,** 13–30.

[21] Hoeffding, W. (1965). *Ann. Math. Statist.,* **36,** 369–408.

[22] Hoeffding, W. (1967). *Proc. 5th Berkeley Symp. Math. Statist. Probab.,* vol. 1, pp. 203–219.

[23] Hoeffding, W. (1971). *J. Approx. Theory,* **5,** 347–356.

[24] Hoeffding, W. (1973). *Ann. Statist.,* **1,** 54–66.

[25] Hoeffding, W. (1974). *J. Approx. Theory,* **11,** 176–193.

[26] Hoeffding, W. (1977). *Ann. Statist.,* **5,** 278–291.

[27] Hoeffding, W. (1977). *Proc. Symp. Statist. Dec. Theor. Rel. Top.,* S. S. Gupta and D. S. Moore, eds. Academic Press, New York, pp. 157–164.

[28] Hoeffding, W. (1984). *J. Amer. Statist. Ass.,* **79,** 712–714.

[29] Hoeffding, W. and Robbins, H. (1948). *Duke Math. J.,* **15,** 773–780.

[30] Hoeffding, W. and Rosenblatt, J. R. (1955). *Ann. Math. Statist.,* **26,** 52–63.

[31] Hoeffding, W. and Shrikhande, S. S. (1955). *Ann. Math. Statist.,* **26,** 439–449.

[32] Hoeffding, W. and Simons, G. (1970). *Ann. Math. Statist.,* **41,** 341–352.

[33] Hoeffding, W. and Wolfowitz, J. (1958). *Ann. Math. Statist.,* **29,** 700–718.

[34] Kotz, S. and Johnson, N. L., eds. (1991). *Breakthroughs in Statistics, Volume 1: Foundations and Basic Theory.* Springer-Verlag, New York.

[35] van Zwet, W. R. (1984). *Z. Wahrsch. Verw. Geb.,* **66,** 425–440.

PRANAB K. SEN

Hotelling, Harold

Harold Hotelling

Born: September 29, 1895, in Fulda, Minnesota, USA.
Died: December 26, 1973, in Chapel Hill, North Carolina, USA.
Contributed to: econometrics, multivariate analysis, statistical inference.

Harold Hotelling, was responsible for much pioneering theoretical work in both statistics and mathematical economics, and did much to encourage and improve the teaching of statistics at U.S. universities. At an early age his father's business forced a move to Seattle, and so, not surprisingly, he attended the University of Washington.

It is interesting that he chose to study journalism as his major, and that he worked on various newspapers in Washington while a student, obtaining his degree in journalism in 1919. This doubtless accounts for the particular interest he always had, even when he was a distinguished leader of the group of theoretical statisticians at the University of North Carolina at Chapel Hill, in ensuring that the achievements of his colleagues received ample attention in the local (and in some cases the national) press.

While majoring in journalism he also took some mathematics courses from Eric Temple Bell, who recognized his analytical abilities and steered him toward mathematics. Thus it was that, now with mathematics as his major subject, he took a Master of Science degree at Washington in 1921, and a Doctorate of Philosophy at Princeton in 1924. His Ph.D. dissertation was in the field of topology, published in 1925 in the *Transactions of the American Mathematical Society.*

After obtaining his doctorate he began work at Stanford University, originally in the Food Research Institute, and his interest in probability and statistics began to take hold; he taught his first courses in statistical theory and practice (a great novelty at that time) and began to publish the first of a long series of scholarly articles. His earliest applications of mathematical ideas concerned journalism and political science; from these he turned to population and food supply, and then to theoretical economics, in which he was one of the initiators of the modern theories of imperfect competition and welfare economics. At the same time he was producing a series of publications in theoretical statistics which were often of such originality and importance that they provoked a considerable amount of later research by many scholars in many lands. In 1931 he published in the *Annals of Mathematical Statistics* what is quite possibly his most important contribution to statistical theory, his paper, "The generalization of Student's ratio."

In 1931 he was appointed Professor of Economics at Columbia University, where he would stay for 15 years. It was while he was there that he was able to assist various refugee scholars from central Europe, including the late Abraham Wald.

During World War II he organized at Columbia University the famous Statistical Research Group, which was engaged in statistical work of a military nature; the group included Wald*, Wallis, and Wolfowitz,* and one of its signal achievements was the theory of sequential procedures.

In 1946 came his final move, to the University of North Carolina at Chapel Hill, where he was almost given carte blanche to create a theoretical statistics department. He rapidly recruited many able scholars, including R. C. Bose,* W. Hoeffding,* W. G. Madow, H. E. Robbins, S. N. Roy, and P. L. Hsu.*

In statistical theory he was a leader in multivariate analysis, being responsible for some of the basic ideas and tools in the treatment of vector-valued random variables. His major contribution to this area has come to be called "Hotelling's generalized T^2." He also played a major role in the development of the notions of principal components and of canonical correlations.

As early as 1927 he studied differential equations subject to error, a topic of general current interest, and published one of the first papers in this field. His papers on rank correlation, on statistical prediction, and on the experimental determination of the maximum of a function also stimulated much further research in succeeding decades.

In economic theory, his papers on demand theory, on the incidence of taxation, and on welfare economics are already regarded as classics that form the basis for much further work that has been done since they were written. In demand theory he was one of a small number of pioneers to revolutionize the basis of that theory and to extend its applications. His work on the incidence of taxation is important still in the literature of public finance.

In 1955 he was awarded an honorary LL.D. by the University of Chicago; in 1963 he was awarded an honorary D.Sc. By the University of Rochester. He was an Honorary Fellow of the Royal Statistical Society and a Distinguished Fellow of the American Economic Association. He served as President of the Econometric Society in 1936–1937 and of the Institute of Mathematical Statistics in 1941. In 1970 he was elected to the National Academy of Sciences, in 1972 he received the North Carolina Award for Science, and in 1973 he was elected to a membership of the Accademia Nazionale dei Lincei in Rome. It is sad to say that, by the time this last honor had come his way, he had already suffered a severe stroke, in May 1972, which led to his death on December 26, 1973.

WALTER L. SMITH

Hsu, Pao-Lu

Born: September 1, 1910 (given as 1909 in ref. [2]), in Beijing, China.
Died: December 18, 1970, in Beijing, China.
Contributed to: multivariate analysis, statistical inference, central limit theorems, design of experiments, teaching of statistics.

Pao-Lu Hsu was born in Beijing, but his family home was in Hangzhou City, Zhejian. He enrolled in the Department of Chemistry in Yan Jing University (later Beijing University) in 1928, but transferred to the Department of Mathematics in Qin Huo University in 1930. He graduated, with a Bachelor of Science degree, in 1933. After teaching for three years in Beijing University he attended University College London from 1936, receiving a Ph.D. degree in 1938 and a D.Sc. In 1940.

At the beginning of 1941, he accepted a professorship at Beijing University. In 1945–1948, he was a visiting professor at Columbia University (New York), the University of North Carolina at Chapel Hill, and the University of California at Berkeley. He then returned to Beijing University, where he remained as a professor for the rest of his life. During this period he was elected a member of the Chinese Academy of Sciences. He never married.

Professor Hsu was an internationally recognized authority in the areas of probability theory and mathematical statistics. He was the first person to obtain world status in modern Chinese history. In connection with the 70th anniversary of his birth, there appeared, in the *Annals of Statistics,* several papers [1, 2, 3, 9] on Hsu's life and work, which contain much detailed information.

During his first two or three years in University College, London, Hsu produced his first papers on mathematical statistics [4–6]. The first of these obtained exact results for distributions of test statistics in the Behrens–Fisher problem—testing the equality of expected values of two normal distributions with unknown variances. The method he used is still referred to, on occasion, as "Hsu's method."

Another early field of study was second-order asymptotic variance estimators in the general linear model. Hsu obtained necessary and sufficient conditions for the Markov estimator to be the best unbiased second-order estimator. This result has been recognized as a fundamental breakthrough.

In multivariate analysis, he obtained an elegant derivation of the Wishart distribution and the distributions of roots of certain determinantal equations in connection with likelihood-ratio tests of multivariate hypotheses on parameters in multinormal general linear models. In the course of this work he developed some new results in matrix theory itself. He also obtained important results in regard to conditions for optimality of likelihood-ratio tests of linear hypotheses.

Hsu also did a considerable amount of research on the distribution of sample variances from nonnormal populations. Not only did he obtain an optimal estimate of the difference between the standardized distribution of sample variance $G_n(n)$ and the standardized normal distribution $\Phi_n(x)$, but also an asymptotic expansion for $G_n(x)$, extending the results of Cramér* and Berry for sample means. His methods are also applicable to distributions of higher-order sample moments, sample correlation coefficients, and Student's t.

A challenging problem in the 1940s was derivation of weak limit distributions for sums of rows of random variables with triangular variance–covariance matrices. Many famous scholars, including P. Lévy,* W. Feller,* A. N. Kolmogorov,* and B. V. Gnedenko,* investigated this problem. Hsu obtained, independently, a set of necessary and sufficient conditions for a weak limit to exist.

Yet another line of Hsu's basic research was on the strong law of large numbers. In 1947, Hsu and H. E. Robbins introduced the concept of complete convergence in this context.

Many of the scientific papers of Hsu have been collected in refs. [7] (Chinese edition) and [8] (English edition). In addition to the topics described above, they include work on Markov processes, design of experiments, and limit theorems for order statistics.

In a more general field, Hsu made many contributions to the development of study and use of modern statistical methods in China. It was largely due to his efforts that in 1956, probability and statistics were chosen as important areas for development in China's long-term science program. Professor Hsu was a leader in these subjects, and the first group of 50 students chosen to study them did so under his direction in Beijing University. This group later had great influence on the development of education and research in probability and statistics in China. At that time, also, Professor Hsu invited experts from Europe and the Soviet Union to visit Beijing and lecture to his group, and arranged for China to send some young people to study abroad. These actions promoted the development of probability theory and mathematical statistics in China.

In 1956, also, the first research institute for the theory of probability and mathematical statistics in China was established at Beijing University, with Professor Hsu as Director, a post he occupied until his death.

Pao-Lu was a dedicated scientist. From 1950 onwards, he was in poor health from chronic tuberculosis, but pursued his research assiduously, continuing to teach from his room when he became unable to move around freely, and advising many students working on their theses. He was an example to all in the amazing will power he evinced in attempting to attain his goals.

ACKNOWLEDGMENT

This entry is based on English translations by Dr. S. T. Lai, Vitreous State Laboratory, The Catholic University of America, Washington, DC 20064, Professor Yi Su, University of Maryland, College Park, MD 20742, and Professor W. L. Pearn, National University Taipei, Taiwan.

REFERENCES

[1] Anderson, R. (1979). Hsu's work on multivariate analysis. *Ann. Statist.*, **7**, 474–478.

[2] Anderson, T. W., Chung, K. L. and Lehmann, E. L. (1979). Pao-lu Hsu 1909–1970. *Ann. Statist.*, **7**, 467–470.

[3] Chung, K. L. (1979). Hsu's work in probability. *Ann. Statist.,* **7,** 479–483.

[4] Hsu, P.-L. (1938). Contributions to the two-sample problems and the theory of Student's *t. Statist. Res. Mem.,* **2,** 1–20.

[5] Hsu, P.-L. (1938). On the best quadratic estimate of the variance, *Statist. Res. Mem.,* **2,** 91–104.

[6] Hsu, P.-L. (1938). Notes on Hotelling's generalized *T, Ann. Math. Statist.,* **9,** 231–243.

[7] Hsu, P.-L. (1981). *Hsu Pao-lu's Collected Papers* (in Chinese). Chinese Sciences, Beijing.

[8] Hsu, P.-L. (1982). *Pao-lu Hsu's Collected Papers.* Springer-Verlag, New York.

[9] Lehmann, E. L. (1979). Hsu's work on inference. *Ann. Statist.,* **7,** 471–473.

CHEN JIADING

Jeffreys, Harold

Born: April 22, 1891, in Co. Durham, U.K.
Died: March 18, 1989, in Cambridge, U.K.
Contributed to: geophysics, astronomy, mathematical physics, probability theory, scientific inference.

Although most of the recognition during Jeffreys' working lifetime came from his work in the first three fields, his important contributions to the last two are what concern us here, where recognition came later.

Jeffreys went up to St. John's College, Cambridge in 1910 and was made a Fellow in 1914, a post he held until his death, a record 75 years. From 1946 until his retirement in 1958, he was Plumian Professor of Astronomy and Experimental Philosophy. He was knighted in 1953. He was a recipient of the Royal Medal of the Royal Society and of a Guy Medal in gold of the Royal Statistical Society. He was a superb science writer whose two major books, *The Earth: Its Origins, History and Physical Constitution* and *Theory of Probability,* describe important, original contributions to their fields. In the first, he developed new ways of analyzing seismic data. These, together with a general interest of the Cambridge of the 1920s in the philosophy of scientific method, led him to develop a general approach to inference and statistics that was then applied to a wide range of scientific situations. The Theory, as he liked to call his book, is remarkable for its blend of theory with practice, and, unlike the other major developers of what is now called Bayesian statistics (de Finetti,* Ramsey,* and Savage*), he used the theory to develop new, operational methods. He disagreed strongly with the views of Popper on scientific method and felt Popper's use of probability was wrong.

The usual model for inference contains data x and parameters θ, linked by probability distributions $\Pr[x|\theta]$ for x given θ. Jeffreys differed from the popular view, exemplified by the work of Neyman* and Fisher,* in also including a probability distribution $\Pr[\theta]$ for θ. The argument is that both types of quantity are, initially, uncertain and that probability is the only suitable mechanism to measure uncertainty. In other words, all statements of uncertainty should combine according to the three rules of the probability calculus: convexity, addition, and multiplication. The first chapter of the Theory explains why this is so, With $\Pr[x|\theta]$ included, inference is accomplished by Bayes' theorem, $\Pr[\theta|x] \propto \Pr[x|\theta]. \Pr[\theta]$ for the uncertainty of θ given data x. Probabilities are interpreted as degrees of belief.

Differing from Neyman and Fisher in the use of $\Pr[\theta]$, he also differed from de Finetti and Savage in adopting an impersonal view of probability. They argued that scientists could legitimately differ in their uncertainties, even on the same data, provided only that they obeyed the probability calculus. According to this subjective view, scientific objectivity only came about through masses of data that drew differing opinions together. Jeffreys felt that rational scientists, on the same evidence, ought to agree on the probability. If this is correct, it should be possible to produce rules to determine the unique probability. One way to do this would be to describe ignorance, say of θ, where there was no knowledge of θ and the evidence E is empty. The general concept, given any evidence, could then be found by Bayes's theorem. An influential contribution of Jeffreys was to develop invariance rules for ignorance. Thus the uniform distribution corresponds to no knowledge of a location parameter. His method was not entirely satisfactory, as he admitted, but it has led to many useful developments, like reference priors, in a field that is active in the 1990s. The term "Jeffreys prior" is widely used.

Another important, original contribution of his was to develop a general scenario for the testing of a scientific hypothesis *H*. According to the thesis described above, this is accomplished by calculating the uncertainty of *H*, expressed through the probability of *H*, $\Pr[H|E]$, on evidence *E*. In agreement with Neyman, but not Fisher, this involved considering alternatives to *H*. His novel approach has a concentration of probability on H, the rest of the probability being spread over the alternatives. Thus a test of $\theta = 0$, where θ is a real number, would typically have $\Pr[\theta = 0] > 0$, with a density of θ over $\theta \neq 0$. Recent work has shown that there are often serious discrepancies between $1 - \Pr[H|E]$ and the tail-area probability popular amongst statisticians and scientists. We see here an example of the fallacy of the transposed conditional, where the probability of *H* given *E* (Jeffreys) is confused with the probability of *E* given *H* (Fisher's tail area). The discrepancies are not so serious, nor so common, in estimation, where the prior distribution does not have a concentration of probability on special values that describe H.

Jeffreys was a poor oral communicator, but this *Theory* is a masterpiece of original, important work written in a beautiful, economical style. Dismissed as outside the mainstream when it appeared in 1939 and for several years after, it is now widely cited. It still has much to teach us and will live as one of the truly important scientific works of the twentieth century. The journal *Chance* **4**(2) (1991) has several articles on aspects of his life and work.

REFERENCES

[1] Jeffreys, H. (1961) *Theory of Probability*. Clarendon Press, Oxford. (The first edition is 1939; this is the third.)

[2] Jeffreys, H. (1924). *The Earth: Its Origin, History and Physical Constitution*. University Press, Cambridge.

[3] Lindley, D. V., Bolt, B. A., Huzurbazar, V. S., Lady Jeffreys, and Knopoff, L. (1991). Sir Harold Jeffreys. *Chance,* **4,** 10–26.

DENNIS V. LINDLEY

Kendall, Maurice George

Born: September 6, 1907, in Kettering, Northamptonshire, England.
Died: March, 29, 1983, in Redhill, Surrey, England.
Contributed to: almost every area of statistical activity, with concentration on social sciences.

The collected works of many mathematicians occupy rather more than one shelf in a typical library bookcase. Fisher's* collected papers on statistics and genetics, together with his books in both these fields, occupy only half of such a shelf. In this respect Fisher is in line with most other reasonably prolific statisticians. The writings for which Maurice Kendall was primarily responsible would fill several library shelves, exceeding in volume those of any mathematician of whom I am aware. The current tenth edition of Kendall's *Advanced Theory of Statistics* runs to five large volumes and covers a truly amazing range of statistical theory by using the device of quoting (with references) many results in the form of "Further exercises" following on from the main expository chapters. A complete set of editions of this work, beginning with the 1943 first edition of vol. 1 and the 1946 first edition of vol. II, would by itself fill one whole shelf, and Maurice's earliest book—the elementary "Yule and Kendall" (1937)—and his *Rank Correlation Methods* (1948), together with his smaller books on the *Geometry of n Dimensions* (1961), *Geometrical Probability* (with P. A. P. Moran, 1962), *Time Series* (1973), *Multivariate Analysis* (1975), and *Studies in the History of Statistics and Probability* [jointly edited with E. S. Pearson* (1970) and with Robin Plackett (vol. II, 1977)], might leave room on the second shelf for *The Statistical Papers of George Udny Yule* (1971, edited with Alan Stuart), but almost certainly not for *Tables of Symmetric and Allied Functions* (1966, edited with F. N. David* and D. E. Barton). The rest of the case would be more than filled with *The Sources and Nature of the Statistics of the United Kingdom,* which he began and edited in 1952 and 1957, and which has continued as a series ever since then: the 1957 *Dictionary of Statistical Terms* (edited with W. R. Buckland, 1957), which also has been extended with many further language glossaries; and the three-volume *Bibliography of Statistical Literature* (with Alison G. Doig, 1962, 1965, and 1968).

Maurice's literary output is remarkable not only for its quality and volume, but also for its wit and style. Fortunately for those who wish to sample its flavor, his publishers, Charles Griffin & Co., encouraged Maurice's closest collaborator, Alan Stuart, to edit under the apt title *Statistics: Theory and Practice* (1984) a selection of Maurice's lectures and papers, along with an obituary and a select bibliography. We statisticians have a reputation, partially deserved, for inability to express our conclusions in terms which the intelligent non specialist finds congenial. It would go a long way to cure this fault if statisticians were to regard this book as enjoyable required reading. Besides discovering the identities of K. A. C. Manderville, Lamia Gurdleneck, and her aunt Sara Nuttal, and enjoying the ballad "Hiawatha Designs an Experiment," we find there model expositions of statistical ideas addressed to economists and other social scientists, while those with more theoretical interests will find a fair sample of Maurice's contributions to statistical theory in his papers on *k*-statistics, time series, rank correlation, paired comparisons, and the like, though not of the music in the style of Handel, composed in successful response to a challenge from his friend, then Chairman of the Royal Opera House, Covent Garden. This last was in support of Maurice's contention that such was not difficult. He may have been unaware that the author of *Erewhon* and *The Way of All Flesh* had spent much of his life trying to do the same.

Son of an engineering worker, Maurice was interested first in languages, but later won a mathematics scholarship to St. John's College, Cambridge, where, after taking both parts of the Mathematical Tripos, in 1930 he passed by examination into the Administrative Class of the UK Civil Service and worked on statistics in the Ministry of Agriculture.

In 1934 he joined the Royal Statistical Society (RRS), whose monthly meetings provided a way of rapidly getting to know other statisticians. A year later a chance encounter with Udny Yule* led to his being invited by Yule to join him in the 11th edition (1937) of the latter's *Introduction to the Theory of Statistics.* By 1939 Kendall was sufficiently well acquainted with Oscar Irwin, Maurice Bartlett, John Wishart, and Egon Pearson* to have agreed with them on the need for an advanced treatise on mathematical statistics. Synopses were drafted, but the declaration of war made collaboration difficult. The only section that appeared during the war was for the most part Maurice's. It was published in 1943 as vol. I of what ultimately become known world-wide as "Kendall's *Advanced Theory of Statistics*."

Soon after joining the RSS, Maurice addressed its Study Group on the mechanical generation of random digits. The resulting papers were published in its *Journal* along with a friendly comment by Udny Yule. The first of his many papers formally read to the Society was on "The geographical distribution of crop productivity" [*J. R. Statist. Soc.*, **102** (1939)]. It foreshadows many of Kendall's later interests—rank correlations, time series, and machine computation. In proposing the usual vote of thanks Mr. H. D. Vigor took great pride in the fact that it was he who had proposed Maurice for Fellowship in the Society. Except for Bowley's* notorious vote of "thanks" to Fisher* and one of the two sets of comments addressed by Fisher to Neyman,* RSS discussions were more polite then than now.

Leon Isserlis, statistician to the British Chamber of Shipping, was prominent among RSS discussants in those days. In 1940, when Isserlis retired, Maurice succeeded him. Early leaving from the upper ranks of the UK Government service is still unusual, and such a move was much less usual then than now. One attraction which might have led Maurice to join the Chamber of Shipping was the fact that his predecessor had clearly been allowed time to engage in mathematical and philosophical speculations having little connection with his day-to-day work. And with submarine warfare at its height, statistics of UK shipping were not required to be published annually, though the data were vital to the war effort. This was the first of a number of changes of employment which helped Maurice make the tremendous contribution to statistics for which he was eventually responsible. It was not that he did not fully carry out his duties to his employers, but that his enormous energy and orderly mind enabled him in his "spare" time to do work which would have kept two or three others fully occupied. His genius for delegation further multiplied the results of his initiatives.

In 1949 Maurice was appointed Professor of Statistics at the London School of Economics. A powerful teacher and, in conjunction with Roy Allen, a strong departmental head, he soon set up a Research Techniques Division with a staff at that time large by comparison with other statistical groups. Much of his prodigious output dates from this period.

The 1950s saw the development of large-scale computing and in 1961 Maurice was persuaded to help set up the UK arm of a computer consultancy which eventually came to be called SCICON. It was typical of his judgement and persuasiveness that he recruited to this organization the young Martin Beale, whose fundamental contributions to integer programming earned him election to the Royal Society of London.

When in 1972 Maurice retired from SCICON the International Statistical Institute undertook to carry through, on behalf of the United Nations, the first World Fertility Survey. It was typical of Maurice to enjoy undertaking virtually alone a task which would daunt three or four others. But the unending travel, coupled with an intense work load, eventually wore down even his extraordinary stamina, and in 1980 the consequences of a heart bypass operation forced retirement. The last time I saw him was at the meeting where the UK Minister for Information Technology presented him with a United Nations Peace Medal for his work on the Fertility Survey. Congratulating him on the medal, and expressing sympathy for the stroke which he had suffered, I was struck by the fury he expressed towards his medical advisers who had failed to warn him of the risk he was running of such an outcome of his surgery. Enforced passivity was the worst blow he could have suffered.

He was awarded the Sc.D. of his old university in 1949. He served as President of the Royal Statistical Society, the Institute of Statisticians, the Operational Research Society and the Market Research Society. The Royal Statistical Society awarded him its highest distinction, the Guy Medal in gold, and both the University of Essex and the University of Lancaster conferred on him their Honorary Doctorates. The British Academy elected him FBA in 1970, and in 1974 he was knighted for services to statistical theory.

He married twice. By his first wife he had two sons and a daughter, and by his second wife he had one son.

G. A. BARNARD

Kitagawa, Tosio

Born: October 3, 1909, in Otaru City, Japan.
Died: March 13, 1993, in Tokyo, Japan.
Contributed to: quality assurance, stochastic processes, design of experiments, information theory.

Professor Tosio Kitagawa, after his high-school education at Sendai, studied in the Department of Mathematics at the University of Tokyo. He graduated there in 1934, and was awarded a Doctorate of Science for his research on functional equations [1, 2] from the University of Tokyo in 1934.

He started his academic career in the Department of Mathematics, Osaka University, and then was appointed an assistant professor at Kyushu University to establish a Department of Mathematics in 1939, where he started education and research in mathematical statistics at universities in Japan. Afterwards he established the Research Association of Statistical Science in 1941, and the Association started to publish the *Bulletin of Mathematical Statistics* in 1941. In 1943, he was promoted to an ordinary professor of mathematical statistics at Kyushu University. He was the first professor of mathematical statistics in Japan. He also held the post of acting director of the Institute of Statistical Mathematics in Tokyo in 1948–1949. During this period he presented the theory and application of weakly contagious stochastic processes [3, 4, 5]. He made gradual efforts to promote education and research in mathematical statistics and also practical statistical applications in Japan [6]. Especially he was interested in the area of statistical quality control, with his colleagues in industry.

After World War II, he introduced, very rapidly, statistical methods of quality control to industrial engineers and was also concerned with designs of sampling surveys for cost-of-living surveys of coal miners, fishery catch volumes, and timber volumes. In the course of this work, he studied several statistical techniques including experimental design and time series analysis. Besides his theoretical and practical activity, he published an introductory book [7] to make clear the historical background of statistical thought and the use of statistical tables [8]. These famous works had a big influence on the Japanese community.

After 1950 he actively developed statistical theories and their underlying statistical methodologies in practice, through domestic and international contacts. He presented many academic papers (see the reference list), mainly in the fields of statistical inference processes, statistical control processes, sample survey theories, and design of experiments. The papers involved a lot of original work, based on his experience and statistical insight. He was awarded the Deming Prize for Statistical Quality Control by the Japanese Union of Scientists and Engineers in 1953, for which he organized a series of sessions on mathematical programming for about ten years. In addition to his research works, he published leading textbooks [23, 26] and dictionaries [13, 22] as an editor.

Consequent to his deep study of the logical foundations of statistics, he developed an interest in information science, and was involved in the establishment of long-range planning of scientific research as a member of the Science Council of

Japan for seventeen years. Also he served as director of the Kyushu University Library 1961–1967, dean of the Faculty of Science 1968–1969, and director of the Research Institute of Fundamental Information Science 1968–1973 at Kyushu University.

He had also many international contacts with outstanding scholars in statistics. He visited many foreign universities and institutes to discuss common research interests, including the Indian Statistical Institute, Iowa State University, Princeton University, and the University of Western Australia, where he could associate with statisticians such as Professors P. C. Mahalanobis,* R. A. Fisher,* T. A. Bancroft, S. S. Wilks,* J. Neyman,* and J. Gani. He was a member of the International Statistical Institute from 1956, and was president of the International Statistical Association for the Physical Sciences. At a meeting of the International Statistical Institute held in Tokyo in 1960, he presented an invited paper [25] summarizing the background and fundamental principles of his various topics developed during the 1950s. He also presented a paper at the Fifth Berkeley Symposium in this connection [28].

In the field of information science, he studied and presented papers related to the logical foundations of information science, creative engineering, methodological considerations in biomathematics, the notion of EIZON, and so on [34–41].

He retired from Kyushu University on reaching the prescribed age, and was made an emeritus professor in 1973. He was appointed a joint director of the International Institute for Advanced Study of Social Information Science (IIAS–SIS), Fujitsu Co. He was there interested in the formation of scientific information systems and the informatic analysis of research activities. In this context he introduced the notions of brainware and research automation [42–50]. In 1975, he was elected President of the Information Processing Society of Japan for a two-year term. He was awarded several honors, including the Honorary Medal of the Japanese Government in 1980, for his scientific achievements.

REFERENCES

[1] Kitagawa, T. (1936). On the associative functional equations and a characterization of general linear spaces by the mixture rule. *Japan J. Math.,* **13,** 435–458.

[2] Kitagawa, T. (1937). On the theory of linear translatable functional equations and Cauchy's series. *Japan. J. Math.,* **13,** 233–332.

[3] Kitagawa, T. (1940). The characterizations of the functional linear operations by means of the operational equations, *Mem. Fac. Sci. Kyushu Univ.,* **A-1,** 1–28.

[4] Kitagawa, T. (1940). The limit theorems of the stochastic contagious processes, *Mem. Fac. Sci. Kyushu Univ.,* **A-1,** 167–194.

[5] Kitagawa, T. (1941). The weakly contagious discrete stochastic process, *Mem. Fac. Sci. Kyushu Univ.,* **A-1,** 37–65.

[6] Kitagawa, T. and Masuyama, M. (1942). *Statistical Tables I,* Res. Assoc. Statist. Sci., Kawadeshobo, Japan.

[7] Kitagawa, T. (1948). *The Foundation for the Recognition of Statistics Methodologies* (in Japanese). Hakuyosya.

[8] Kitagawa, T. (1950). *Tables of Poisson Distribution with Supplementary Note on Statistical Analysis of Rare Events* (in Japanese). Baifukan.

[9] Kitagawa, T. (1950). Successive process of statistical inference (1). *Mem. Fac. Sci. Kyushu Univ.,* **A5,** 139–180.

[10] Kitagawa, T. (1951). Successive process of statistical inference (2). *Mem. Fac. Sci. Kyushu Univ.,* **A6,** 54–95.

[11] Kitagawa, T. (1951). Successive process of statistical inference (3). *Mem. Fac. Sci. Kyushu Univ.,* **A6,** 131–155.

[12] Kitagawa, T. (1951). Successive process of statistical inference (4). *Bull. Math. Statist.,* **5,** 35–50.

[13] Kitagawa, T. (1951). *Dictionary of Statistics.* Toyo Keizai Shimposha.

[14] Kitagawa, T. (1952). Successive process of statistical controls (1). *Mem. Fac. Sci. Kyushu Univ.,* **A7,** 13–26.

[15] Kitagawa, T. and Mitome, M. (1953). *Tables for the Design of Factorial Experiments* (in Japanese). Baifukan.

[16] Kitagawa, T. (1953). Successive process of statistical inference (5). *Mem. Fac. Sci. Kyushu Univ.,* **A7,** 81–106.

[17] Kitagawa, T. (1953). Successive process of statistical inference (6). Mem. Fac. Sci. Kyushu Univ., A8, 1–29.

[18] Kitagawa, T. (1956). *Lectures on Design of Experiments I, II* (in Japanese). Baifukan.

[19] Kitagawa, T. (1954). Some contributions to the design of sample surveys. Sankhyá, 14, 317–362.

[20] Kitagawa, T. (1956). Some contributions to the design of sample surveys. Sankhyá, 17, 1–36.

[21] Kitagawa, T. (1957). Successive process of statistical inference associated with an additive family of sufficient statistics. *Bull. Math. Statist.,* **7,** 92–112.

[22] Kitagawa, T. (1957). *Systematic Dictionary of Statistics.* Toyo Keizai Shimposha.

[23] Kitagawa, T. (1959). *Statistical Inference I, II.* Iwanami.

[24] Kitagawa, T. (1959). Successive process of statistical controls (2). *Mem. Fac. Sci. Kyushu Univ.,* **A13,** 1–16.

[25] Kitagawa, T. (1960). Successive process of statistical controls (3). *Mem. Fac. Sci. Kyushu Univ.,* **A14,** 1–33.

[26] Kitagawa, T. (1960). *General Course of Statistics.* Kyoritsu Shuppann.

[27] Kitagawa, T. (1961). A mathematical formulation of the evolutionary operation program. *Mem. Fac. Sci. Kyushu Univ.,* **A15,** 21–71.

[28] Kitagawa, T. (1961). Successive process of optimizing procedures. *Proc. 4th Berkeley Symp. Math. Statist. Probab.,* **1,** 407–434.

[29] Kitagawa, T. (1961). The logical aspect of successive processes of statistical inferences and controls. *Bull. Int. Statist. Inst.,* **38,** 151–164.

[30] Kitagawa, T. (1963). The relativistic logic of mutual specification in statistics. *Mem. Fac. Sci. Kyushu Univ.,* **A17,** 76–105.

[31] Kitagawa, T. (1963). Automatically controlled sequence of statistical procedures in data analysis, *Mem. Fac. Sci. Kyushu Univ.,* **A17,** 106–129.

[32] Kitagawa, T. (1963). Estimation after preliminary test of significance, *Univ. Calif. Publ. Statist.,* **3,** 147–186.

[33] Kitagawa, T. (1965). Automatically controlled sequence of statistical procedures. In *Bernoulli, Bayes, Laplace Anniversary Volume.* Springer-Verlag, Berlin, pp. 146–178.

[34] Kitagawa, T. (1968). Information science and its connection with statistics. *Proc. 5th Berkeley Symp. Math. Probab.,* vol. 1, pp. 491–530.

[35] Kitagawa, T. (1969). *The Logic of Information Science* (in Japanese). Kodansya.

[36] Kitagawa, T. (1969). Information science approaches to scientific information systems and their implication to scientific researches. *Res. Inst. Fund. Inf. Sci. Kyushu Univ. Res. Rep.,* **3,** 1–51.

[37] Kitagawa, T. (1970). *Viewpoint of Information Science—Search for a New Image of the Sciences* (in Japanese). Kyoritsu Shuppan.

[38] Kitagawa, T. (1971). A contribution ot the methodology of biomathematics information science approach to biomathematics. *Math. Biosci.,* **12,** 25–41.

[39] Kitagawa, T. (1972). Three coordinate systems for information science approaches, *Inf. Sci.,* **5,** 157–169.

[40] Kitagawa, T. (1973). Dynamical systems and operators associated with a single neuronic equation. *Math. Biosci.,* **18,** 191–244.

[41] Kitagawa, T. (1973). Environments and eizonsphere (in Japanese). *Soc. Inf. Sci. Ser. Gakushu Kenkyu Sha,* **16**(1), 132–199.

[42] Kitagawa, T. (1974). The logic of information sciences and its implication for control process in large system. *2nd FORMATOR Symp. Math. Methods for Analysis of Large Scale Systems,* Prague, pp. 13–31.

[43] Kitagawa, T. (1974). Brainware concept in intelligent and integrated system of information. *Res. Inst. Fund. Inf. Sci. Kyushu Univ. Res. Rep.,* **39,** 1–20.

[44] Kitagawa, T. (1974). *Historical Image of Human Civilization—a Prolegomenon to Informative Conception of History* (in Japanese). *Soc. Inf. Sci. Ser.* **17**(1). Gakushu Kenkyu Sha.

[45] Kitagawa, T. (1974). Cell space approaches in biomathematics. *Math. Biosci.,* **19,** 27–71.

[46] Kitagawa, T. (1975). Dynamical behaviours associated with a system of neuronic equations with time lag, *Res. Inst. Fund. Inf. Sci. Kyushu Univ. Res. Rep.,* **46,** 1–59; 49, 1–87; 54, 1–97.

[47] Kitagawa, T. (1978). An informatical formulation of generalized relational ecosphere on the basis of paired categories. *Int. Conf. Cybern. Systems,* pp. 322–327.

[48] Kitagawa, T. (1979). The logics of social information science. *Soc. Sci. Ser., Gakushu Kenkyu Sha,* **18**(1), 26–137.

[49] Kitagawa, T. (1979). Generalized artificial grammars and their implications to knowledge engineering approaches. *Int. Inst. Adv. Study Soc. Inf. Sci. Res. Rep.,* **6,** 1–29.

[50] Kitagawa, T. (1980). Some methodological consideration on research automation: (1) objectivization and operatorization. *Res. Inst. Fund. Inf. Sci. Res. Rep.,* **96,** 1–31.

[51] Kitagawa, T. (1982). From cosmos to chaos: my scientific career in mathematics, statistics, and informatics. *The Making of Statisticians,*. J. Gani, ed. Applied Probability Trust, Springer-Verlag, pp. 165–178.

CHOOICHIRO ASANO

Neyman, Jerzy

Born: April 16, 1894, in Bendery, Russia.
Died: August 5, 1981, in Berkeley, California.
Contributed to: mathematical statistics, probability theory, testing hypotheses, confidence intervals, generalized chi-square, stochastic models, statistics in substantive fields.

Jerzy Neyman in 1938.

Jerzy Neyman was one of the great founders of modern statistics. He made fundamental contributions to theoretical statistics and also to the innovative yet precise use of statistics in a wide spectrum of substantive fields ranging from agriculture and astronomy through biology and social insurance to weather modification. He tackled new areas with great interest and enthusiasm, especially when the problem was of societal importance, because he wanted to "find out" and because he knew that real problems are the source of interesting mathematical–statistical questions. It was not just finding the answer that attracted him; his attention centered on "how to find what we need to know" and even on how to pose the question of what we want to study.

Jerzy Neyman was born of Polish parents in Bendery, Moldavia, then in Russia and now in Moldova. Both his grandfathers were landowning gentry (as reflected in his full surname, Spława-Neyman) who participated in the Polish uprising of 1863 and thereafter had their lands confiscated and their families exiled to Siberia and to Tashkent. But Jerzy Neyman's father Czeslaw was only five years old at this time, and was allowed to stay in Ukraine on condition that he was not to live near Warsaw. He became a lawyer and was a judge when Jerzy was born. However, Czeslaw had a greater interest in archaeology, and Jerzy remembered going with his father on the digs in the Crimea.

Jerzy Neyman's early education was provided by governesses, alternately French and German, who contributed to his proficiency in many languages. He attended school in Simferopol and then in Kharkov, where his family moved after the death of his father when Jerzy Neyman was twelve. In 1912, he entered the University of Kharkov to study physics and mathematics. One of his lecturers was the Russian probabilist S. N. Bernstein,* who called his attention to the ideas in Karl Pearson's* *Grammar of Science* [28], which Neyman described as influencing his development. Nevertheless, his major interest was in new research in the measure theory of Lebesgue, and this was the area of his early papers. From 1917 to 1921, Neyman was studying graduate mathematics, tutoring, and teaching. Life was very difficult in Kharkov in those years of war and revolution. When reestablished Poland and Russia started fighting over their boundary, Neyman found himself apprehended as an enemy alien, together with the rector of the University of Kharkov, who also happened to be a Pole. After some weeks in jail, they were released because their teaching was needed in the university. In 1921, in an exchange of prisoners of war, Neyman went to Poland for the first time, at the age of 27. Raised and educated in "eastern Poland," Neyman was always fond of being Polish and appreciative of his heritage, although sometimes critical of his country's governance.

In Warsaw, Neyman visited the Polish mathematician W. Sierpiński, who was interested in his research and helped him get some of his papers published (those where Neyman had not been anticipated by others while he was isolated in Russia). A position was tentatively promised Neyman in Warsaw for the fall, so now the problem was to earn his living during the summer. With the help of Sierpiński, Neyman obtained a position as a statistician at the National Institute of Agriculture in Bydgoszcz. One of his duties was making meteorological observations, but his main duty was to assist in agricultural trials. Neyman applied himself—obtaining funds for a month of study in Berlin, since these problems were unknown in Poland—and started to publish. His 1923 doctoral thesis from the University of Warsaw was on probabilistic problems in agricultural experimentation. Beginning in 1923, Neyman lectured at the Central College of Agriculture in Warsaw and at the Universities of Warsaw and Cracow, commuting by train each week.

In 1924 he obtained a postdoctoral fellowship to study under Karl Pearson* at University College, London. As Neyman described the situation, the mathematicians in Poland wanted to know whether the statistics he was writing made any sense, that is, whether Karl Pearson would publish it in *Biometrika.* Actually, some of Neyman's statistical papers had been translated into English and were known abroad. During this early London period. Neyman had contacts with Pearson and his son Egon* and several other statisticians including R. A. Fisher* and W. S. Gosset,* who turned out to be the first statistician Neyman met at University College. Attired in formal morning coat, Neyman had gone at noon to call on Professor Pearson at his office—the correct dress and the correct hour for a Polish gentleman—but Karl Pearson was out to lunch. Thus their friendship started, before Neyman even knew Gosset's identity—"Student." In retrospect, these early contacts do not seem to have been deep. Neyman's papers, accepted by Karl Pearson, were similar to what he had been doing in Poland. His interest in set theory remained. With the collapse of the Polish currency, Neyman's government fellowship was essentially worthless. He obtained a Rockefeller fellowship and spent the next year in Paris, hearing lectures from Lebesgue and Hadamard and meeting Borel.

From 1928 to 1934, Neyman was busy in Poland. His activities were broad, not only agricultural experimentation but also problems in the chemical industry, problems in social health insurance and other socioeconomic questions, and especially statistical research in the behavior of bacteria and viruses, leading to his appointment as head of the Statistical Laboratory of the Nencki Institute of Experimental Biology. He was also working in mathematical statistics with his students and with Egon Pearson. His collaboration with Pearson started in 1925 and at first was carried out largely by correspondence except for a few brief meetings in France. By 1934 Karl Pearson had retired, and his department was divided between his son and Fisher. Egon invited Neyman to return to University College, first as senior lecturer and then as reader (associate professor); now cooperation in person was possible and Neyman had a secure position that allowed him time to develop his own research.

The decade 1928–1938 was a fruitful period for Neyman. In their fundamental 1928 paper, Neyman and Pearson [17] put the theory of testing hypotheses on a firm basis, supplying the logical foundation and mathematical rigor that were missing in the early methodology. An important component was the formulation of statistical hypotheses by the construction of careful stochastic models of the phenomena under questions. Whether a proposed statistical test is optimum or even whether it is valid for testing a hypothetical model must be judged in comparison with the other possible modtion by confidence sets [5]. Estimation by intervals that had endpoints determined by the observations and had size dependent on the probability that the interval will cover the true point had long been a problem in statistics. Using the relations with the theory of testing hypotheses, Neyman adapted the results from testing theory to the problem of finding a confidence interval such that he could guarantee that the probability of covering the true value of the parameter to be estimated was at least equal to a preassigned value called the confidence coefficient. Further, the probability of covering a false value, given the first property for the true value, was minimized. Thus the confidence interval was shortest conditionally in a probability sense. Later statisticians constructed confidence intervals that are shortest conditionally in a geometric sense. The concepts were quickly extended to confidence sets of several dimensions.

At, first Neyman thought that confidence sets would coincide with the fiducial intervals of Fisher if those intervals were also required to be shortest in the same sense. In most examples, the two kinds of intervals do coincide, and one can understand this if care is taken to distinguish between the random variable and particular values that it may take on and also to distinguish between the random variable and the true unknown parameter to which the random variable is related. Once the observations are made, the interval is no longer random, and there is no longer any probability involved. Either the observed interval covers the true value or it does not. If a series of confidence intervals are constructed by the Neyman method, for the same or different observers, the proportion of confidence intervals that cover correctly will tend to the confidence coefficient. There are cases where the confidence interval and the fiducial interval produce different results; the Behrens–Fisher problem is an example [9].

The uses of confidence intervals soon appeared in many textbooks and works on statistical methodology. Again there usually would be no reference to Neyman as originator. The logic and the rigor behind confidence intervals was appealing, as was the easy relation between confidence-interval estimation and testing hypotheses. In this lectures, Neyman used to say that this relation "brought order out of chaos."

The characteristic pattern of Neyman's research evident in all three of these fundamental research efforts is to take a rather vague statistical question and make of it a precisely stated mathematical problem for which one can search for an optimal solution. His papers established a new climate in statistical research and new directions in statistical methodology. Neyman, his students, and others vastly extended the new theories. Neyman's ideas quickly went into many fields. As David Kendall reports, "We have all learned to speak statistics with a Polish accent."

Neyman made a six-week tour of the United States in 1937 to lecture at the Department of Agriculture in Washington, and at several universities. He was offered a position as professor at one of those universities and also at the University of California at Berkeley, where professors in several departments had joined in an effort to have a strong statistics unit. Many person have asked Jerzy Neyman why he moved from a secure position at University College, then the statistical center of the universe, to the faraway University of California, where he had never been and where there was no statistics unit. Two points always appeared in his response: one point is the fact that Neyman was being asked to build a statistical unit at Berkeley where there was none already. The other point was that he realized that World War II was coming fast and he feared what Hitler might do to Poland and to the rest of Europe. He anticipated that Poland would be destroyed again and that he would be interned in Britain as an enemy alien. Neyman had survived a war and revolution with vivid difficulties; he wanted to move himself and his family as far away as possible. There was no suggestion of pressure from English statisticians, in particular from R. A. Fisher. It seems unjust to both individuals to suggest pressure. In any case, it is not in Neyman's character to yield to pressure. Neyman was aware of Fisher's attacks but generally did not respond [12]: "It would not be appropriate." They met and talked at international meetings and at Cambridge where Neyman spent a sabbatical year in 1957–1958. In 1961, there was a response [12], however, at the suggestion of Japanese statisticians after a strong series of attacks; it was entitled "Silver Jubilee of my dispute with Fisher."

Jerzy Neyman arrived in Berkeley in August 1938 to spend nearly half of his life at the University of California. When he reached the age of 67, he continued as director of the Statistical Laboratory and as professor of statistics, being recalled each year to active duty. He was going forward all the time. He liked Berkeley and worked with great enthusiasm to build the best department of statistics and to vitalize the university toward stronger yet more flexible programs.

The Statistical Laboratory was founded during his first year at Berkeley, which provided some autonomy, but Neyman felt that a completely independent Department of Statistics was important (cf. Reid [29]). Reaching out to the other departments and to other centers, Neyman established yearly summer sessions and then regular (every five years) Berkeley Symposia on Mathematical Statistics and Probability with participants from all over the world, the leaders in each new direction. Much efforts was involved and many persons had to be convinced, especially in the early years, but Neyman persisted, with the results that Berkeley soon became an exciting place to be and the Berkeley Statistics Department came into existence in 1955, blossoming out as the best. Neyman struggled to obtain more university positions for statistics and then to bring the best possible faculty and students and to support them academically and financially. The University of California has a large measure of faculty governance, but this is very time-consuming, especially during times of stress such as the Year of the Oath (1950. When the Regents imposed a special loyalty oath that interfered with academic freedom) and the student movements in the sixties when more academic governance was demanded by students. Neyman was at the forefront of protecting the rights of the protestors.

With his arrival in Berkeley, Neyman turned more toward the use of statistics in large problems. During World War II, research on directed topics, typically multiple bomb aiming, absorbed all of this time outside of a heavy teaching schedule. As the war was ending, the first Berkeley symposium took place. Neyman presented an important paper [11] which was written before 1945, using a class of procedures that he called best asymptotically normal (BAN) estimates, for estimation and also for testing. Neyman showed that by minimizing various appropriate expressions, one can obtain classes of estimates that are asymptotically equivalent, all producing limiting normal distributions that have minimum variance. The wide choice in the expressions to be minimized, ranging through likelihood ratio and minimum chi-square, each with possible modifications, accompanied with a wide choice of estimators for the unknown parameters, provided flexibility that allowed simpler expressions and much easier computation.

Important and useful concepts also appeared in this seminal paper, including what are now called restricted chi-square tests. In order to show that the various test procedures are asymptotically optimal and also asymptotically equivalent, Neyman considered the large-sample performance of these tests by considering "nearby alternatives" that approach the hypothesis at the rate $1/n^{1/2}$. This method is now widely used in asymptotic studies: it was introduced by Neyman in 1937 in an unusual paper on the "smooth" test for goodness of fit [6]. The theory of BAN estimation is now widely used in a manner similar to the use of least squares, as its asymptotic equivalent.

Neyman intensified his interest in how one studies stochastic problems. His method consisted in constructing a stochastic model consistent with the knowledge available and checking the consequences of the model against possible observations. Usually, certain aspects were of particular interest; he would determine the kinds of observations that would lead to sensitive responses and try to arrange for a cooperative study whenever new observations were needed. The range of topics was broad indeed. He extended [22] his clustering models beyond what are now called Neyman type A, type B, and so forth, which originally described how larvae crawl out from where the eggs are clustered, to other entities in other fields, e.g., physical particles and clusters of galaxies [8]. These clustering models were widely adopted in many areas of biology, ecology, and physics by researchers trying to get a better understanding of the processes involved rather than applying interpolation formulas. Neyman, mostly with Scott and with Shane, published a long series of articles on the clustering of galaxies in space as derived from the apparent clustering on the photographic plate [21]. Several cosmological questions are of continuing interest, such as the evolution of galactic types from elliptical to spiral or the reverse.

Neyman studied many other stochastic processes, including the mechanism and detection of accident proneness, with Bates, where more realistic and more general models were set up and tested, thus deriving new information about the process [1]. The same vein underlies all of these studies of stochastic processes: catching schools of fish [11], spread of epidemics [24], and carcinogenesis, in a long series of papers [13, 25] including recent studies where the mechanism inducing the cancer involves high-energy radiation [20]. In his studies of the relation between pollutants and health, the direction had to be more diverse, as it also had to be in a long series of studies on the effectiveness of weather modification on which he was working at the time of his death and which he started at the request of the state of California, where lack of rainfall is often a serious difficulty [14, 27].

Jerzy Neyman saw that the models he was using were too complex for the application of his optimum tests developed in the thirties. He turned to asymptotic methods [10], such as his BAN estimates, and developed tests that are locally optimal in the presence of nuisance parameters. These are called *C*-alpha tests, in honor of H. Cramér* and because they are asymptotically similar, even though no similar test exists. These optimal tests are not difficult to construct [30] from the logarithmic derivatives of the densities; they solve a large class of problems that cannot be solved by straight substitution of observed values for unknown parameters.

There were other situations where the conflicting hypotheses are not identifiable or the probability of making a correct decision was very tiny. These difficulties arise not only in widely used normal theory [4], but also in applied problems such as competing risks, when one wants to disentangle which of the possible causes of death (or which combination) actually is the cause; or competition between species, where one wants to predict which of two competing species will be the survivor when both will survive if kept separate [16]. What additional information must be supplied to allow the study of models of relapse and recovery, for example, to be complete?

Jerzy Neyman wanted to ensure that science was not obscured by political expedience or by commercialism. He turned to the scientific societies to which he belonged, of which there were many, for help in enforcing a strict code of ethics. Several organizations have taken action; others are moving. The problems are not easy, but Neyman had the courage to speak out for honesty [15, 23].

Neyman was a superb teacher using his version of the Socratic method. It was not easy to be sent to the chalkboard but the training was invaluable. Also, he was an inspiring research leader who shared his ideas and granted full credit to his students and young colleagues, always in strict alphabetical order! He was always a winner in number and quality of doctoral students. He maintained a steady interest in all of their activities and liked to follow their progress through the years. He dedicated the 1952 edition of *Lectures and Conferences* to the memory of his Polish students and early co-workers lost in World War II, listing for each the grisly cause of death—nearly all of his Polish students were killed. But he kept his faith in the new students who came from all over the world, helping them in every possible way: "I like young people!"

Neyman strove to strengthen statistics and to build statistical groups. One reason was to make it easier to publish good papers without prejudiced editing, a difficulty that he and other members of his generation had faced. He worked to strengthen the Institute of Mathematical Statistics, organizing many sessions and serving on the council and as president. He felt that the International Statistical Institute should be widened to open the program and the elected membership to academic statisticians and to young researchers. He found a way to do this by establishing, with the help of many colleagues, a new section of the ISI, the Bernoulli Society, whose members could attend ISI sessions. The Society had freedom to organize part of the ISI program. Never mind that in 1958 Neyman agreed to accept the name International Association for Statistics in the Physical Sciences. The name was not a restriction on its activities, and it was very active not only at ISI sessions but also in organizing satellite symposia on a wide range of topics. By 1975 the name Bernoulli Society became official, with a strengthening of its world-wide role. Neyman became an honorary president of the ISI.

Jerzy Neyman was interested in having stronger science and more influence of scientists in general decision making. He spoke out strongly against inequities and worked hard year after year to establish fair treatment, especially for black people. Perhaps his scientific hero was Copernicus. He gave a special talk at the 450th anniversary of Copernicus while he was still in Bydgoszcz. At Copernicus' 500th anniversary in Berkeley, Neyman was even more active. In addition to talks and articles, he edited an unusual book, *The Heritage of Copernicus: Theories "Pleasing to the Mind,"* in which scientists from many fields described neo-Copernican revolutions in thought. Neyman received a shower of Copernican medals and plaques.

Jerzy Neyman's achievements received wide recognition. In addition to being elected to the International Statistical Institute and the International Astronomical Union, he was elected to the U.S. National Academy of Sciences, and made a foreign member of the Swedish and Polish Academies of Science and of the Royal Society. He received many medals, including the Guy Medal in gold from the Royal Statistical Society (London) and, in 1968, the U.S. National Medal of Science. He received honorary doctorates from the University of Chicago, the University of California, Berkeley, the University of Stockholm, the University of Warsaw, and the Indian Statistical Institute.

Neyman was greatly esteemed, and he was greatly loved. He gave of his own affection, his warmth, and his talents in such a way that they became a part of science to be held and handed on.

REFERENCES

[1] Bates, G. E. and Neyman, J. (1952). *Univ. Calif. Publ. Statist., 1,* 215–254, 255–276. (Theory of accident proneness: true or false contagion.)

[2] Hacking, I. (1965). *Logic of Statistical Inference.* Cambridge University Press. Cambridge. England.

[3] Neyman, J. (1934). *J. R. Statist. Soc.,* **97,** 558–625. Also in *A Selection of Early Statistical Papers of J. Neyman* (1967). University of California Press, Berkeley, No. 10. [Spanish version appeared in *Estadistica* (1959). **17,** 587–651.] (Fundamental paper on sampling, optimal design, confidence intervals.)

[4] Neyman, J. (1935). *J. R. Statist. Soc. Suppl.,* **2,** 235-242. Also in *A Selection of Early Statistical Papers of J. Neyman* (1967). University of California Press, Berkeley, No. 15. (Difficulties in interpretation of complex experiments.)

[5] Neyman, J. (1937). *Philos. Trans. R. Soc. Lond. A,* **236,** 333–380. Also in *A Selection of Early Statistical Papers by J. Neyman* (1967). University of California Press, Berkeley,

No. 20. (Fundamental paper on theory of estimation by confidence sets. See also ref. [7].)

[6] Neyman, J. (1937). *Skand. Aktuarietidskr.,* **20,** 149–199. Also in *A Selection of Early Statistical Papers of J. Neyman* (1967). University of California Press, Berkeley, No. 21. (Unusual paper on "smooth" test for goodness of fit.)

[7] Neyman, J. (1938, 1952). *Lectures and Conferences on Mathematical Statistics and Probability.* Graduate School, U.S. Dept. of Agriculture, Washington, DC. [Spanish version published by InterAmerican Statistical Institute (1967)).] (The revised and enlarged second edition provides an interesting account of many of Neyman's ideas.)

[8] Neyman, J. (1939). *Ann. Math. Statist.,* **10,** 35–57. (First model of "contagious" distributions, including Neyman type A clustering.) Also in *A Selection of Early Statistical Papers of J. Neyman* (1967). University of California Press, Berkeley No. 25.

[9] Neyman, J. (1941). *Biometrika,* **32,** 128–150. Also in *A Selection of Early Statistical Papers of J. Neyman* (1967). University of California Press, Berkeley, No. 26. (Investigation of relation between confidence intervals and Fisher's fiducial theory.)

[10] Neyman, J. (1949). *Proc. Berkeley Symp. Math. Statist. Probab.* (of 1945), pp. 239–273. Also in *A Selection of Early Statistical Papers of J. Neyman* (1967). University of California Press, Berkeley, No. 28. (Seminal paper on restricted chi-square tests, BAN estimation, and asymptotically optimal and asymptotically equivalent procedures.)

[11] Neyman, J. (1949). *Univ. Calif. Publ. Statist.,* **1,** 21–36.. (Catching schools of fish, a study of the decrease in sardine catches.)

[12] Neyman, J. (1961). *J. Operat. Res. Soc. Jpn.,* **3,** 145–154. (Neyman conducts himself "not inappropriately" in controversy with R. A. Fisher.)

[13] Neyman, J. (1961). *Bull. Int. Inst. Statist.,* **38,** 123–135. (Modeling for a better understanding of carcinogenesis. Summary paper in a series on carcinogenesis.)

[14] Neyman, J. (1977). *Proc. Natl. Acad. Sci. U.S.A.,* **74,** 4714–4721. (Invited review paper on a statistician's view of weather modification technology.)

[15] Neyman, J. (1980). *Statistical Analysis of Weather Modification Experiments,* E. Wegman and D. DePriest, eds. Marcel Dekker, New York, pp. 131–137. (Comments on scientific honesty in certain experiments and operations.)

[16] Neyman, J., Park, T., and Scott, E. L. (1956). *Proc. 3rd Berkeley Symp. Math. Statist. Probab.,* vol. 4, pp. 41–79. (Struggle for existence: Tribolium model.)

[17] Neyman, J. and Pearson, E. S. (1928). *Biometrika,* **20-A,** 175–240, 263–294. Also in *Joint Statistical Papers of J. Neyman and E. S. Pearson* (1967). University of California Press, Berkeley. No. 1 and 2. (Fundamental paper on testing hypotheses, in two parts.)

[18] Neyman, J. and Pearson, E. S. (1933). *Phil. Trans. R. Soc. Lond. A,* **231,** 289–337. Also in *Joint Statistical Papers of J. Neyman and E. S. Pearson* (1967). University of California Press, Berkeley, No. 6.

[19] Neyman, J. and Pearson, E. S. (1933). *Proc. Camb. Phil. Soc.,* **29,** 492–510. Also in *Joint Statistical Papers of J. Neyman and E. S. Pearson* (1967). University of California Press, Berkeley, No. 7.

[20] Neyman, J. and Puri, P. S. (1982). *Proc. R. Soc. Lond. B,* **213,** 139–160. (Models of carcinogenesis for different types of radiation.)

[21] Neyman, J. and Scott, E. L. (1952). *Astrophys. J.,* **116,** 144–163.. (Theory of spatial distribution of galaxies: first paper in a long series.)

[22] Neyman, J. and Scott, E. L. (1957). *Proc. Cold Spring Harbor Symp. Quant. Biol.,* **22,**

109–120. [Summary paper on populations as conglomerations of clusters. See also the paper (1959) in *Science,* **130,** 303–308.]

[23] Neyman, J. and Scott, E. L. (1960). *Ann. Math. Statist.,* **31,** 643-655. (Correction of bias introduced by transformation of variables.)

[24] Neyman, J. and Scott, E. L. (1964). In *Stochastic Models in Medicine and Biology,* J. Gurland, ed. University of Wisconsin Press, Madison, pp. 45–83. (Stochastic models of epidemics.)

[25] Neyman, J. and Scott, E. L. (1967). *Proc. 5th Berkeley Symp. Math. Statist. Probab.,* vol. 4, pp. 745–776. (Construction and test of two-stage model of carcinogenesis: summary.)

[26] Neyman, J. and Scott, E. L. (1967). *Bull. Int. Inst. Statist.,* **41,** 477–496. (Use of C-alpha optimal tests of composite hypotheses. Summary paper with examples.)

[27] Neyman, J. and Scott, E. L. (1967). *Proc. 5th Berkeley Symp. Math. Statist. Prob.,* vol. 5, pp. 327–350. (Timely summary paper on statistical analysis of weather modification experiments; one of 48 published papers and many reports in this field.)

[28] Pearson, K. (1892, 1937). *The Grammar of Science,* 3rd ed., revised, enlarged. E. P. Dutton, New York. (A paperback edition was published in 1957 by Meridian.)

[29] Reid, C. (1982). *Neyman—From Life,* Springer-Verlag, New York. (A sensitive and knowledgeable biography, beautifully rendered.)

[30] Wald, A. (1950). *Statistical Decision Functions.* Wiley, New York (Fundamental book on decision functions.)

BIBLIOGRAPHY

Kendall, D. G., Bartlett, M. S., and Page, T. I. (1982). *Biogr. Mem. Fellows R. Soc. Lond.,* **28,** 378–412. (This interesting and extensive biography is in three parts and contains complete bibliography through 1982.)

Klonecki, W. and Urbanik, K. (1982). *Prob. Math. Statist., Polish Acad. Sci.,* vol. 2, pp. 1–111. (Neyman viewed by his countrymen.)

LeCam, L. and Lehmann, E. L. (1974). *Ann. Statist.,* **2,** vii–xiii. (Review of Neyman's scientific work, on the occasion of his eightieth birthday.)

ELIZABETH L. SCOTT

Pearson, Egon Sharpe

Born: August 11, 1895, in Hampstead (London), England.
Died: June 12, 1980, in Midhurst, Sussex, England.
Contributed to: applications of statistical techniques, statistical theory, quality control, operations research, statistical education.

Egon S. Pearson

Egon Sharpe Pearson (E. S. P.), the only son of the British statistician Karl Pearson, was born in 1895, a time when the latter's interest in statistical methods, kindled by the work of Francis Galton,* was growing rapidly. Following school education at the Dragon School, Cambridge (1907–1909) and Winchester (1909–1914), he entered Trinity College, Cambridge to read mathematics in 1914. His undergraduate studies were interrupted first by a severe bout of influenza (August-December 1914) and then by war work (1915–1918). After finally obtaining his first degree in 1919, he continued with graduate studies in astronomy, attending, inter alia, lectures by A. S. Eddington on the theory of errors and F. J. M. Stratton on combinations of observations. He also worked with F. L. Engledow and G. U. Yule* at that time.

In the fall of 1921, he took a position in the Department of Applied Statistics at University College, London. This department, the first of its kind, had been established through the efforts of Karl Pearson in 1911 and was still very much under his control. The next five years were an apprenticeship; Egon was not required (or permitted) to teach until 1926, when his father was suffering from cataract. By this time, he felt the need to work out his own philosophy on the use of statistical inference and scientific method generally if he were to develop an independent career. He was not fully satisfied with the methods of the Pearsonian school, with their considerable reliance on a wide variety of "indices," and he was ripe to receive fresh external influences. In fact, two such influences of some importance entered his life about that time. While they exerted considerable interplay on each other (and were not, of course, exhaustive), it is convenient to regard them as representing the more theoretical and the more practical interests in his statistical work.

The first, and perhaps better known, was the collaboration with Jerzy Neyman,* whom he met when the latter was a visitor in the department in 1925–1926. The second was a sustained correspondence with W. S. Gosset ("Student")* which started in 1926. Both episodes lasted about the same time, till Neyman's departure for the United States in 1938 and Gosset's death in 1937, respectively.

As already noted, this dichotomy, although a convenient simplification, omits much that is relevant and important. Egon Pearson never regarded the corpus of methods associated with the name Neyman–Pearson as a *theory,* but rather as a collection of principles representing an approach to statistical—or more broadly scientific—problems in general. The essentials of this approach are consideration of the results to be expected from applying statistical procedures, usually expressed in probabilistic terms. The approach is not limited to the traditional fields of hypothesis testing and estimation, but has been applied in other contexts, notably, in discriminant analysis (wherein a property of major interest is the probability of correct classification).

One of the basic concepts in the Neyman–Pearson approach—that of an alternative hypothesis, which might be valid if a hypothesis under test were not valid—was in fact mentioned by Gosset in a letter to Egon Pearson in 1926. In November of that year, the latter wrote to Neyman, now returned to Warsaw, mentioning this concept. At first Neyman suggested that some form of inverse-probability argument might provide an appropriate technical basis, but it was soon recognized that sufficient knowledge of prior probabilities would be available but rarely. A period of intensive, though mostly long-range, discussions ensued. The remarkable story of the development of this collaboration, whose members were usually some thousand miles apart until 1934, is told in some detail in David [1] and Reid [23]. Progress was accelerated by occasional meetings, starting with ten days in Paris in April 1927 and continuing in Warsaw in the summer of 1929 and January 1932. Consequent on the first meeting, the first general outline of the approach appeared in two long papers in 1928 [14, 15]. From the 1929 discussions in Warsaw came the basic papers [16, 17] on two- and k-sample problems, published in 1930–1931; in 1933 there appeared what may be regarded as a definitive summary [18] of this phase of the Neyman–Pearson approach to hypothesis testing.

We now return to the other major influence and describe some contemporaneous developments associated with the correspondence with Gosset. At an early stage, the latter had drawn Pearson's attention to the important topic of robustness—lack of sensitivity of statistical procedures to departures from assumptions (e.g., of independence, normality, etc.) on which their "standard" properties are assessed. The tables of random numbers produced by L. H. C. Tippett while working in the department in University College (1925–1927) made it possible to investigate robustness by means of sampling experiments (later termed *Monte Carlo methods* and—currently and more generally—*simulation*). The results of some of this work were published in 1928–1931 [3, 4, 11, 12]. (The first published experimental estimate of power of a test, an important concept in the Neyman–Pearson approach, appeared in Pearson and Adyanthaya [12].) The topic of robustness remained of perennial interest and was reflected in a publication as late as 1975 [22]. A little later in the correspondence, Gosset drew attention to the application of statistical methods in quality control of industrial production, then beginning to flourish in the United States under the enthusiastic guidance of W. A. Shewhart and F. W. Winters. Pearson visited the United States in 1931 and met Shewhart and a number of other prominent American statisticians, giving an address on Probability and Ways of Thinking at the annual meeting of the American Statistical Association. In this talk, he set out some basic ideas of the then very novel Neyman–Pearson approach.

Later, he played an active role in the formation of the Industrial and Agricultural Research Section of the Royal Statistical Society in 1932–1933 and its initial progress. (Its later development into the Research Section, and its organ's development from the *Supplement* into *Series B of the Journal of the Royal Statistical Society*, was accompanied by a marked increase in emphasis on theory. The regional and specialized sections of the Royal Statistical Society, and *Applied Statistics* or *Section C* of the *Journal,* have greater claims to be regarded as true heirs of the original organization.)

At this time there also began a long association with the British Standards Institution. In 1936, the monograph in ref. [5] appeared. It was the first official British standard on the use of statistical methods in industrial quality control. The association continued for some 40 years, and Pearson was attending meetings of committees of the International Standards Organization as late as 1976.

On Karl Pearson's retirement in 1933, his department at University College, London was split into the Department of Eugenics and the Department of Applied Statistics. Egon was appointed head of the latter. He soon found it possible (in 1934) to invite Neyman to take a position on the staff—at first temporary but, before long, permanent. It might have been expected, and must have been hoped, that this would lead to more rapid further developments in the Neyman–Pearson approach to the philosophy of statistical procedures. In fact, this did not happen, for several reasons. There were new preoccupations—Pearson now had new administrative responsibilities and had married in 1934. On his father's death in 1936, he became managing editor of *Biometrika,* a post he held, creating a growing reputation for care and thoroughness, both for the journal and himself, for the next 30 years. During this time also he began his work on successive revisions of Karl Pearson's *Tables for Statisticians and Biometricians* in collaboration first with L. J. Comrie and later with H. O. Hartley.* He was personally concerned in calculating many new tables, notably those of the distribution of range in normal samples, percentage points of Pearson curves, and distribution of skewness and kurtosis coefficients. Although further joint work [19–21] with Neyman was published in the short-lived but influential departmentally produced journal *Statistical Research Memoirs* in 1936 and 1938, by 1936 Pearson felt he had gone as far as he wished to go along the path of further elaboration of the Neyman–Pearson approach. In his later work, there is, in general, more emphasis on specific applications.

This shift in emphasis received powerful support with the beginning of World War II in 1939. Pearson was head of a group of statisticians working on weapons assessment with the Ordnance Board. There is an interesting general account of some phases of this work in ref. [8]. During this period there was a very rapid increase in the use of statistical methods in quality control—already one of Pearson's major interests. There were related developments in formation of the new discipline of *operations research*—Pearson was a founding member of the (British) Operational Research Club (later Society) in 1948.

This very full life continued until Pearson retired as head of the department at University College in 1960. Thereafter, Pearson's activity decreased only gradually. With the relaxation of administrative burdens, time was again found for more scholarly work—especially in the field of frequency curves, in which his father was pioneering at the time of his birth. His last statistical paper [13] included wide-ranging comparisons among no fewer than eight systems of frequency curves.

His last major work was the fulfillment of a promise made to his father—production of a scholarly annotated version [10] of Karl Pearson's lectures on the early history of statistics and University College in 1921–1933, based on Pearson's own lecture notes, but considerably enriched by illuminating editorial comments.

REFERENCES

References [14–21] are included in *Joint Statistical Papers of J. Neyman and E. S. Pearson* (University of California Press, Berkeley, 1967). References [3], [4], [6–8], [11], and [12] are in *The Selected Papers of E. S. Pearson* (University of California Press, 1966).

[1] David, F. N., ed., (1966). In *Research Papers in Statistics. Festschrift for J. Neyman.* Wiley, New York, pp. 1–23.
[2] Moore, P. G. (1975). *J. R. Statist. Soc.,* **138A,** 129–130.
[3] Pearson, E. S. (1929). *Biometrika,* **21,** 337–360.
[4] Pearson, E. S. (1931). *Biometrika,* **23,** 114–133.
[5] Pearson, E. S. (1936). *The Application of Statistical Methods to Industrial Standardization and Quality Control,* Brit. Stand. 600. British Standards Institution, London.
[6] Pearson, E. S. (1947). *Biometrika,* **34,** 139–167.
[7] Pearson, E. S. (1956). *J. R. Statist. Soc.,* **119A,** 125–146.
[8] Pearson, E. S. (1963). *Proc. 8th Conf. Des. Exper. Army Res. Dev.,* pp. 1–15.
[9] Pearson, E. S. (1968). *Biometrika,* **55,** 445-457.
[10] Pearson, E. S., ed. (1979). *The History of Statistics in the Seventeenth and Eighteenth Centuries.* Macmillan, New York (Lectures by Karl Pearson given at University College, London, during the academic sessions 1921–1933.)
[11] Pearson, E. S. and Adyanthaya, N. K. (1928). *Biometrika,* **20A,** 356–360 [21, 259–286 (1929)].
[12] Pearson, E. S. and Adyanthaya, N. K. (1928). *Biometrika,* **20A,** 259–286.
[13] Pearson, E. S., Johnson, N. L., and Burr, I. W. (1979). *Commun. Statist. B,* **8,** 191–229.
[14] Pearson, E. S. and Neyman, J. (1928). *Biometrika,* **20A,** 175–240.
[15] Pearson, E. S. and Neyman, J. (1928). *Biometrika,* **20A,** 263–294.
[16] Pearson, E. S. and Neyman, J. (1930). *Bull. Acad. Pol. Sci.,* 73–96.
[17] Pearson, E. S. and Neyman, J. (1931). *Bull. Acad. Pol. Sci.,* 460–481.
[18] Pearson, E. S. and Neyman, J. (1933). *Philos. Trans. R. Soc. Lond.,* **231A,** 289–237.
[19] Pearson, E. S. and Neyman, J. (1936). *Statist. Res. Memo.,* **1,** 1–37.
[20] Pearson, E. S. and Neyman, J. (1936). *Statist. Res. Memo.,* **2,** 25–57.
[21] Pearson, E. S. and Neyman, J. (1936). *Statist. Res. Memo.,* **1,** 113–137.
[22] Pearson, E. S. and Please, N. W. (1975). *Biometrika,* **62,** 223–241.
[23] Reid, C. (1982). *Neyman from Life.* Springer-Verlag, New York.

PEARSON, KARL

Born: March 27, 1857, in London, England.
Died: April 27, 1936, in Coldharbour, Surrey, England.
Contributed to: anthropology, biometry, eugenics, scientific method, statistical theory.

Karl Pearson was educated at University College School and privately, and at Kings College, Cambridge, where he was a scholar. He took his degree in 1879, being the Third Wrangler in the Mathematical Tripos. In 1885 he was appointed to the Chair of Applied Mathematics at University College, London, where he stayed for the whole of his working life, moving to the newly instituted Chair of Eugenics in 1911. Both his parents were of Yorkshire stock, and he thought of himself as a Yorkshireman. He was married twice, and there were three children, Sigrid, Egon, and Helga, by the first marriage.

A man of great intellect, Pearson, when young, had many and varied interests. His ability as a mathematician was unquestioned, but he also studied physics and metaphysics, law, folklore, and the history of religion and social life in Germany. The echoes of this last are found among his first publications: *The Trinity: A Nineteenth Century Passion Play* (1882), *Die Fronica* (1887), and the *Ethic of Freethought* (1888). Throughout the intellectual ferment of these early years, however, his interest in the sciences persisted and emerged in the happy mingling of science, history, and mathematics in his completion (1886) of Todhunter's *History of the Theory of Elasticity* and his own production *The Grammar of Science* (1892).

The appointment to the Chair of Applied Mathematics (1885) did not check the outpourings of this fertile mind, although it inevitably gave them some direction. His principal professorial duty was teaching mathematics to engineering students, which influenced the way in which he subsequently thought about observational problems, but the real impetus in his thinking, both then and for many years, came from his friendship with Francis Galton* and with Weldon. The latter came to University College in 1890. He was interested in Galton's *Natural Inheritance* (1889) and had been collecting data on shrimps and shore crabs; he appealed to Pearson for the mathematical tools to carry out his analyses with the idea of studying evolution through morphological measurements. Pearson clearly found his lifework and in the process of answering his friends' pleas for help laid the foundations of modern mathematical statistics. For the fact that foundations must be buried does not mean that they are not there. In an astonishing output of original papers, we get the now familiar apparatus of moments and correlation, the system of frequency curves, the probable errors of moments and product moments reached by maximizing the likelihood function, and the χ^2 goodness-of-fit test, all in the space of 15 years when he was also lecturing many hours a week. It was not until 1903 that a grant from the Draper's Company enabled "biometric" research to be carried on with Pearson under less pressure. He was elected to the Royal Society in 1896 and was the Darwin Medallist in 1898. Among the students during these years were G. U. Yule* and L. N. G. Filon.

The concept of *ancestral inheritance* put forward by Pearson and Weldon was not received kindly by many biologists—chief among them, William Bateson—and inevitable controversies ensued. From these quarrels, arising from critical prepublication reviews, sprang the statistical journal *Biometrika,* which Pearson edited from its inception in 1901 until his death in 1936. The effect of the controversies on Pearson was to make him cut himself off from participation in meetings of learned societies. Thus, for example, he was never a Fellow of the Royal Statistical Society. He must have felt even more isolated by the movement of Weldon to a chair at Oxford and by Weldon's death in 1906.

Francis Galton was supportive of Pearson both in friendship and financial aid throughout the period 1885 until his death in 1911. He helped with the founding of *Biometrika.* When eventually the Biometrics Laboratory was started, he helped with the funding, and in his will money was left for a Chair of Eugenics, which Pearson was the first to occupy. The Biometrics Laboratory and Galton's Eugenics Laboratory both became part of the new Department of Applied Statistics with Pearson at the head.

During the years after Weldon left for Oxford, there was a lull in Pearson's mathematical–statistical activity. Endless data of all kinds were collected and analyzed, many controversies—mostly in the eugenics field—were pursued, many graduate students were instructed in the new biometric field, and it was from one of these that the next stage in statistical methods originated. W. S. Gosset* ("Student") was a student of Pearson's in 1906, bringing with him the small-sample problem. This was to occupy the attention of statisticians for many years, although it did not affect the course of Pearson's work unduly. He was always concerned with the collection and analysis of large quantities of data, with the mathematics very definitely ancillary to the main purpose—a tool to aid the understanding and not an end in itself. This led him in later years into many mathematical controversies in which he was wrong in the exact sense, but close enough in the practical sense.

But if at this time and in the later years his research seems to have been directed toward the working out of the ideas, both mathematical and practical, of a former era, his teaching side flourished. In 1915 the University of London instituted an undergraduate honors degree in statistics, and after a brief interlude of war service, Pearson drew up courses and gave lectures. In addition to the students, professors from all over the world came to these classes and went back after their sabbaticals were over to teach and research in their turn. These postwar years, remembered by many for the bitter controversies with R. A. Fisher,* are more important historically for this dissemination by teaching of how to analyze data and for the elevation of statistics both in England and abroad to the status of a respectable field of university study.

Pearson resigned in 1933. His department was split into two, one part as the Department of Eugenics with R. A. Fisher as Galton Professor, and the other as the Department of Statistics under E. S. Pearson,* with the latter responsible for the undergraduate teaching. Pearson moved to rooms in the Department of Zoology, where he continued to edit *Biometrika* until his death. The goal for the statistician, which he always taught his students, may be summed up in some words which he wrote in 1922: "The imagination of man has always run riot, but to imagine a thing is not meritorious unless we demonstrate its reasonableness by the laborious process of studying how it fits experience."

F. N. DAVID

Pitman, Edwin James George

Born: October 29, 1897, in Melbourne, Australia.
Died: July 21, 1993, in Kingston, near Hobart, Tasmania.
Contributed to: theory of estimation, tests of significance, probability theory.

Edwin James George Pitman was educated at the University of Melbourne, graduating with B.Sc. and M.A. degrees. He was acting professor of mathematics at the University of New Zealand 1922–1923, then tutor in mathematics and physics at residential colleges of the University of Melbourne 1924–1925. In 1926 he was appointed professor of mathematics at the University of Tasmania, a position he held until his retirement at the end of 1962.

As part of his duties at the University of Tasmania, Pitman was required to offer a course in statistics. He was also consulted by experimentalists of the Tasmanian Department of Agriculture about the analysis of their data. These circumstances led to his studying the then new statistical methods and ideas being developed by R. A. Fisher.* As Fisher's ideas were not then widely understood, Pitman set about mastering his work, in particular, *Statistical Methods for Research Workers.*

Pitman became interested in testing hypotheses about means without any accompanying specification of form of population distributions. He examined the nonnull distribution of the test statistic for what would now be called the permutation test, first for the test of equality of two means, and then for tests of several means (analysis of variance) and of correlation. The results are given in three papers [2, 3].

This work on distribution-free methods culminated in the lecture notes [6] developed during a visit to the United States in 1948 and 1949. These notes, produced for the lectures given at the University of North Carolina, were never published but were widely circulated in mimeographed form. They were frequently cited and undoubtedly provided a starting point for much subsequent work in this field.

However, Pitman's first published papers dealt with basic theoretical questions arising from the work of R. A. Fisher. His first paper [1] not only discussed the applicability of the concept of intrinsic accuracy but also established the result that families of distributions admitting a nontrivial sufficient statistic are of exponential type.

In two major papers [4, 5] on inference about location and scale parameters, Pitman systematically developed the ideas of Fisher [14] on estimation conditional on a sample "configuration." He applied the concept of invariance, previously used by Hotelling* [15], essentially as a practical restriction on the class of estimators to be considered. The restrictions on simultaneous estimation of location and scale parameters were clearly spelled out. The powerful methods developed were applied to show the unbiasedness of Bartlett's test for homogeneity of variances, as against Neyman* and Pearson's* related test.

Concern that much of the basic mathematics of statistical inference was unnecessarily unattractive led Pitman to reexamine some of the theory of the subject; the result was the monograph [11]. As he states in the preface: "The book is an attempt to present some of the basic mathematical results required for statistical inference with some elegance as well as precision. . . ."

In his later years, Pitman turned his attention to topics in probability theory. The most important results are those [8, 9] on the behavior of the characteristic function of a probability distribution in the neighborhood of the origin; this behavior determines the nature of the limiting distribution of sums.

His most recent contribution [12] was to subexponential distributions, for which he gave new results and methods.

Pitman was remarkable as a statistician in that, though he worked for many years in virtual isolation, he was able to contribute to the solution of central problems of statistical inference and to originate some important fundamental ideas. His first contact with statisticians outside Australia was in 1948–1949, when he was invited to visit Columbia University, the University of North Carolina, and Princeton University to lecture on nonparametric inference.

Pitman contributed several important concepts in statistics. Perhaps the most useful and widely known is that of asymptotic relative efficiency [13, pp. 337–338], which provides a means of comparing alternative tests of a hypothesis.

Pitman was much in demand as a visiting lecturer. He was visiting professor of statistics at Stanford in 1957. After his retirement from the University of Tasmania he visited Berkeley, Johns Hopkins, Adelaide, Melbourne, Chicago, and Dundee.

He was honored by election as a fellow of the Australian Academy of Science (FAA) in 1954, a fellow of the Institute of Mathematical Statistics in 1948, president of the Australian Mathematical Society in 1958 and 1959, and a vice-president in 1960 of the International Statistical Institute (to membership of which he was elected in 1956). He was elected an honorary fellow of the Royal Statistical Society in 1965, an honorary life member of the Statistical Society of Australia in 1966, and an honorary life member of the Australian Mathematical Society in 1968.

In honor of his seventy-fifth birthday à collection of essays by colleagues and former students [16] was published. In honor of his eightieth birthday in 1977, and in recognition of his outstanding contributions to scholarship, the University of Tasmania conferred on him the Honorary Degree of Doctor of Science. In 1977 the Statistical Society of Australia instituted the Pitman medal, a gold medal to be awarded to a member of the Society for high distinction in statistics; the first such medal was awarded to Pitman himself.

In 1981 the American Statistical Association invited Pitman, along with other leading statisticians, to record a lecture on videotape for historical purposes.

At his home in Hobart, Tasmania, Edwin Pitman and his wife Elinor provided warmhearted hospitality and kept "open house" over the past 40 years to visiting statisticians, mathematicians, and others of the academic community.

SELECTED REFERENCES OF E. J. G. PITMAN

[1] (1936). *Proc. Camb. Phil. Soc.,* **32,** 567–579.

[2] (1937). *J. R. Statist. Soc. Suppl.,* **4,** 119–130; 225-232.

[3] (1938). *Biometrika,* **29,** 322–335.

[4] (1939). *Biometrika,* **30,** 391–421.

[5] (1939). *Biometrika,* **31,** 200–215.

[6] (1949). *Lecture Notes on Non-parametric Statistical Inference* (unpublished).

[7] (1949b—with Herbert Robbins). *Ann. Math. Statist.,* **20,** 552–560.

[8] (1960). *Proc. 4th Berkeley Symp. Math. Statist. Prob.,* University of California Press, Berkeley, Calif., vol. 2, pp. 393–402.

[9] (1968). *J. Aust. Math. Soc.,* **8,** 423–443.

[10] (1978). *Aust. J. Statist.,* **20,** 60–74.

[11] (1979). *Some Basic Theory for Statistical Inference.* Chapman & Hall, London.

[12] (1980). *J. Aust. Math. Soc.,* **A29,** 337–347.

OTHER REFERENCES

[13] Cox, D. R. and Hinkley, D. V. (1974). *Theoretical Statistics.* Chapman & Hall, London.

[14] Fisher, R. A. (1934). *Proc. Roy. Soc.,* **A144,** 285–307.

[15] Hotelling, H. (1936). *Biometrika,* **28,** 321–377.

[16] Williams, E. J. (1974). *Studies in Probability and Statistics: Papers in Honour of Edwin J. G. Pitman.* North-Holland, Amsterdam.

E. J. WILLIAMS

Savage, Leonard Jimmie

Born: November 20, 1917, in Detroit, Michigan.
Died: November 1, 1971, in New Haven, Connecticut.
Contributed to: probability theory, foundations of statistics, Bayesian statistics.

Leonard Jimmie Savage was the most original thinker in statistics active in America during the period from the end of World War II until his premature death. Statistics, previously a collection of useful but somewhat disconnected ideas, was given firm foundations by him: foundations that led to new ideas to replace the old. Although almost all his published work lies in the fields of statistics and probability, he was interested in and informed about so many things that he made a superb statistical consultant for those who were not satisfied with a glib response but would benefit from a thorough understanding of the situation. He was a good writer and lecturer.

He was raised in Detroit and eventually, after some difficulties largely caused by his poor eyesight, graduated in mathematics from the University of Michigan. His statistical work began when he joined the Statistical Research Group at Columbia University in 1943. Later he joined the University of Chicago, and it was there that he wrote his great work, *The Foundations of Statistics* [8]. In 1960 he moved to Michigan and in 1964 to Yale, where he remained until his death. A collection of his principal works has appeared [11]; this book also contains four personal tributes and an account of his technical achievements.

In the late 1940s and early 1950s, statistics was dominated by the work of R. A. Fisher* [3] and by decision-theory ideas due to Neyman* [4], Pearson* [5, 6], and Wald* [12]. These proved to be practically useful, but there was no general underlying theory, although Wald had gone some way to providing one. Savage had briefly worked with von Neumann and had seen the power of the axiomatic method in constructing a general system. Using basic probability ideas, von Neumann had shown that the only sensible decision procedure was maximization of expected utility. Savage conceived and later executed the idea of widening the axiomatic structure and deducing, not only the utility, but the probability aspect. This he did in the first seven chapters of his first book [8].

Essentially he showed that the only satisfactory description of uncertainty is probability and of worth is utility, and that the best decision is the one that maximizes expected utility. It is the first of these that was both important and new. Now the foundations were secure; he felt he could establish the Fisherian and other results as consequences of the basic results. In the remainder of the book he attempted to do this, and gloriously failed (He explains this in the second edition, which differs from the first only in minor details except for a new, illuminating preface [9].) Instead, gradually, over the years he developed new ideas that replaced the old ones, constructing what is now usually called Bayesian statistics. That he correctly continued to enjoy and respect the enduring values in the older work is beautifully brought out in his marvelous work on Fisher [10]. Nevertheless, he had created a new paradigm.

Savage had the attribute, unfortunately not always present in contemporary scientists, of being a true scholar, always respecting the repository of knowledge above personal aggrandizement. Having constructed new foundations, he was able to understand the work of Ramsey* [7] and de Finetti* [1], who earlier had developed arguments leading to essentially the same results. He spent some time in Italy with de Finetti and was largely responsible for making de Finetti's brilliant ideas available to English-speaking statisticians and for extending them. Many of the papers he wrote between 1954 and his death are concerned with his increasing understanding of the power of Bayesian methods, and to read these in sequence is to appreciate a great mind making new discoveries. His concern is usually with matters of principle. Only in his consulting work, the bulk of which was never published, does he attempt to develop techniques of Bayesian statistics to match the brilliant ones put forward by Fisher.

Savage also did important work in probability, much of it in collaboration with Lester Dubins [2]. The basic problem they considered was this: you have fortune f, $0 < f < 1$, and may stake any amount $s \leq f$ to win s with probability $\frac{1}{2}$ or lose it with probability $\frac{1}{2}$. Once this has been settled, you may repeat the process with $f' = f \pm s$ so long as $f' > 0$: How should the stakes be selected to maximize the probability of reaching a fortune of 1 or more? They showed that bold play is optimal: that is, stake f, or at least enough to get $f' = 1$. Generalizations of these ideas provided a rich store of results in stochastic processes. The treatment is remarkable in using finite rather than sigma additivity.

Proper judgement of the importance of Savage's work can only come when the Bayesian paradigm is established as the statistical method or when it is shown to be defective. Even if the latter happens, Savage can be credited with creating a method that led to important, new ideas of lasting value. He was a true originator.

REFERENCES

[1] de Finetti, B. (1974–1975). *Theory of Probability*. Wiley, New York, two volumes.

[2] Dubins, L. E. and Savage, L. J. (1965). *How to Gamble If You Must: Inequalities for Stochastic Processes*. McGraw-Hill, New York.

[3] Fisher, R. A. (1950). *Contributions to Mathematical Statistics*. Wiley, New York.

[4] Neyman, J. (1967). *Early Statistical Papers*. University of California Press, Berkeley.

[5] Neyman, J. and Pearson, E. S. (1967). *Joint Statistical Papers*. Univ. of California Press, Berkeley.

[6] Pearson, E. S. (1966). *Selected Papers*. University of California Press, Berkeley.

[7] Ramsey, F. P. (1950). *The Foundations of Mathematics and Other Logical Essays*. Humanities Press, New York.

[8] Savage, L. J. (1954). *The Foundations of Statistics*. Wiley, New York.

[9] Savage, L. J. (1972). *The Foundations of Statistics*, 2nd rev. ed. Dover, New York.

[10] Savage, L. J. (1976). *Ann. Statist.*, **4**, 441–500.

[11] Savage, L. J. (1981). *The Writings of Leonard Jimmie Savage—A Memorial Selection.* Amer. Statist. Assoc. and Inst. Math. Statist., Washington, DC.

[12] Wald, A. (1950). *Statistical Decision Functions.* Wiley, New York.

DENNIS V. LINDLEY

Sverdrup, Erlang

Born: February 23, 1917, in Bergen, Norway.
Died: March 15, 1994, in Oslo, Norway.
Contributed to: Neyman–Pearson theory, multiple decision theory, inference in time-continuous Markov processes, graduation by moving averages, probabilistic life-insurance mathematics.

Erlang Sverdrup was a pioneer in actuarial and statistical science in Norway. He was born in Bergen, Norway, on February 23, 1917, son of Professor Georg Johan, and Gunvor (Gregusson) Sverdrup. He was 23 years old and a student of actuarial mathematics at the University of Oslo (UO) on the outbreak of World War II. He voluntarily reported for duty and commenced a five-year-long service as intelligence officer in the Norwegian free forces, with special responsibility for the cipher service, first at the legations in Helsinki, Moscow, Tokyo, and Washington, and from 1942 in London. After the conclusion of peace in 1945 he obtained the actuarial degree. He was then recruited to the econometrics group at UO and came to work with such personalities as Ragnar Frisch and Trygve Haavelmo, a reflection of whose scientific views can be seen as a background for those of Sverdrup. In 1948 he was appointed research assistant at the Seminar of Insurance Mathematics at UO and, holding the only scientific position there, he also became head of the Seminar. In 1949–1950 he visited the University of California at Berkeley, the University of Chicago, and Columbia University in New York.

He received his doctoral degree from UO in 1952 with a dissertation on minimax procedures, which together with the follow-up work [3] constitutes an important early contribution to the development of the general theory of admissible and minimax procedures and to the Neyman–Pearson theory of testing statistical hypotheses. A particularly notable contribution is an elegant proof, based on analytical continuation, for the general version of the Neyman–Pearson result about the construction of most powerful unbiased tests. A survey of the field is given in ref. [5].

In 1953 he was appointed professor of insurance mathematics and mathematical statistics at UO. At that time statistics was not a separate discipline in any Norwegian university, and Sverdrup devoted himself to the demanding task of building a modern educational program in mathematical statistics. The core of the study program he launched is presented in a two-volume textbook [6]. On this basis he also organized an up-to-date actuarial program with a scientific orientation on a par with other studies in mathematics and science. The teaching material he worked out for the instruction in life-insurance mathematics and general risk theory was based on stochastic models and was in that respect ahead of every contemporary textbook in the field. Relevant references here are [2] and [4]. The latter, which treats statistical inference in continous-time Markov models, also provides evidence of Sverdrup's pioneering role in the Scandinavian school in life-history analysis. Sverdrup's strong interests in general scientific methodology and in particular the logic of mode-based statistical inference are clearly pronounced in his contributions to multiple decision theory, e.g. refs. [7] and [8]. They clarify the rationale of testing a null hypothesis, which he prefers to call the null state: "It is chosen not because we have any a priori confidence in it, or are interested in the truth of it, but because we are interested in certain effects that are contrasts to the hypothesis. The test is just a clearance test, performed initially to determine whether or not to search further for significant contrasts/departures from the hypothesis."

Sverdrup was a fellow of the Institute of Mathematical Statistics, an elected member of the Norwegian Academy of Sciences, the Royal Norwegian Scientific Society, and the International Statistical Institute, and an honorary member of the Norwegian Actuarial Society and the Norwegian Statistical Society.

He was assigned a number of administrative tasks at and outside the University. He was chairman of the Norwegian Insurance Council in 1970–1974 and the Norwegian Mathematics Council in 1974–1976. He was a visiting professor at Columbia University in 1963–1964 and at Aarhus University in 1969. As editor of *Skandinavisk Aktuarietidsskrift,* 1968–1982, he was instrumental in reshaping it into the present *Scandinavian Actuarial Journal.*

A more complete biography and a bibliography are published in [1].

REFERENCES

[1] Norberg, R., ed. (1989). *Festschrift to Erling Sverdrup.* Scandinavian University Press, Oslo.

[2] Sverdrup, E. (1952). *Skand. Aktuarietidskr.,* **35,** 115–131.

[3] Sverdrup, E. (1953). *Skand. Aktuarietidskr.,* **36,** 64–86.

[4] Sverdrup, E. (1965). *Skand. Aktuarietidskr.,* **48,** 184–211.

[5] Sverdrup, E. (1966). *Rev. Int. Statist. Inst.,* **34,** 309–333.

[6] Sverdrup, E. (1967). *Laws and Chance Variations,* vols., I–II. North-Holland, Amsterdam.

[7] Sverdrup, E. (1977). *Bull. Int. Statist. Inst.,* **XLVII,** 573–606.

[8] Sverdrup, E. (1986). *Scand. Actuarial J.,* **1986,** 13–63.

RAGNAR NORBERG

Thiele, Thorvald Nicolai

Thorvald Nicolai Thiele

Born: December 24, 1838, in Copenhagen, Denmark.
Died: September 26, 1910, in Copenhagen, Denmark.
Contributed to: numerical analysis, distribution theory, statistical inference, time series, astronomy, actuarial mathematics.

Thorvald Thiele belonged to a prominent family of book printers, instrument makers, and opticians. His father, Just Mathias, was for many years private librarian to King Christian VIII; his duties included acting as secretary of the Royal Academy of Fine Arts. He was also a well-known dramatist, poet and folklorist, and art historian.

Thorvald was named after the famous sculptor Bertel Thorvaldsson—one of his godfathers. After attending the Metropolitan School, he was admitted to the University of Copenhagen in 1856. Even while still at school, his interest in mathematical subjects had been stimulated by Professor Christian Jörgenson, who, however, urged him to choose astronomy as his main subject at the University, to avoid the imputed "one-sidedness" involved in the study of pure mathematics.

In 1859, Thiele was awarded the Gold Medal of the University for his astronomical thesis, and in 1860 he completed his M.Sc. Degree, with astronomy as his major subject. From 1860 to 1870 he was an assistant to Professor H. L. d'Arrest at the Copenhagen Observatory. In 1862, he accompanied d'Arrest on a trip to Spain to observe a total solar eclipse. His work on double-star systems earned him a doctorate in 1866. Apart from this, during the early 1860s he played a part in establishing the Hafnia Life Insurance Company, for which he worked as an actuary and, for nearly 40 years, manager.

In 1875, Thiele became Professor of Astronomy and Director of the Copenhagen Observatory, with which he was associated until his retirement in 1907. From 1895 onwards, he was a member of the University Senate and served as rector of the University from 1900 to 1906. He was also a founder of the Danish Society of Actuaries in 1901, and served as its president until his death in 1910.

Thiele married Maria Martin Trolle in 1867. They had six children. An unmarried daughter, Margarethe (1868–1928), was a school teacher and also a translator for the Royal Danish Society and co-author of the first Danish–French dictionary, which appeared posthumously in 1937. Thiele's wife died, at the early age of 48, in 1889.

Thiele was a versatile mathematician, with special interests in applied mathematics—notably in numerical analysis (wherein he constructed a well-known representation of Newton's divided-difference interpolation formula in the form of a continued fraction) and actuarial mathematics. His contributions to astronomy include the determination of orbits of double stars—the Thiele–Innes method (R. Innes arranged some of Thiele's formulas for machine computation), the reduction of errors of observation, and the three-body problem.

In actuarial mathematics, his most important contribution is his differential equation for net premium reserve, derived in 1875. He was an expert in calculation, similarly to his fellow countryman Harald Westergaard.*

In a paper by the Danish mathematical statistician and historian of statistics Anders Hald [5], the contributions of T. N. Thiele to statistics are classified into the following categories: skew distributions, cumulants, estimation theory, linear models, analysis of variance, and time series. Thiele's work on skew distributions related to Gram–Charliar Type A distributions, and transformations. According to Gram [4], he fitted a lognormal distribution to data on marriage age of females as early as 1875.

It is perhaps for his work on cumulants and their uses that Thiele has had the most enduring influence on statistical theory and methods. He introduced them in ref. [8], but gave the classical definition, via moment generating functions, in the form (with an obvious notation)

$$\exp\left(\sum_{j=1}^{\infty} \frac{\kappa_j t^j}{j!}\right) \equiv 1 + \sum_{j=1}^{\infty} \frac{\mu'_j t^j}{j!},$$

only ten years later, in ref. [9]. Karl Pearson [7] included a somewhat cursory reference to Thiele's "half-invariants." R. A. Fisher [2] does not include a reference to Thiele's work, though in ref. [3] he remarked that cumulants". . . seem to have been later [after Laplace] discovered independently by Thiele (1889)." Thiele is not mentioned in Cornish and Fisher [1].

In estimation theory, Thiele contributed an important generalization of the method of moments, and the construction of k-statistics. In the theory of linear models, he introduced what is now called the canonical form of a general linear hypothesis (orthogonalization of linear models), and also developed one-way and two-way analyses of variance (though not using that name) in his book *General Theory of Observations,* published in Danish in 1889, and a more popular version, *Elementary Theory of Observations,* published in 1897. The book was translated into English in ref. [10]. Remarkably, it was reprinted in full, in 1931, in the *Annals of Mathematical Statistics,* providing a somewhat belated, though welcome, recognition of the importance (and preceding neglect) of Thiele's work. This neglect was, indeed, but one sad example of lack of communication between the continental and English schools of statistics during the period around the turn of the century. However, the faults were not all on one side. As Hald [3] perceptively noted, "Thiele had the bad habit of giving imprecise references or no references at all to other authors; he just supposed that his readers were fully acquainted with the literature." It was thus ironic that he was himself neglected, for instance, by K. Pearson* and R. A. Fisher.*

We conclude with a quotation from an obituary for Thiele, written by J. P. Gram [4], who has already been mentioned in this article, and was closely associated with Thiele at the Hafnia Insurance Company from 1872 to 1875. It provides a vivid, balanced summary of Thiele's personality.

He thought profoundly and thoroughly on any matter which occupied him and he had a wonderful perseverance and faculty of combination. But he liked more to construct his own methods than to study the methods of other people. Therefore his reading was not very extensive, and he often took a one-sided view which had a restrictive influence on the results of his own speculations. . . . Thiele's importance as a theoretician lies therefore more in the original ideas he started than in his formulations, and his ideas were in many respects not only original but far ahead of his time. Therefore he did not get the recogniton he deserved, and some time will elapse before his ideas will be brought into such a form that they will be accessible to the great majority.

REFERENCES

[1] Cornish, E. A. and Fisher, R. A. (1937). Moments and cumulants in the specification of distributions. *Rev. Int. Statist. Inst.,* **4,** 1–14.

[2] Fisher, R. A. (1928). Moments and product moments of sampling distributions. *Proc. Lond. Math. Soc.* **2,** 30, 199–238.

[3] Fisher, R. A. (1932). *Statistical Methods for Research Workers,* 4th ed. Oliver and Boyd, Edinburgh.

[4] Gram, J. P. (1910). Professor Thiele som Aktuar, *Dansk Forsikringsårbog,* **7,** 26–37.

[5] Hald, A. (1981). T. N. Thiele's contributions to statistics. *Int. Statist. Rev.,* **49,** 1–20.

[6] Heegaard, P. (1942) and Thiele, T. N. In *Dansk Biografisk Leksikon,* Paul Engelstaft, ed. Schultz, Copenhagen, vol. 23, pp. 503–506.

[7] Pearson, K. (1895). Contributions to the mathematical theory of evolution II, Skew variation in homogeneous material. *Phil. Trans. R. Soc. Lond. A,* **186,** 343–414.

[8] Thiele, T. N. (1889). *Foreloesninger over Almindelig Iagttagelses Poere.* Reitzel, Copenhagen.

[9] Thiele, T. N. (1899). *Om Iagttagelseslaerens Halvinvarianter.* Oversigt Videnskabernes Selskab Forhandlinger, pp. 135–141.

[10] Thiele, T. N. (1903). *Theory of Observations.* Layton, London. [Reprinted in *Ann. Math. Statist,* **2,** 165–308 (1931).]

[11] Thiele, T. N. (1909). *Interpolationsrechnung.* Leipzig, Germany. (Contains the author's succinct definition: "Interpolation ist der Kunst zwischen den Zeilen einer Tafeln zu lesen.")

Wald, Abraham

Born: October 31, 1902, in Cluj, Hungary (now Romania).
Died: December 13, 1950, in Travancore, India.
Contributed to: decision theory, sequential analysis, geometry, econometrics.

Abraham Wald

Abraham Wald was born in Cluj on October 31, 1902. At that time, Cluj belonged to Hungary, but after World War I it belonged to Romania. Menger [2] states that when he first met Wald in Vienna in 1927, Wald spoke "fluent German, but with an unmistakable Hungarian accent." A short time after Wald emigrated to the United States, he spoke fluent English, with an accent.

Wald would not attend school on Saturday, the Jewish sabbath, and as a result he did not attend primary or secondary school, but was educated at home by his family. On the basis of this education, he was admitted to and graduated from the local university. He entered the University of Vienna in 1927 and received his Ph.D. in mathematics in 1931. Wald's first research interest was in geometry, and he published 21 papers in that area between 1931 and 1937. Two later papers, published in 1943 and 1944, were on a statistical generalization of metric spaces, and are described in [4]. A discussion of Wald's research in geometry by Menger [2] describes it as deep, beautiful, and of fundamental importance.

During the 1930s, economic and political conditions in Vienna made it impossible for Wald to obtain an academic position there. To support himself, he obtained a position as tutor in mathematics to Karl Schlesinger, a prominent Viennese banker and economist. As a result of this association, Wald became interested in economics and econometrics, and published 10 papers on those subjects, plus a monograph [6] on seasonal movements in time series. Morgenstern [3] describes this monograph as developing techniques superior to all others. Wald's first exposure to statistical ideas was a result of his research in econometrics. This research in economics and econometrics is described by Morgenstern [3] and Tintner [5]. Once again, his contributions to these areas are characterized as of fundamental importance.

Austria was seized by the Nazis early in 1938, and Wald came to the United States in the summer of that year, as a fellow of the Cowles Commission. This invitation from the Cowles Commission probably saved Wald's life, for almost all of the members of his family in Europe were murdered by the Nazis. In the fall of 1938, Wald became a fellow of the Carnegie Corporation and started to study statistics at Columbia University with Harold Hotelling.* Wald stayed at Columbia as a fellow of the Carnegie Corporation until 1941, lecturing during the academic year 1939–1940. In 1941, he joined the Columbia faculty and remained there for the rest of his life. During the war years, he was also a member of the Statistics Research Group at Columbia, doing war-related research. The techniques he developed for estimating aircraft vulnerability are still used and are described in a reprint [16] published in 1980.

Wald's first papers in statistics were published in 1939, and one of them [7] is certainly one of his most important contributions to statistical theory. J. Wolfowitz, who became a close friend and collaborator of Wald soon after Wald arrived in New York, in [22] describes this 1939 paper by Wald as probably Wald's most important single paper. In it, Wald points out that the two major problems of statistical theory at that time, testing hypotheses and estimation, can both be regarded as simple special cases of a more general problem—known nowadays as a "statistical decision problem." This generalization seems quite natural once it is pointed out, and the wonder is that nobody had thought of it before. Perhaps it needed a talented person with a fresh outlook to see it.

Wald does much more than merely point out the generalization in this paper. He defines loss functions, risk functions, *a priori* distributions, Bayes decision rules, admissible decision rules, and minimax decision rules, and proves that a minimax decision rule has a constant risk under certain regularity conditions. It is interesting that in this paper Wald states that the reason for introducing *a priori* distributions on the unknown parameters is that it is useful in deducing certain theorems on admissibility and in the construction of minimax decision rules: That is, he is not considering the unknown parameters to be random variables. This paper did not receive much attention at first, but many other papers, by Wald and others, have extended statistical decision theory. Wald's 1950 book [13] contains most of the results developed up to that year. This book is accessible only to those with a strong background in mathematics. The 1952 paper [14] summarizes the basic ideas and is more easily read.

Wald's other great contribution to statistical theory is the construction of optimal statistical procedures when sequential sampling is permitted. (Sequential sampling is any sampling scheme in which the total number of observations taken is a random variable.) Unlike statistical decision theory, the concept of sequential sampling is not due to Wald and is not included in the 1939 paper on decision theory. Just who first thought of sequential sampling is apparently not known. In ref. [11] it is stated that Captain G. L. Schuyler of the U.S. Navy made some comments that alerted M. Friedman and W. Allen Wallis to the possible advantages of sequential sampling, and Friedman and Wallis proposed the problem to Wald in March 1943. Wald's great contribution to sequential analysis was in finding optimal sequential procedures. He started by considering the problem of testing a simple hypothesis against a simple alternative using sequential sampling and conjectured that the subsequently famous sequential probability ratio test is optimal for this problem, in the sense that among all test procedures with preassigned upper bounds on the probabilities of making the wrong decision, the sequential probability ratio test minimizes the expected sample size under both the hypothesis and the alternative. Early in his investigation, he was able to show that this conjecture is at least approximately true, but it was not until his 1948 paper with Wolfowitz [21] that a proof that it is exactly true was given. This proof was based on the ideas that Wald had developed in his work on statistical decision theory and thus united his two major contributions to statistical theory. The 1947 book [11] describes the results on sequential analysis known up to that time in an elementary manner and is accessible to anyone with a knowledge of elementary probability theory. The 1950 book [13] incorporates sequential sampling into statistical decision theory.

Besides statistical decision theory and sequential analysis, Wald made many other fundamental contributions to statistical theory, some of which will be described briefly. In [9] he derived the large-sample distribution of the likelehood-ratio test under alternatives to the hypothesis being tested and proved the asymptotic optimality of the test. In collaboration with Wolfowitz, he made fundamental contributions to nonparametric statistical inference in [17–20]. He wrote a pioneering paper on the optimal design of experiments [8], a field which became very active a few years later. In [10] Wald generalized a theorem of von Neumann on zerosum two-person games. In collaboration with Mann [1], he developed statistical inference for stochastic difference equations. Reference [12] contains a new proof of the consistency of maximum likelihood estimators.

Wald was an excellent teacher, always precise and clear. He was a master at deriving complicated results in amazingly simple ways. His influence on the teaching of statistics extended far beyond the students who actually attended his classes, because, with his permission, the Columbia students reproduced the notes they took in his classes. These reproduced notes were only supposed to be circulated to other Columbia students, but they had a much wider circulation than that.

In late 1950, Wald was in India, at the invitation of the Indian Government, lecturing on statistics. He was accompanied by his wife. On December 13, 1950, both were killed in a plane crash.

A fuller account of Wald's research and a list of 104 of his publications are contained in ref. [15]. This list is complete except for the 1980 reprint [16].

REFERENCES

[1] Mann, H. B. and Wald, A. (1943). *Econometrica,* **11,** 173–220.

[2] Menger, K. (1952). *Ann. Math. Statist.,* **23,** 14–20.

[3] Morgenstern, O. (1951). *Econometrica,* **19,** 361–367.

[4] Schweizer, B. and Sklar, A. (1983). *Probabilistic Metric Spaces.* North-Holland, New York.

[5] Tintner, G. (1952). *Ann. Math. Statist.,* **23,** 21–28.

[6] Wald, A. (1936). *Berechnung und Ausschaltung von Saisonschwankungen.* Springer, Vienna.

[7] Wald, A. (1939). *Ann. Math. Statist.,* **10,** 299–326.

[8] Wald, A. (1943). *Ann. Math. Statist.,* **14,** 134–140.

[9] Wald, A. (1943). *Trans. Amer. Math. Soc.,* **54,** 426–482.

[10] Wald, A. (1945). *Ann. Math.,* **46,** 281–286.

[11] Wald, A. (1947). *Sequential Analysis.* Wiley, New York.

[12] Wald, A. (1949). *Ann. Math. Statist.,* **20,** 595–601.

[13] Wald, A. (1950). *Statistical Decision Functions.* Wiley, New York.

[14] Wald, A. (1952). *Proc. Int. Congress of Mathematicians.* Harvard University Press, Cambridge, Mass.

[15] Wald, A. (1955). *Selected Papers in Statistics and Probability.* McGraw-Hill, New York.

[16] Wald, A. (1980). *A Method of Estimating Plane Vulnerability Based on Damage of Survivors.* Center for Naval Analyses, Washington D. C.

[17] Wald, A. and Wolfowitz, J. (1939). *Ann. Math. Statist.,* **10,** 105–118.

[18] Wald, A. and Wolfowitz, J. (1940). *Ann. Math. Statist.,* **11,** 147–162.

[19] Wald, A. and Wolfowitz, J. (1943). *Ann. Math. Statist.,* **14,** 378–388.

[20] Wald, A. and Wolfowitz, J. (1944). *Ann. Math. Statist.,* **15,** 358–372.

[21] Wald, A. and Wolfowitz, J. (1948). *Ann. Math. Statist.,* **19,** 326–339.

[22] Wolfowitz, J. (1952). *Ann. Math. Statist.,* **23,** 1–13.

L. WEISS

Yule, George Udny

George U. Yule

Born: February 18, 1871, in Beech Hill, near Haddington, Scotland.
Died: June 26, 1951, in Cambridge, England.
Contributed to: correlation theory, distribution theory, stochastic processes, statistics of literary vocabulary.

George Udny Yule was a member of a Scottish family steeped in literary and administrative traditions. After schooling at Winchester, he proceeded at the age of 16 to study engineering at University College, London. His first published work was on research into electric waves under Heinrich Hertz during his sojourn at Bonn 1892. He wrote six papers on electromagnetic theory, but, after taking up a demonstratorship offered him in 1893 by Karl Pearson* (then a professor of applied mathematics at University College), he turned his attention to statistical problems, starting with his 1895 paper, "On the Correlation of Total Pauperism with Proportion of Outrelief." There is a fascinating discussion of this paper in Selvin [3], describing how Yule introduced correlation coefficients in studying two-way tables in the earlier volumes of the monumental work of Booth [2].

For Yule, Pearson was an inspiring teacher, and Yule made fundamental contributions in 1897 and 1899 to the theory of statistics of regression and correlation. In 1899, Yule left University College for a post at the City and Guilds of London Institute. Between 1902 and 1909, he also gave the Newmarch lectures in statistics at University College. These lectures formed the basis of his famous *Introduction to the Theory of Statistics,* which, during his lifetime, ran to 14 editions [the 11th (1937), 12th (1940), 13th (1945), and 14th (1950) being joint with M. G. Kendall]. He continued to publish papers on association and correlation (1900, 1901, 1903), and was awarded the Guy Medal in Gold in 1911 by the Royal Statistical Society. His theoretical works were accompanied by contributions to various economic and sociological subjects (1906, 1907, 1910).

In 1912, Yule was appointed a lecturer in statistics at Cambridge University (later raised to the status of a readership). The years 1920–1930 were the most productive in his career. During this period he introduced the correlogram and laid the foundations of the theory of autoregressive series. He was president of the Royal Statistical Society from 1924 to 1926. In one of his last publications in the *Journal of the Royal Statistical Society,* in late 1933, he showed that German authorities had overestimated the number of Jews entering Germany from Poland and Galicia during and after World War I by a factor of about 5.

In 1931, he retired from his post at Cambridge University. However, he "felt young enough to learn to fly. Accordingly, he went through the intricacies of training, got a pilot's license, and bought a plane. Unfortunately, a heart attack cut short his flying and, to a considerable degree, his scholarly work" (Bates and Neyman [1]). The death of Karl Pearson in 1936 affected him deeply. However, according to

M. G. Kendall, "the publication of the revised Introduction in 1937 gave him a new lease of life." In the later years, his main work was related to frequency of occurrence of words (particularly nouns) in various texts. This research found expression in his last book (1944) on *The Statistics of Literary Vocabulary.*

A great deal of Yule's contributions to statistics resides in the stimulus he gave to students, in discussion with his colleagues, and advice he generously tendered to all who consulted him. His work on correlation and regression is now so standard that only history buffs would consult the original sources; he invented the correlogram and the autoregressive series; he also paved the way for Fisher's derivation of the distributions of partial correlations. The terms *Yule process* and *Yule distribution* are now firmly established in the literature.

REFERENCES

[1] Bates, G. E. and Neyman, J. (1952). *Univ. California Publ. Statist.,* **1,** 215–254.

[2] Booth, C. E., ed. (1889–1893). *Life and labour of the People in London.* Macmillan, London, England (first 4 of 17 volumes).

[3] Selvin, H. C. (1976). *Archives Europ. J. Sociol.,* **17,** 39–51.

WORKS BY G. U. YULE

(1895). *Econ. J.,* **5,** 477–489.

(1897). *Proc. R. Soc. Lond.,* **60,** 477–489.

(1899). *J. R. Statist. Soc.,* **62,** 249–286.

(1900). *Philos. Trans. R. Soc. Lond.,* **194A,** 257–319.

(1901). *Philos. Trans. R. Soc. Lond.,* **197A,** 91–133.

(1903). *Biometrika,* **2,** 121–134.

(1906). *J. R. Statist. Soc.,* **69,** 88–132.

(1907). *J. R. Statist. Soc.,* **70,** 52–87.

(1910). *J. R. Statist. Soc.,* **73,** 26–38.

(1933). *J. R. Statist. Soc.,* **96,** 478–480.

(1944). *The Statistics of Literary Vocabulary.* Cambridge University Press, London.

A list of Yule's publications is included in the obituary

Kendall, M. G. (1952). *J. R. Statist. Soc. A,* **115,** 156–161.

A list of publications is also given in

Stuart, A. and Kendall, M. G., eds. (1971). *Statistical Papers of George Udny Yule.* Hafner, New York.

SECTION **3**

Statistical Theory

Aitken, Alexander Craig

Born: April, 1, 1895, in Dunedin, New Zealand.
Died: November 3, 1967, in Edinburgh, Scotland.
Contributed to: statistics, numerical methods, algebra.

Alexander Craig Aitken was born in Dunedin, New Zealand and died in Edinburgh, Scotland, where he had spent his working life. Dunedin is a largely Scottish community on the southern tip of New Zealand. His father was a farmer. Aitken's extraordinary memory, musical gift, attractive personality, and work and teaching talents are well described in the obituary articles [1, 2].

After attending Otago Boys High School, he studied classical languages for two years at Otago University. In April 1915 he enlisted in the New Zealand Infantry and served in Gallipoli, Egypt, and France in World War I, as movingly described in a manuscript written while he was recovering from wounds in 1917 but turned into a book [3] only in 1962, his last publication. Aitken had total recall of his past, and this section of it always gave him great pain. His platoon was all but wiped out in the battle of the Somme, along with all records. He was able to remember and write down all the information in the records of all these men. Upon recovery, he returned to Otago University. He could not study mathematics there, though it was realized that he had a gift for it. He then taught languages in his old school for three years. Upon graduating in 1920 he married a fellow student and later had a son and daughter.

Fortunately, in 1923 he was given a scholarship to study mathematics under E. T. Whittaker in Edinburgh. The first edition of Whittaker & Robinson's *Calculus of Observations* [4] (W&R) appeared in 1924, but Whittaker had earlier considered *graduation* or *smoothing of data* as a statistical problem, his motivation was mainly actuarial. So arose—almost—what we now call *splines*. The function chosen to be minimized was the sum of the squares of the differences between the observed u_n and "true" values u_n' plus a multiple of the sum of squares of the *third* differences of the "true" values. How to execute this was Aitken's Ph.D. problem. Whittaker was so pleased with Aitken's results that he was awarded a D.Sc. instead and a staff appointment. This method is given in W&R. In a preceding section in W&R, they speak of the *method of interlaced parabolas,* in which they fit a cubic polynomial to each successive four graduated values u_n'—this allows interpolation. Had the second term in the minimand been the integral of the square of the third derivative of the interpolating function, they would have invented the modern method of getting a spline.

The Mathematics Department in Edinburgh had, over many years, very broad interests that spanned all of applied mathematics—Whittaker was a towering figure in many fields. In particular, it was then the only place in Britain that taught determinants and matrices, and these immediately appealed to Aitken. In 1932 he published, with H. W. Turnbull, *An Introduction to the Theory of Canonical Matrices* [5]. Its last chapter gives, among other applications, some to statistics. So by then he had shown his real interests—algebra, numerical analysis, and statistics. Aitken succeeded Whittaker in the Chair in Mathematics in 1946, holding it until he retired in 1965.

Aitken was a renowned teacher. In 1939 he published the first two volumes in the Oliver & Boyd series (of which he was a joint editor with D. E. Rutherford) University Mathematical Texts. They were *Statistical Mathematics* [6] and *Determinants and Matrices* [7]. These two books were my first acquaintance with his work. When I was an undergraduate there were very few books on these topics with any style to them. Around 1949, as a graduate student I became aware of his research, largely written in the preceding ten years, in statistics, especially of his matrix treatment of least squares (see e.g. [8])—idempotents like $\mathbf{X}(\mathbf{X}'\mathbf{X})^{-1}\mathbf{X}'$, the use of the trace operator (e.g. $E[\mathbf{x}'\mathbf{A}\mathbf{x}] = \text{Tr}\, E\,[\mathbf{A}\mathbf{x}\mathbf{x}']$), etc. This has now become standard. Indeed, with the linear model $\mathbf{y} = \mathbf{X}\boldsymbol{\beta} + \mathbf{u}$ where the random error vector has mean 0 and nonsingular covariance matrix $\mathbf{V}$, the estimating equations $\mathbf{X}'\mathbf{V}^{-1}\mathbf{y} = \mathbf{X}'\mathbf{V}^{-1}\mathbf{X}\boldsymbol{\beta}$ are known as the *Aitken equations.* But he never mentioned vector spaces, although he would have been aware of the backgroud geometry. He wrote many papers about least squares and especially about the fitting of polynomials and other statistical topics—we have only singled out three areas to mention here.

In the late thirties he gave a student from New Zealand, H. Silverstone, a Ph.D. topic—the optimal estimation of statistical parameters—which he had apparently already worked out for a scalar parameter. See Aitken and Silverstone (1941) [9]. In a 1947 paper [10] Aitken completed the work for many parameters. But he has never received credit for this work, which seems unfair to me. *For the time at which it was written,* the formulation is correct, as are the answers. But instead of a direct proof of a minimum, they simply give the "Euler equation" that the calculus of variations (CofV) throws up, though with the full knowledge that that method has difficulties. This was very natural, as the study of minimal surfaces was then of great interest and much use was made of the CofV in mathematical physics.

To take the simpler problem with his notation, let $\Phi(x, \theta)$ be the density of the vector $\mathbf{x}$ which ranges over a region which is the same for all θ and be uniformly differentiable with respect to θ. Suppose there exists $t(\mathbf{x})$, a minimum-variance estimator of θ. Then $\int t\, \partial\Phi/\partial\theta\, dx = 1$, and $\int (t-\theta)^2\Phi\, dx$ must be a minimum. Writing $I(t) = \int (t-\theta)^2\Phi\, dx - 2\lambda \int t\, \partial\Phi/\partial\theta\, dx$, consider any other estimator $t + h$. Then

$$\begin{aligned} I(t+h) &= I(t) + \int h\left(2(t-\theta)\Phi - \frac{2\lambda\partial\Phi}{\partial\theta}\right) dx + \int h^2\Phi\, dx \\ &\geq I(t) \end{aligned}$$

if and only if $(t-\theta) = \lambda\, \partial\Phi/\partial\theta$, where λ may be a function of θ. This is the "Euler" equation they give (and, I would guess, the proof they had) from which they correctly draw all the now well-known conclusions. It took, however, many other statisticians many years to clarify these questions and to find what other assumptions are necessary. M. S. Bartlett has published some correspondence [11] with Aitken.

Several other features of Aitken's life are his love of music, his awesome memory, and his arithmetical ability. He played the violin well and for a time was leader of the Edinburgh University Musical Society Orchestra, which was sometimes conducted by his close friend Sir Donald Tovey. His violin was particularly important to him when in the Army, and it now is displayed in his old school. He said that 75% of the time his thoughts were musical. On occasion he would demonstrate his arithmetical feats. He wrote a book [12] against the decimalization of the English coinage—the use of twelve, which has so many factors as a basis appealed to him. He was able to dictate rapidly the first 707 digits of π.

Among his honors, he was a Fellow of the Royal Societies of Edinburgh and London and of the Royal Society of Literature.

REFERENCES

[1] Copson, E. T., Kendall, D. G., Miller, J. C. P., and Ledermann, W. (1968). Obituary articles. *Proc. Edinburgh. Math. Soc.,* **16,** 151–176.

[2] Whittaker, J. M. and Bartlett, M. S. (1968). Alexander Craig Aitken. *Biogr. Mem. Fell. R. Soc. Lond.,* **14,** 1–14.

[3] Aitken, A. C. (1962). *Gallipoli to the Somme.* Oxford University Press, London, 177 pp.

[4] Whittaker, E. T. and Robinson, G. (1924). *The Calculus of Observations.* Blackie & Son, London & Glasgow (sixth impression, 1937).

[5] Turnbull, H. W. and Aitken, A. C. (1932). *An Introduction to the Theory of Canonical Matrices.* Blackie & Son, London and Glasgow, 192 pp.

[6] Aitken, A. C. (1939). *Statistical Mathematics.* Oliver & Boyd, Edinburgh, 153 pp.

[7] Aitken, A. C. (1939). *Determinants and Matrices.* Oliver & Boyd, Edinburgh, pp. 144

[8] Aitken, A. C. (1934–1935). On least squares and linear combinations of observations. *Proc. Roy. Soc. Edinburgh.,* **55,** 42–48.

[9] Aitken, A. C. and Silverstone, H. (1941). On the estimation of statistical parameters. *Proc. Roy. Soc. Edinburgh,* **A61,** 186–194.

[10] Aitken, A. C. (1947). On the estimation of many statistical parameters. *Proc. Roy. Soc. Edinburgh,* **A62,** 369–377.

[11] Bartlett, M. S. (1994). Some pre-war statistical correspondence. In *Probability, Statistics and Optimization,* F. P. Kelly, ed. Wiley, Chichester and New York, pp. 297–413.

[12] Aitken, A. C. (1962). *The Case against Decimalization.* Oliver & Boyd, Edinburgh, 22 pp.

GEOFFREY S. WATSON

Anderson Oskar Nikolaevich

Born: August 2, 1887, in Minsk, Russia.
Died: February 12, 1960, in Munich, Federal Republic of Germany.
Contributed to: correlation analysis, index numbers, quantitative economics, sample surveys, time-series analysis, nonparametric methods, foundations of probability, applications in sociology.

The work and life of Oskar Anderson received a great deal of attention in the periodical statistical literature following his death in 1960. In addition to the customary obituary in the *Journal of the Royal Statistical Society (Series A),* there was a relatively rare long appreciative article—written by the famous statistician and econometrist H. O. A. Wold,* whose biography also appears in this volume—in the *Annals of Mathematical Statistics,* with an extensive bibliography, and a remarkable obituary and survey of his activities in the *Journal of the American Statistical Association,* written by the well-known statistician G. Tintner, also containing a bibliography. Furthermore there was detailed obituary in *Econometrica*. The first entry in the *International Statistical Encyclopedia* (J. M. Tanur and W. H. Kruskal, eds.) contains a rather detailed biography and an analysis of Anderson's contributions.

Some of this special interest may be associated with unusual and tragic events the first 40 years of Anderson's life, as aptly noted by Wold (1961):

> The course of outer events in Oskar Anderson's life reflects the turbulence and agonies of a Europe torn by wars and revolutions. [In fact Fels (1961) noted that "a daughter perished when the Andersons were refugees; a son a little later. Another son fell in the Second World War."]

Both German and Russian economists and statisticians compete to claim Oskar Anderson as their own. His father became a professor of Finno-Ugric languages at the University of Kazan (the famous Russian mathematician I. N. Lobachevsky, who was born in Kazan, was also a professor at the University). The Andersons were ethnically German. Oskar studied mathematics at the University of Kazan for a year, after graduating (with a gold medal) from gymnasium in that city in 1906. In 1907, he entered the Economics Department of the Polytechnical Institute at St. Petersburg. From 1907 to 1915 he was a assistant to A. A. Chuprov* at the Institute, and a librarian of the Statistical-Geographical "Cabinet" attached to it. He proved himself an outstanding student of Chuprov's, whose influence on Anderson persisted throughout his life. Also, during the years 1912–1917, he was a lecturer at a "commercial gymnasium" in St. Petersburg, and managed to obtain a law degree. Among his other activities at that time, he organized and participated in an expedition in 1915 to Turkestan to carry out an agricultural survey in the area around the Syr Darya river. This survey was on a large scale, and possessed a representativity ahead of contemporary surveys in Europe and the USA. In 1917 he worked as a research economist for a large cooperative society in southern Russia.

In 1917, Anderson moved to Kiev and trained at the Commercial Institute in that city, becoming a docent, while simultaneously holding a job in the Demographic Institute of the Kiev Academy of Sciences, in association with E. Slutskii.*

In 1920, he and his family left Russia, although it was said that Lenin had offered him a very high position in the economic administration of the country. It is possible that his feelings of loyalty to colleagues who were in disfavor with the authorities influenced this decision. For a few years he worked as a high-school principal in Budapest, and then for many years (1924–1943) except for a two-year gap he lived in Bulgaria, being a professor at the Commercial Institute in Varna from 1924 to 1933, and holding a similar position at the University of Sofia from 1935 to 1942. (In the period 1933–1935 he was a Rockefeller Fellow in England and Germany, and his first textbook on mathematical statistics was published.) While in Bulgaria he was very active in various sample surveys and censuses, utilizing, from time to time, the methodology of purposive sampling. In 1940 he was sent to Germany by the Bulgarian government to study rationing, and in 1942, in the midst of World War II, he accepted an appointment at the University of Kiel.

After the war, in 1947, Anderson became a professor of statistics in the Economics Department of the University of Munich, and he remained there till his death in 1960. His son Oskar Jr. served as a professor of economics in the University of Mannheim.

Anderson was a cofounder—with Irving Fisher and Ragnar Frisch—of the Econometric Society. He was also a coeditor of the journal *Metrika*. In the years 1930–1960 he was one of the most widely known statisticians in Central and Western Europe, serving as a link between the Russian and Anglo-American schools in statistics, while working within the German tradition exemplified by such statisticians as Lexis* and von Bortkiewicz.* He contributed substantially to the statistical training of economists in German universities. His main strengths lay in systematic coordination of statistical theory and practice; he had good intuition and insight, and a superb understanding of statistical problems. His second textbook, *Probleme der Statistischen Methodenlehre,* published in 1954, went through three editions in his lifetime; a fourth edition appeared posthumously in 1962. He was awarded honorary doctorates from the Universities of Vienna and Mannheim and was an honorary Fellow of the Royal Statistical Society.

His dissertation in St. Petersburg, "On application of correlation coefficients in dynamic series," was a development of Chuprov's ideas on correlation. He later published a paper on this topic in *Biometrika,* and a monograph in 1929. While in Varna, he published a monograph in 1928 (reprinted in Bonn in 1929), criticizing the Harvard method of time-series analysis, and developed his well-known method of "variate differences" (concurrently with and independently of W. S. Gosset*). This method compares the estimated variances of different orders of differences in a time series to attempt to estimate the appropriate degree of a polynomial for a local fit. In 1947, Anderson published a long paper in the *Schweizerische Zeitschrift für Volkswirtschaft und Statistik,* devoted to the use of prior and posterior probabilities in statistics, aiming at unifying mathematical statistics with the practices of statistical investigators. Anderson was against abstract mathematical studies in economics, and often criticized the so-called "Anglo-American school," claiming that the Econometric Society had abandoned the goals originally envisioned by its founders.

During the last period of his life, he turned to nonparametric methods, advocating, *inter alia,* the use of Chebyshev-type inequalities, as opposed to the "sigma rule" based on assumptions of normality. Some of his endeavors were well ahead of his time, in particular, his emphasis on causal analysis of nonexperimental data, which was developed later by H. Wold,* and his emphasis on the importance of elimination of systematic errors in sample surveys. Although he had severed physical contact with the land of his birth as early as 1920, he followed with close attention the development of statistics in the Soviet Union in the thirties and forties, subjecting it to harsh criticism in several scorching reviews of the Marxist orientation of books on statistics published in the USSR.

REFERENCES

[1] Anderson, O. N. (1963). *Ausgewáhlte Schriften,* H. Strecker and H. Kellerer, eds. Mohr Verlag, Tübingen, Germany. 2 vols. (These collected works, in German, of O. N. Anderson include some 150 articles, originally written in Russian, Bulgarian, and English as well as German. They are supplemented by a biography.)

[2] Fels, E. (1961). Oskar Anderson, 1887–1960. *Econometrica,* **29,** 74–79.

[3] Fels, E. (1968. Anderson, Oskar N. In *International Encyclopedia of Statistics,* J. M. Tanur and W. H. Kruskal, eds. Free Press, New York.

[4] Rabikin, V. I. (1972). O. Anderson—a student of A. A. Chuprov, *Uchenye Zapiski po Statistike,* **18,** 161–174.

[5] Sagoroff, S. (1960). Obituary: Oskar Anderson, 1887–1960, *J. R. Statist. Soc. A,* **123,** 518–519.

[6] Tintner, G. (1961). The statistical work of Oscar Anderson. *J. Amer. Statist. Ass.,* **56,** 273–280.

[7] Wold, H. (1961). Oskar Anderson: 1887–1960, *Ann. Math. Statist.,* **32,** 651–660.

Bol'shev, Login Nikolaevich

Born: March 6, 1922, in Moscow, USSR.
Died: August 29, 1978, in Moscow, USSR.
Contributed to: transformations of random variables, asymptotic expansions for probability distributions, tables of mathematical statistics, mathematical epidemiology and demography.

L. N. Bol'shev was one of the leading Soviet experts in the field of mathematical statistics during the period 1950–1978.

He was born in 1922 in the family of a white-collar worker. His father was a military man who distinguished himself in World War I, and his grandfather a well-known military topographer of his time.

Bol'shev graduated in 1951 from the Mechanical–Mathematical Faculty of the Moscow State University. Based on A. N. Kolmogorov's* recommendations and initiative, he started his tenure at the Probability Theory and Mathematical Statistics Branch of the prestigious Steklov Mathematical Institute in Moscow, where he worked throughout his whole life, and where A. Ya. Khinchine, N. V. Smirnov,* and Yu. V. Prohorov were his colleagues. Later, Bol'shev combined his work at the Institute with lecturing at the Moscow State University.

In 1955 Bol'shev defended his master's dissertation entitled "On the problem of testing composite statistical hypotheses" at the Institute, and in 1966 he was awarded the Doctor of Science degree by the Scientific Council of the Institute for his research on "Transformation of random variables."

The theory of transformation of random variables and the methods of asymptotic expansions for the basic probability distributors utilized in statistics (in particular Pearson, Poisson, and generalized hypergometric) which Bol'shev developed in his thesis are widely used in statistical theory and practice [1, 4–7].

L. N. Bol'shev was highly successful in developing and implementing the construction of tables of mathematical statistics which are required for probabilistic and statistical calculations [12–14, 15]. These activities were originated in the USSR by E. E. Slutskii* and N. V. Smirnov.* Jointly with Smirnov, he compiled in 1965 the well-known *Tables of Mathematical Statistics* [13], which are the most advanced and authoritative international contribution to this area.

Additional contributions of L. N. Bol'shev which have attracted international attention are in the areas of statistical parameter estimation [2, 8, 11], testing for outliers [3, 9], and epidemiology and demography [10].

An excellent and popular teacher, Bol'shev attracted many students who later became well-known Soviet statisticians. He was often called as an expert in various scientific discussions connected with the application of statistical methodology in industry, geology, biology, medicine, and sports refereeing.

In 1974 he was elected a corresponding member of the Academy of Sciences of the USSR, and the following year a member of the International Statistical Institute. He served with great distinction in the Red Army during World War II and was awarded the Order of the Red Banner of Labor and other medals.

REFERENCES

[1] Bol'shev, L. N. (1959). On transformation of random variables. *Theory Probab. Appl.*, **4,** (2), 136–149.

[2] Bol'shev, L. N. (1961). A refinement of the Cramér–Rao inequality, *Theory Probab. Appl.*, **6**(3), 319–326.

[3] Bol'shev, L. N. (1961). On elimination of outliers, *Theory Probab. Appl.*, **6**(4), 482–484.

[4] Bol'shev, L. N. (1963). Asymptotic Pearsonian transformations, *Theory Probab. Appl.*, **8**(2), 129–155; Corr., 473.

[5] Bol'shev, L. N. (1964). Distributions related to hypergeometic. *Theory Probab. Appl.*, **9**(4), 687–692.

[6] Bol'shev, L. N. (1964). Some applications of Pearson transformations. *Rev. Int. Statist. Inst.*, **32,** 14–16.

[7] Bol'shev, L. N. (1965). On characterization of Poisson distribution and its statistical applications. *Theory Probab. Appl.*, **10**(3), 488–499.

[8] Bol'shev, L. N. (1965). On construction of confidence intervals, *Theory Probab. Appl.*, **10**(1), 187–192.

[9] Bol'shev, L. N. (1966). Testing of outliers in the case of least squares estimation (in Russian). *Abstracts of Scientific Papers at the International Congress of Mathematicians, Section 11,* Moscow, pp. 30–31.

[10] Bol'shev, L. N. (1966). A comparison of intensities of simple streams. *Theory Probab. Appl.*, **11**(3), 353–355.

[11] Bol'shev, L. N. and Loginov, E. A. (1966). Interval estimation in the presence of nuisance parameters. *Theory Probab. Appl.*, **11**(1), 94–107.

[12] Bol'shev, L. N. and Smirnov, N. V. (1962). *Tables for Calculation of Bivariate Normal Distribution Functions.* Fizmatgiz, Moscow.

[13] Bol'shev, L. N. and Smirnov, N. V. (1965). *Tables of Mathematical Statistics.* Fizmatigiz, Moscow. (Translated into English.)

[14] Bol'shev, L. N., Gladkov, B. V., and Shcheglova, V. (1961). Tables for calculation of functions of B- and Z-distributions. *Theory Probab. Appl.*, **6**(4), 446–455.

[15] Bol'shev, L. N., Bark, L. S., Kuznetsov, P. I., and Cherenkov, A. P. (1964). *Tables of the Rayleigh–Rice Distribution,* Fizmatgiz, Moscow.

S. A. AIVAZIAN (TRANSLATED AND EDITED BY S. KOTZ)

Bortkiewicz, Ladislaus Josephowitsch Von

Born: August 7, 1868, in St. Petersburg, Russia.
Died: July 15, 1931, in Berlin, Germany.
Contributed to: statistics, economics.

Bortkiewicz graduated from the University of St. Petersburg Faculty of Law in 1890, and subsequently studied under Lexis* in Göttingen, where he defended his doctoral thesis in 1893. Privatdozent in Strasbourg from 1895 to 1897, Bortkiewicz then returned to Russia until 1901, when he was appointed to a professorial position at the University of Berlin, where he taught statistics and economics until his death.

Best known for his modeling of rare-event phenomena by the Poisson distribution, Bortkiewicz also made numerous other contributions to mathematical statistics, notably the statistical analysis of radioactivity, the theory of runs, and the distributional properties of extreme values. His work on population theory, actuarial science, and political economy was also noteworthy; for the latter, see Gumbel [2].

Bortkiewicz's monograph *Das Gesetz der kleinen Zahlen* (The Law of Small Numbers) [1] is unquestionably his best-known work and was in large part responsible for the subsequent popularity of the Poisson distribution. Contrary to popular belief, Bortkiewicz meant by the expression "law of small numbers" not the Poisson distribution itself, but the tendency of data in binomial sampling with small and variable success probability to appear as though the success probability were constant when the sample size was large. The Lexis ratio was advocated as a means of detecting the presence of variable success probabilities.

A meticulous scholar, Bortkiewicz wrote in a difficult style that may have lessened his influence. In England, one of the few to appreciate his work was John Maynard Keynes, who devoted a chapter of *A Treatise on Probability* [3] to describing the results of Lexis* and Bortkiewicz.

Literature

Thor Andersson's long obituary in the *Nordic Statistical Journal,* 1931 (**3,** 9–26), includes an essentially complete bibliography. The biography by E. J. Gumbel in the *International Encyclopedia of Statistics* (1978, Free Press, New York) contains a useful list of secondary literature; see also the entry by O. B. Sheynin in the *Dictionary of Scientific Biography* (1970, Scribner's, New York). Bortkiewicz's contributions to dispersion theory and the Lexis ratio are discussed within a historical perspective by C. C. Heyde and E. Seneta in *I. J. Bienaymé: Statistical Theory Anticipated* (1977, Springer-Verlag, New York), pp. 49–58. Other historical comments, including a description of Bortkiewicz's applications of the Poisson distribution and the controversies with Gini and Whittaker, may be found in Frank Haight's *Handbook of the Poisson Distribution* (1967, Wiley, New York), Chap. 9.

Bortkiewicz was a member of the International Statistical Institute and participated in its meetings; William Kruskal and Frederick Mosteller describe his criticism of A. N. Kiaer's* concept of "representative sample" at one such meeting in their "Representative Sampling IV: The History of the Concept in Statistics," *International Statistical Review,* 1980 (**48,** August).

REFERENCES

[1] Bortkiewicz, L. von (1898). *Das Gesetz der kleinen Zahlen.* Teubner, Leipzig.

[2] Gumbel, E. J. (1978). In *International Encyclopedia of Statistics.* Free Press, New York.

[3] Keynes, J. M. (1921). *A Treatise on Probability.* Macmillan, London.

SANDY L. ZABELL

Bose, Raj Chandra

Born: June 19, 1901, in Hoshangabad, Madhya Pradesh, India.
Died: October 30, 1987, in Fort Collins, Colorado, USA.
Contributed to: coding theory, design and analysis of experiments, geometry, graph theory, multivariate analysis distribution theory.

Raj Chandra Bose was educated at Punjab University, Delhi University, and Calcutta University. He received M. A. degrees from the latter two institutions in 1924 (in applied mathematics) and 1927 (in pure mathematics), respectively. At Calcutta he came under the influence of the outstanding geometer S. Mukhopadhyaya.

There followed a lectureship at Asutosh College, Calcutta. This involved very heavy teaching duties, but in his limited spare time, Bose continued to produce research, including joint publications with Mukhopadhyaya, in multidimensional and nonEuclidean geometry.

In 1932, this came to the attention of Mahalanobis,* who was in the process of forming the Indian Statistical Institute. He needed a research worker who could apply geometrical techniques to distributional problems in multivariate analysis, along the lines pioneered by Fisher,* and invited Bose to join the young institute.

Initially, Bose worked, as expected, on applications of geometrical methods in multivariate analysis, in particular on the distribution of Mahalanobis' D^2-statistic. Much of this work was in with Roy (e.g., ref. [6]). Later, as a consequence of attending seminars given by Levi (appointed Hardinge Professor of Pure Mathematics at Calcutta University in 1936), he developed an interest in the application of geometrical methods to the construction of experiment designs. His monumental paper [2] on the construction of balanced incomplete block designs established his reputation in this field. Further development of these ideas constitutes a major part of Bose's work. It includes the now well-known concept of partially balanced incomplete block designs (see ref. [4]) and also the demonstration (with S. S. Shrikhande and E. T. Parker) of the falsity of Euler's conjecture (ref. [7], foreshadowed by ref. [1]) on the nonexistence of $n \times n$ Graeco–Latin squares for values of n other than a prime or a power of a prime. Reference [3] contains a broad summary of this work, including treatment of confounding and fractional factorial designs.

Bose took a part-time position in the Department of Mathematics at Calcutta University in 1938, and moved to the Department of Statistics upon its formation in 1941. He was head of the department from 1945 to 1949, and received a D.Litt. Degree in 1947. In 1949, he became a professor in the Department of Statistics at the University of North Carolina, where he stayed until his "retirement" in 1971.

During this period, Bose's interests came to include, with growing emphasis, coding theory—in particular, the use of geometry to construct codes. Among the outcomes of this interest there resulted, from collaboration with Ray-Chaudhuri, the well-known Bose–Chaudhuri (BCH) codes [5].

After his "retirement," Bose accepted a position, jointly with the Departments of Mathematics and Statistics (later, Mathematics alone), at Colorado State University in Fort Collins. His interests continued to change, gradually returning to more purely mathematical topics, including graph theory. He finally retired in 1980, but retained an active interest in research.

REFERENCES

[1] Bose, R. C. (1938). *Sankhyā,* **3,** 323–339.

[2] Bose, R. C. (1938). *Ann. Eugen. (Lond.),* **9,** 358–399.

[3] Bose, R. C. (1947). *Sankhyā,* **8,** 107–166.

[4] Bose, R. C., Clatworthy, W. H., and Shrikhande, S. S. (1954). Tables of Partially Balanced Designs with Two Associate Classes. *Tech. Bull. 107,* North Carolina Agricultural Experiment Station, Raleigh, N. C. [An extensive revision, by W. H. Clatworthy, with assistance from J. M. Cameron and J. A. Speakman, appeared in 1963 in National Bureau of Standards (U. S.) *Applied Mathematics Series,* vol. 63.]

[5] Bose R. C. and Ray-Chaudhuri, D. K. (1960). *Inform. Control,* **3,** 68–79.

[6] Bose, R. C. and Roy, S. N. (1936). *Sankhyā,* **4,** 19–38.

[7] Bose, R. C., Shrikhande, S. S., and Parker, E. T. (1960). *Canad. J. Math.,* **12,** 189–203.

Chuprov (or Tschuprow), Alexander Alexandrovich

Born: February 18 (n.s.), 1874, in Mosal'sk, Russia.
Died: April 19, 1926, in Geneva, Switzerland.
Contributed to: mathematical statistics, demography, rural economics.

The formative years of Chuprov's education were heavily influenced by his father, A. I. Chuprov (1842–1908), for many years a professor of political economy and statistics at Moscow University. Alexander graduated from the physico mathematical faculty of Moscow University in 1896 with a dissertation on probability theory as a basis for theoretical statistics, then traveled to Germany to study political economy. During a semester at the University of Berlin, he established a lasting friendship with L. Bortkiewicz,* who introduced him to W. Lexis* in Göttingen. The years 1897–1901 were spent at Strasbourg University, where among his teachers were Bortkiewicz and G. F. Knapp, who supervised his doctoral dissertation "Die Feldgemeinschaft, eine morphologische Untersuchung," published in 1902. In 1902, in order to gain a teaching position in Russia, Chuprov completed master's examinations at the University of Moscow, concentrating on theoretical economics and the application of mathematical methods.

He was offered a position in 1902 in the newly formed Economics Section of the St. Petersburg Polytechnic Institute, where he was in charge of the organization and teaching of statistics until 1916. In this area he developed a remarkably broad and modern-style pedagogical approach to the subject and a fine library, and later produced a number of notable disciples, including O. N. Anderson* and N. S. Chetverikov (or Tschetwerikoff). In 1909, his work, *Ocherki po Teorii Statistiki* [2], for which he was awarded a doctor's degree by Moscow University, was published and ran into a second edition within a year. This work had enormous influence for a number of years in Russia due to its stress on the logical and mathematical approach to statistics. It includes principles of probability theory with strong emphasis on the frequency interpretation of objective probability on the basis of the law of large numbers; and an account of the stability theory of statistical series of trials (dispersion theory) of Lexis and Bortkiewicz. A response in 1910 to the work by A. A. Markov* brought Chuprov into awareness of Markov's work, and marked the beginning of an intense correspondence between the two [8] that resulted in a focusing of Chuprov's researches on theory.

In May 1917, he went to Scandinavia for a vacation but never returned to Russia, because of the revolution. Most of Chuprov's work in mathematical statistics was published during his emigré period. The initial thrust of it is in the framework of dispersion theory. If X_{ij} is the jth observation ($j = 1, \dots, n$) in the ith set of observations ($i = 1, \dots, m$), then interest within this theory focused on the *dispersion coefficient L*. Chuprov showed, *inter alia,* that in the case of general i.i.d. variables X_{ij}, indeed $EL = 1$, but that this could still hold in the case of $N = mn$ variables that are not i.i.d. The notable conceptual advances here consist of an extension to general variables, and the investigation (under Markov's influence) of the moment structure of a sample statistic. Clearly in evidence also are ideas of analysis of variance.

The work on dispersion theory led to an extensive investigation of expectations of sample moments in terms of population moments (i.e., the "method of moments") under, ultimately, very general conditions embracing samples of correlated observations [10]. An obvious application of such results to sample-survey theory, in which Chuprov had always been interested, anticipated several results of J. Neyman,* especially the well-known formula for optimal allocation $\{n_h\}$ among t strata of a sample of fixed size n: $n_h = nN_hS_h/\sum_{r=1}^{t} N_rS_r$, $h = 1, \ldots, t$, where N_h and S_h are the population size and standard deviation of stratum h [10, Chap. V, Sec. III]. His investigations in correlation theory are presented in ref. [11].

Work of this general nature led to his propagation of the modern "stochastic" view of statistical theory, based on the notion of empirical observations as manifestations of random variables following a probability distribution, and leading to a sampling distribution for sample statistics, with statistical inference based on conceptual repetitions of samples to accord with the law of large numbers. From the law of large numbers, he was also led to the notion of consistency of an estimator; and in his writings a limiting Gaussian form of a sampling distribution is adduced from a consideration of moments. Indeed, his lasting contribution to mathematical statistics would seem to rest on an early recognition of such fundamental ideas and their lucid and extensive popularization on the continent. He is often credited, on the one hand, with unifying several streams of statistical thinking, i.e., the Russian probabilistic stream, the German stream typified by Lexis and Bortkiewicz, and that of the English biometric school led by Karl Pearson,* although his technical contributions were soon overshadowed by those of R. A. Fisher.*

On the other hand, he is also often regarded as the main proponent of the "Continental direction" of statistics, which sought to develop statistical methods free of those extensive assumptions about underlying distributions of the kind later developed by the English school.

In the area of demography, Chuprov presented to the International Statistical Institute, during his St. Petersburg period, the results of an extensive data-analytic consideration of the problem of decrease in the surplus of male births.

Chuprov was a high-minded ascetic. That part of his emigré period which he spent in Dresden, which began in mid 1920, was spent in solitary and intense productivity without steady income. He appears to have been apolitical, maintaining extensive professional contacts with his homeland and publishing in the Russian journal *Vestnik Statistiki,* while sporadically participating in Russian emigré–professional–academic organizations and publishing in their journals. Following his Dresden period, he took up an appointment with the Russian College in Prague in 1925. There soon followed a serious decline in his health, leading to his death.

He exerted a profound influence on the evolution of statistics, in Scandinavia in particular, and many of his papers were published in *Nordisk Statistisk Tidskrift (N.S.T.),* founded in 1922. His work was known to most of the notable English statisticians of the time and was championed by L. Isserlis in particular.

The most comprehensive account of Chuprov's life and work is ref. [5]; more readily accessible is the obituary [9]; and vol. 5 of *N.S.T.* (1926) contains much relevant material, including an obituary [7]. Several other obituaries are derivative of refs. [7] and [9].) Volume 18 of *Biometrika* (1926) has a portrait of Chuprov. Russian versions of many of his papers originally published in *N.S.T.* are collected in ref. [3]; refs. [2] and [11] were reprinted in the USSR.

REFERENCES

[1] Chetverikov, N. S., ed. (1968). *O Teorii Dispersii. Statistika,* Moscow. (Presents evolution of dispersion theory through a series of key papers, in Russian versions, including Lexis', Bortkiewicz's, and Chuprov's.)

[2] Chuprov, A. A. (1910). *Ocherki po Teorii Statistiki* (1st ed., 1909). St. Petersburg. (Reprinted by Gosstatizdat, Moscow, 1959.)

[3] Chuprov, A. A. (1960). *Voprosy statistiki.* Gosstatizdat, Moscow. (Introductory remarks, and translations into Russian, by B. I. Karpenko and N. S. Chetverikov.)

[4] Heyde, C. C. and Seneta, E. (1977). *I. J. Bienaymé: Statistical Theory Anticipated.* Springer-Verlag, New York. (Chapter 3 gives an account of dispersion theory from a modern standpoint.)

[5] Karpenko, B. I. (1957). *Uch. Zap. Statist.,* **3,** 282–317 (in Russian). (Written with the cooperation of N. S. Chetverikov. Contains the most complete listing of Chuprov's writings.)

[6] Kendall, M. G. and Doig, A. (1968). *Bibliography of Statistical Literature, vol. 3: Pre-1940 with Supplements.* Oliver & Boyd, Edinburgh. (Page 281 lists most of Chuprov's papers published during his emigré period.)

[7] Kohn, S. (1926). *Nord. Statist. Tidskr.,* **5,** 171–194.

[8] Ondar, Kh. O., ed. (1977). *O. Teorii Veroiatnostei i Matematicheskoi Statistike,* Nauka, Moscow. (Annotated correspondence between Chuprov and Markov, Nov. 1910–Feb. 1917.)

[9] Tschetwerikoff, N. S. (1926). *Metron,* **6,** 314–320.

[10] Tschuprow, A. A. (1923). *Metron,* **2,** 461–493, 646–680.

[11] Tschuprow, A. A. (1925). *Grundbegriffe und Grundprobleme der Korrelationstheorie.* Teubner, Leipzig–Berlin. (Published in Russian version: *Osnovnie Problemi Teorii Korrelatsii,* M. and S. Sabashnikov, Moscow, 1926; and in English translation: *Principles of the Mathematical Theory of Correlation,* Hodge, London, 1939. Russian version reprinted by Gosstatizdat, 1960.)

[12] Vinogradova, N. M. (1957). *Uch. Zap. Statist.,* **3,** 318–324 (in Russian).

E. SENETA

Cochran, William Gemmell

Born: July 15, 1909, in Rutherglen, Scotland
Died: March 29, 1980, on Cape Cod, Massachusetts, USA.
Contributed to: the development of statistical methods and their dissemination, through 121 papers and 5 books.

Cochran brought the statistical expertise of Rothamsted to the USA in 1939. This, plus his wisdom and good humor, soon made him the leading statistical consultant in the USA. This was later reinforced by his influential books *Experimental Designs* (with Gertrude Cox, 1950) [6] *Sampling Techniques* (1935) [4] and *Statistical Techniques* (sixth edition of the classic by Snedecor*) [8]. He wrote a number of insightful reviews (e.g. refs. [3] and [5] on chi-square) which also still read well. A more detailed biography and a complete list of his writings may be found in ref. [9]. To this we can now add his collected papers [1], whose introduction gives a further commentary.

Cochran entered Glasgow University in 1927, graduating in mathematics with prizes including a scholarship to St. John's College, Cambridge University, where he became a Wrangler in 1933. The next step was a Ph.D. His first paper was on fluid dynamics, followed by the "Cochran's Theorem" paper [2] suggested by J. Wishart, who gave the only statistics course offered in Cambridge. This caught his fancy. Yates* had taken Fisher's position as head of statistics at Rothamsted when Fisher went to the Galton chair of eugenics at University College, London. Yates offered Cochran a job as his assistant. Since the Depression was at its height, Cochran accepted. Ph.D. degrees were then rather scorned, anyway. It also allowed him to see much of Fisher,* who kept his house in Harpenden. In his six years there he wrote 29 papers on topics that interested him all his life, and he became a well-known and distinguished statistician. There also he married an entomologist, Betty I. M. Mitchell, by whom he later had three children.

After a visit to Ames, Iowa, Cochran accepted a position there in 1939. At no other teaching institution then were sampling and the design and analysis of agricultural data pursued more vigorously. During World War II he worked with the Princeton Statistical Research Group, led by S. S. Wilks.* In 1946 Gertrude Cox* formed the Institute of Statistics in North Carolina with H. Hotelling* as the leader of the theoretical group in Chapel Hill and Cochran as the leader of the applied group in Raleigh. This formidable threesome built a distinguished team. The books [6] and [4] were largely written there.

Cochran had many Ph.D. students, to whom he was very kind and encouraging. He had much influence on all who knew him and was universally liked. The Cochrans were very hospitable. As a student of his in North Carolina, I recall many kindnesses, personal and intellectual, and the square dances at their house.

In January 1949, Cochran moved to the Department of Biostatistics (which had an illustrious history) in the School of Public Health of Johns Hopkins University. He didn't find this very congenial. However, at Hopkins Cochran first faced the problems of sampling human populations and analyzing observational data rather than data obtained from designed experiments. This became an added interest for the rest of his career. In his time there he wrote a study [7] of the Kinsey Report with F. Mosteller and J. W. Tukey. He moved to Harvard in 1957 when it established its Department of Statistics. At Harvard he took a further interest in a sequential method for estimating the median effective dose.

Cochran was president of the Institute of Mathematical Statistics, the American Statistical Association, the Biometric Society, and the International Statistical Institute. He was elected to the American Academy of Arts and Sciences and the National Academy of Science.

REFERENCES

[1] Cochran, B. I. M., compiler (1982). *Contributions to Statistics of W. G. Cochran.* Wiley, New York.

[2] Cochran, W. G. (1934). The distribution of quadratic forms in a normal system. *Proc. Cambridge Phil. Soc.,* **30,** 178–189.

[3] Cochran, W. G. (1952). The chisquare test of goodness-of-fit. *Ann. Math. Statist.,* **23,** 315–345.

[4] Cochran, W. G. (1953). *Sampling Techniques.* Wiley, New York.

[5] Cochran, W. G. (1954). Some methods for strengthening the common chisquare tests. *Biometrics,* **10,** 427–451.

[6] Cochran, W. G. and Cox, G. M. (1950). *Experimental Designs.* Wiley, New York.

[7] Cochran, W. G., Mosteller, F., and Tukey, J. W. (1954). *Statistical Problems of the Kinsey Report.* American Statistical Association, Washington, D.C.

[8] Snedecor, G. W. and Cochran, W. G. (1967). *Statistical Methods.* Iowa State University Press, Ames.

[9] Watson, G. S. (1982). William Gemmell Cochran. *Ann. Statist.,* **10,** 1–10.

GEOFFREY S. WATSON

Edgeworth, Francis Ysidro

Born: February 8, 1845, in County Lonford, Ireland.
Died: February 13, 1926, in London, England.
Contributed to: index numbers, laws of error, estimation, economics, agricultural statistics.

Francis Ysidro Edgeworth (1845–1926) remains a relatively unheralded economist and statistician even though he is responsible for many novel and important contributions to both fields. Edgeworth's original contributions to mathematical or analytical economics include the indifference curve, the contract curve (and the related construct known as the Edgeworth box), the law of diminishing returns (which he also used in his editorial capacity to encourage brevity in journal submissions [3]), and the determination of economic equilibria. His statistical contributions include works on index numbers, the law of error, the theory of estimation, correlation, goodness of fit, and probability theory.

Edgeworth was born February 8, 1845, in County Lonford, Ireland, a descendent of a family of Anglo-Irish gentry. He received a classical education common to his time: he attended Trinity College, Dublin at age 17, entered Oxford as a scholar of Magdalen Hall in 1867, and was admitted to Balliol College a year later. He was awarded first-class honors in Literis Humanioribus in 1869, and took his B.A. degree in 1873. Edgeworth was apparently primarily a "self-taught" mathematician, having received little formal training in advanced mathematics [2].

Edgeworth's earliest known publication is *New and Old Methods of Ethics,* published by Parker and Company of Oxford in 1877, in which he investigated quantitative problems arising in utilitarianism. His only book, *Mathematical Psychics* (published in 1881), represents his first contribution to economics, and contains his treatment of contracts in a free market (and his beginning work on mathematical economics generally).

He was made Lecturer in Logic at King's College, London in 1880, and in 1890 succeeded Thorold Rogers as Tooke Professor of Economic Science and Statistics. A year later he succeeded Thorold Rogers as Drummond Professor of Political Economy at Oxford, where he remained until retiring with the title of Emeritus Professor in 1922.

He began his publications on probability and statistics while at King's College. Edgeworth was concerned with *a priori* probabilities, which he fitted into a frequency theory of probability. (His views of "inverse probability" are distinct from Fisher's.) He wrote at length on the law of error, emphasizing the prevalence of the normal law in nature and studying the distribution of averages. His work on index numbers for prices, begun in 1883, drew heavily from his work on averages and error. In 1892, he published his first paper on correlation, in which he attempted to provide a framework for the concept of multiple correlation. Edgeworth published approximately 40 statistical papers between 1893 and 1926 [5].

A large part of Edgeworth's last 35 years was occupied with the editorship of the *Economics Journal,* which was begun under his editorship in 1891. Although most of his service was devoted to economics, including serving as president of the Economic Section of the British Association for the Advancement of Science (1889 and 1922), Edgeworth also served as president of the Royal Statistical Society from 1912 to 1914. He was made a Fellow of the British Academy in 1903.

Edgeworth was a life-long bachelor who led the life of an ascetic. He was extremely shy, with a self-effacing manner [4]. He owned few possessions, not even bothering to collect books, but preferring public libraries. (Keynes claimed that the only two material objects he knew that Edgeworth privately owned were red tape and gum.) While at Oxford, Edgeworth lived in a Fellow's room at All Soul's College. He maintained the same two barely furnished rooms on the outskirts of London for the last 50 years of his life. He died on February 13, 1926 [3].

Edgeworth's contributions to the field of mathematical statistics (74 papers and 9 reviews in all) are summarized in ref. [1].

REFERENCES

[1] Bowley, A. L. (1928). *F. Y. Edgeworth's Contributions to Mathematical Statistics.* (Pamphlet published by the Royal Statistical Society. Contains an annotated bibliography.)

[2] Creedy, J. (1983). *J. R. Statist. Soc. A,* **146,** 158–162. (Speculates on mathematical connections in Edgeworth's life.)

[3] Keynes, J. M. (1951). *Essays in Biography.* Norton, New York.

[4] Schumpeter, J. A. (1954). *History of Economic Analysis.* Oxford University Press, New York.

[5] Spiegel, H. W., ed. (1952). *The Development of Economic Thought.* Wiley, New York.

BIBLIOGRAPHY

Edgeworth, F. Y. (1925). *Papers Relating to Political Economy,* 3 vols. Published on behalf of the Royal Economic Society by Macmillan.

Edgeworth, F. Y. (1881). *Mathematical Psychics.* Kegan Paul, Long. (Reprinted by London School of Economics, 1932.)

Henderson, J. M. and Quandt, R. E. (1971). *Microeconomic Theory: A Mathematical Approach.* McGraw-Hill, New York.

Kendall, M. G. (1968). *Biometrika,* **55,** 269–276.

Stigler, S. M. (1978). *J. R. Statist. Soc. A,* **141,** 287–322. (Contains a detailed discussion of Edgeworth's contributions in probability and statistics, including early work on an analysis of variance for a two-way classification.)

Stigler, S. M. (1986). *The History of Statistics: The Measurement of Uncertainty before 1900.* Belknap Harvard University Press, Cambridge, Mass. (A clear and thorough exposition. Comprehensible for general readers.)

RAMONA L. TRADER

Gumbel, Emil Julius

Born: July 18, 1891, in Munich, Germany.
Died: September 10, 1966, in Brooklyn, New York, USA.
Contributed to: extreme-value theory and its applications.

In statistical circles, Emil Gumbel's name is indissolubly linked with the development of extreme-value theory and applications of extreme-value distributions in many fields. In the broader area of public policy, however, he made a mark as a tireless worker against authoritarianism, first as a republican and later as a pacifist and anti-Nazi. These political positions led him into some difficult situations in the earlier part of his career, as we will see. The information in this essay is derived mainly from refs. [1–3], which will be acknowledged at appropriate points.

Gumbel studied at the University of Munich from 1910 to 1914, obtaining a Ph.D. degree with a thesis on population statistics. Following military service during World War I, he returned to civilian life from 1919 to 1923. During this period he published a political work *Vier Jahre Lüge* (Four Years of Lies), which earned him, on March 14, 1919, a raid on his dwelling in Berlin—at that time under martial law—by an officer and ten heavily armed soldiers searching for evidence of "Spartacist intrigue." (There was a further visit by a soldier the next day, claiming that he had "forgotten" something, who left with a bag containing not only documents, but some of Gumbel's best clothing [3].)

In 1923 Gumbel joined the faculty at the University of Heidelberg as a *Privatdozent,* becoming an *ausserordentlicher Professor* (reader) in 1930. During this period, his scientific interests remained primarily in population statistics—in particular, the construction and use of life tables. His political interests turned to exposing the activities of nationalists aimed at destabilizing the Weimar Republic. Quoting from [1]:

> In eight books, Professor Gumbel revealed the extent of the assassinations, for political motives, by various "clandestine" para-military organizations, undermining the democratic efforts of the Weimar Republic. Most notable among these are "Vier Jahre Politischer Mord," Malik, Berlin, 1922; "Verraeter Verfallen der Feme," Malik, Berlin, 1929; and "Les crimes politiques en Allemagne, 1919–1929, Nouvelle Revue Francaise, Paris, 1931. These writings, as well as a number of sharply worded speeches, caused an organized upheaval of national-socialist groups and student organizations who forced the expulsion of Professor Gumbel from the University of Heidelberg as early as 1932.

Together with his wife and son, he went into exile in France, working as a visiting professor at the Institut Henri Poincaré in Paris in 1933, and as a research professor at the University of Lyon from 1934 to 1940. During this period his work on extreme-value distributions flourished, both in theory and in applications, especially in regard to return periods of flood flows. Among his published papers at this time was a definitive study, appearing in 1935, of the distribution which bears his name (*Ann Inst. H. Poincaré,* **5,** 115–158). Although this distribution was already known—as one of three possible limiting distributions of extreme (maximum or minimum) order statistics from random samples as the sample size increases [shown by R. A. Fisher and L. H. C. Tippett (1928) and M. Fréchet (1927)]—Gumbel's exposition was so thorough that the distribution is now almost universally known under his name.

In 1940, however,

> Under pressure of the "surrender on demand" clause of the French armistice, Professor Gumbel and his family found refuge in the United States of America. This third period of his life proved to be a period of great scientific activity, of personal tragedy caused by the death of his wife in New York in 1952, as well as of personal and scientific fulfillment. [1]

He worked in the New School for Social Research from 1940, and was an adjunct professor at Columbia University, New York, from 1953 until the summer of 1966, when he ceased professional activities. After 1940, his political activities lessened, though he wrote a report (1944–1945) for the Office of Strategic Services on "The Situation in the German Reich 1928–1940."

Gumbel's scientific publications (of which there is a list in ref. [2]) clearly reflect the developments of his interests. Especially noteworthy is his prolific output while in France between 1932 and 1940. This includes determination of the limiting distributions of the second, third, etc. order statistics in random samples as the sample size increases. After 1940 the list contains work on functions of order statistics—notably range, midrange, and extremal quotient. His work on extreme distributions culminated and was consolidated in *Statistics of Extremes* (Columbia University Press, New York, 1958). Later, we find articles on the construction of multivariate distributions with specified marginals [*Bull. Inst. Int. Statist.*, **37**, 363–373 (1960); *J. Amer. Statist. Assoc.*, **55**, 698–707 (1960)]. This topic is one of special interest to the authors of this present article.

> During this period, he lectured in many international universities such as the Free University of West Berlin; the Imperial College of Science and Technology, London; the International Course in Hydraulic Engineering, Delft; Institut Henri Poincaré, Paris; and the Universities of Tokyo, Kyoto, Hamburg, Lisbon, Bonn, Tuebingen, Munich and Zurich. He was elected a member of the International Statistical Institute in 1953. [1]

SELECTED SOURCES

[1] Gumbel, H. (1967). Emil J. Gumbel 1891–1966. Obituary note. *Int. Statist. Inst. Rev.*, **35,** 104–105.

[2] Kogelschatz, H. (1993). Emil Julius Gumbel—Würdigung seines Wissenschaftlichen Werkes, *Allg. Statist. Arch.*, **77,** 433–440.

[3] *Mitteilungen des Bundes Neues Vaterland* (September 1920). Berlin.

Hájek, Jaroslav

Jaroslav Hájek

Born: February 4, 1926, in Poděbrady, Czechoslovakia.
Died: June 10, 1974, in Prague, Czechoslovakia.
Contributed to: sample surveys, rank tests, stochastic processes, statistical education.

Much of this essay is based on Zbynek Šidák's obituary [1] and a personal communication from Professor Šidák [2]. Quoting from [1]:

> He lost his father at an early age and the family was rather badly off then. At that time and also in subsequent years, he found a second father in his uncle. Some five years later his mother remarried but in 1944 she became a widow again. Nevertheless, in spite of her slender resources she let her son study because as early as a boy he showed much talent for mathematics.

From [2]:

> His mother never had to regret her sacrifice. For the rest of her life he was standing by her and helping her, and, moreover, he supported also his younger sister to enable her to attain also high education. Similarly as he was an exemplary son and brother, he was also an excellent husband and father. He married in 1954, and he was not only a strong personality in his family but, what must be more appreciated, he was a prudent adviser to all, and he unified in this way his family. He patiently tried to teach his nearest relation—his wife Betty, daughters Betty and Helen, and his sister Jarmila—his life prudence, gentle relations to other people, and the art of overcoming obstacles. All of them accepted his advice gladly because he was himself an example of such a behavior, both in his family and among his collaborators and students in his work.

Again from [1]:

> After his studies at the grammar school, J. Hájek studied statistical engineering at the College of Special Sciences at the Czech Technical University and graduated in 1950 as Doctor of Statistical Engineering (in the sense of the word of that time). After his military service he spent the years from 1951 to 1954 as graduate student in the Institute of Mathematics of the Czechoslovak Academy of Sciences and obtained his CSc. Degree there in 1955. From 1954 to 1966 he served as a research worker in that Institute. In 1963 he got his Doctor of Physico-Mathematical Sciences degree for his thesis on statistical problems in stochastic processes.

He started his teaching activity as early in 1948–1949 as an assistant lecturer in the College of Special Sciences. He also lectured at the Economical University in Prague. In 1963 he habilitated at the Faculty of Mathematics and Physics of the Charles University and started teaching there. In 1964 he became external head of the Department for many years. Finally, in 1966, he was appointed full professor at the Faculty and joined it for good.

In 1973, shortly before the end of his life, Hájek was awarded the prestigious State Prize for his work on the asymptotic theory of rank tests.

Hájek was active in national and international scientific organizations, and was an elected Fellow of the Institute of Mathematical Statistics and member of the International Statistical Institute. He received many invitations to work and lecture in foreign countries—during his final visit to the USA, in 1970, he received invitations to lecture at no fewer than 45 universities.

An early and continuing major scientific interest of Hájek's was probability sampling. He wrote a book [3] on this topic, and had nearly completed a monograph on *Theory of Probability Sampling* at the time of his death. (This work was never published.)

A later-developing interest was the theory of rank tests, with special reference to asymptotics, in which field he rapidly gained a high international reputation. The results of this work (up to 1965) were consolidated in a three-hundred-page monograph [3]. Since then the book [4] with Šidák has become a classic in the field.

Hájek's work on stochastic processes includes the Feldman–Hájek theorem on dichotomy—that the probability measures of any two Gaussian processes are either equivalent or mutally singular.

Despite the theoretical nature of much of Hájek's original work, he was intensely interested in practical applications. Quoting, once more, from [1]:

> All Hájek's scientific activity was closely associated with applications of mathematical statistics. The results obtained by him are of considerable importance in this respect and are taken advantage of in practice (sample surveys, application of rank tests, stationary processes etc.). Moreover, Prof. Hájek himself cooperated in many problems of practice in various fields. Let us mention e.g. his extensive cooperation in sample surveys concerned with the condition of teeth of the population, with food, in certain anthropometric surveys etc.
>
> Prof. Hájek was an ardent supporter of mathematico-statistical methods because "statistics increases the culture and productivity of human thinking, for it is able to distinguish between justified judgements and the hasty ones, to fix the border beyond which simpler models should be replaced by more complex ones, and to determine the size of data necessary for an appropriate decision" (Hájek's own quotation of 1970).

Professor Zbynek Šidák [2] has given the following assessment of Hájek's personality:

> As a person, Hájek was interested in many sides of cultural life, e.g. in modern painting (for some time he intended to build some mathematical theory of aesthetics), music etc. In particular, he loved concert guitar and played this instrument quite well. In our younger years, there were occasional parties organized in the Institute, and several times we played there together: he played guitar, myself either mandoline or flute. Another remark: He was a friendly person, cheerful, having full comprehension for jokes and fun. Returning to the parties in the Institute, he also used to write some funny

scenes (little dramatic funny plays) for these parties and to perform them before the audience.

For relaxation and rest he used to return to Poděbrady, his birthplace, a city in central Bohemia with a spa. There he devoted himself either to fishing or gardening in his little garden in the outskirts of the city. He used to say that in these activities he could relax, find peace in his mind, and also (possibly) give birth to new ideas in statistics.

And, finally:

He lived on 48 years, and the end of his life was as follows: He had a serious kidney disease (hereditary in his family). For some time he had even both kidneys removed, he lived completely with no kidneys, he had to go twice a week, or later three times a week, for dialysis to the hospital. At last, he had a transplant of a kidney, for some time everything looked quite sucessful, but at the end of three months period after the transplant (doctors said three months are critical) he unfortunately and almost immediately died.

References [5–7] contain further information on Hájek's life and work.

REFERENCES

[1] Šidák, Z. (1975). Jaroslav Hájek, 1926–1974. *Czechoslovak Math. J.,* **25,** 491–501.

[2] Šidák, Z. (1996). Personal communication.

[3] Hájek, J. (1960). *Theory of Probability Sampling with Applications to Sample Surveys* (in Czech). NC SAV, Prague.

[4] Hájek, J. and Šidák, Z. (1967). *Theory of Rank Tests.* Academia, Prague and Academic Press, New York. (There is also a Russian translation.)

[5] Šidák, Z. (1974). Prof. Jaroslav Hájek awarded the State Prize. *Czechoslovak Math. J.,* **24,** 167–169.

[6] Dupac, V. (1975). Jaroslav Hájek, 1926–1974. *Ann. Statist.,* **3,** 1031–1037.

[7] Dalenius, T. (1979). Some personal recollections of Jaroslav Hájek. In *Contributions to Statistics, Jaroslav Hájek Memorial Volume,* J. Jurečková, ed. Reidel, Dordrecht, The Netherlands.

Hartley, Herman Otto

Born: April 13, 1912, in Berlin, Germany.
Died: December 30, 1980, in Durham, North Carolina, USA.
Contributed to: analysis of variance, numerical analysis, sampling distributions and theory, statistical methods.

Herman O. Hartley

H. O. Hartley was one of the most prominent and influential statisticians of the twentieth century. He was born in Germany and in 1933 obtained his Ph.D. in mathematics in Berlin. In 1934 he emigrated to England, where his collaboration with Egon Pearson* after the war led to the extensive and well-known *Biometrika Tables for Statisticians* [11, 12].

In 1953, Hartley visited the United States, and stayed to become Research Professor at Iowa State College; from 1963 to 1977 he founded and administered the Institute of Statistics at Texas A & M University, moving in "retirement" in 1979 to a full-time position at Duke University, where he served until his death.

Hartley's versatile and penetrating mind produced significant contributions in many branches of statistics over a period of 40 years. These include data processing, numerical analysis, and tabulation (see refs. [4–6]). They also include innovative results in the analysis of variance [2, 4, 8], estimation [8–10], sampling theory [3, 7], the study of sampling distributions [1, 4], hypothesis testing [2], variance components [4], sample surveys [9], and the development of statistical methods in fields of application such as carcinogenicity experiments [10]. The references given here are only an illustrative cross section of his research publications.

Professor Hartley was a fellow of the American Statistical Association and in 1979 served as its president (see ref. [6]). He also served as president of the Eastern North American Region of the Biometric Society and was a fellow of the Institute of Mathematical Statistics; among many honors he was the recipient in 1973 of the S. S. Wilks* medal.

REFERENCES

[1] David, H. A., Hartley, H. O., and Pearson, E. S. (1954). *Biometrika,* **41,** 482–493.

[2] Hartley, H. O. (1950). *Biometrika,* **37,** 308–512.

[3] Hartley, H. O. (1966). *J. Amer. Statist. Ass.,* **61,** 739–748.

[4] Hartley, H. O. (1967). *Biometrics,* **23,** 105–114.

[5] Hartley, H. O. (1976). In *On the History of Statistics and Probability.* D. B. Owen, ed. Marcel Dekker, New York, pp. 419–442.

[6] Hartley, H. O. (1980). *J. Amer. Statist. Ass.,* **75,** 1–7. (Hartley's presidential address to

the American Statistical Association: a thought-provoking essay for anybody concerned with statistics as a science and as a profession.)

[7] Hartley, H. O. and Rao, J. N. K. (1962). *Ann. Math. Statist.,* **33,** 350–374.

[8] Hartley, H. O. and Rao, J. N. K. (1967). *Biometrika,* **54,** 93–108.

[9] Hartley, H. O. and Rao, J. N. K. (1968). *Biometrika,* **55,** 547–557.

[10] Hartley, H. O. and Sielken, R. L., Jr. (1977). *Biometrika,* **33,** 1–30.

[11] Pearson, E. S. and Hartley, H. O. (1966). *Biometrika Tables for Statisticians,* vol. 1, 3rd ed. Cambridge University Press, Cambridge.

[12] Pearson, E. S. and Hartley, H. O. (1972). *Biometrika Tables for Statisticians,* vol. 2. Cambridge University Press, Cambridge.

Langevin, Paul

Born: January, 23, 1872, in Paris, France.
Died: December 19, 1946, in Paris, France.
Contributed to: theory of magnetism, stochastic differential equations.

A brief account of the life and work of this distinguished physicist is given in ref. [2]. As a young man he was in close contact with Jean Perrin, Pierre and Marie Curie, J. J. Thompson, E. Rutherford, and others concerned with atomic events which display some randomness. Perrin showed that cathode rays were a stream of particles (electrons) and used Brownian motion to show that atoms had a real rather than theoretical existence. Perrin's book [6] was the inspiration for a recent book by Mandelbrot [5] on random phenomena.

In 1905, Langevin [3] used the then new ideas of atomic structure to explain para- and diamagnetism. For the former he supposed each molecule to have a magnetic moment **m** which, in the absence of an external field, will be oriented at random due to thermal agitation. When a field of strength **H** is applied, Boltzmann's method gives a probability density for the orientation of the magnetic moments proportional to

$$\exp(\mathbf{m} \cdot \mathbf{H}/kT),$$

where T is the absolute temperature and k a constant. Writing $\mathbf{m} \cdot \mathbf{H} = mH \cos\theta$, $\kappa = mH/kT$, he arrived at the density

$$\frac{\kappa}{4\pi \sin \kappa} \exp(\kappa \cos \theta), \tag{1}$$

much used in the analysis of directional data in three dimensions. For x and μ unit vectors in a space of q dimensions, the generalization of (1) is

$$(2\pi)^{-q/2} \kappa^{-q/2-1} I_{q/2-1}^{-1}(\kappa) \exp(\kappa \mu \cdot x), \tag{2}$$

the *Langevin distribution.*

In 1908, Langevin [4] described the motion of a particle in a fluid due to Brownian motion by stochastic differential equations, a then novel technique. Thus if υ_i is the velocity component in the direction of coordinate i, $i = 1, 2, 3$, he set

$$\dot{\upsilon}_i = -\alpha \upsilon_i + \sigma \xi_i, \tag{3}$$

where the term $-\alpha\upsilon_i$ is the retarding force of Stokes' law and $\sigma\xi$ is the component of a rapidly oscillating random force due to molecular collisions. This latter force is now idealized to be a "white noise." Langevin went on to derive directly Einstein's results on diffusion. Langevin's technique of stochastic differential equations has become very important (see, e.g., Arnold [1] and Schuss [7]) and the basic form (3) is known as the *Langevin equation.*

REFERENCES

[1] Arnold, L. (1974). *Stochastic Differential Equations*. Wiley-Interscience, New York, p. 228.

[2] Gillespie, C. C., ed. (1973). *Dictionary of Scientific Biography,* vol. 8. Scribner's, New York.

[3] Langevin, P. (1905). *Ann. Chim. Phys.,* **5,** 70–127.

[4] Langevin, P. (1908). *C. R. Acad. Sci. Paris,* **146,** 530–533.

[5] Mandelbrot, B. B. (1977). *Fractals, Form, Chance and Dimension.* W. H. Freeman, San Francisco, p. 365.

[6] Perrin, J. (1923). *Atoms.* Constable, London, p. 231.

[7] Schuss, Z. (1980). *Theory and Applications of Stochastic Differential Equations.* Wiley, New York, p. 321.

[8] Zakusilo, O. K. (1981). Langevin Equations with Poisson Perturbances. *Tech. Rep. No. ONR 81-02,* Statistical Laboratory, University of California, Berkeley, Calif.

G. S. WATSON

Lévy, Paul-Pierre

Born: September 1, 1886, in Paris, France.
Died: December 15, 1971, in Paris, France
Contributed to: functional analysis, calculus of probability, study of stochastic processes.

Paul-Pierre Lévy was educated in the Lycée Louis le Grand and Lycée Saint-Louis Paris and then at the Ecole Polytechnique (1904–1906), where he had Henri Poincaré as a professor. He left this school as a mine engineer and then converted to research and academic teaching. He was a professor in the Mine School of Saint Etienne, Mine School of Paris, and Ecole Polytechnique until 1959. He gave lectures at the Collège de France and at the Sorbonne. He was a member of the Institut de France (Académie des Sciences). His scientific work started with functional analysis (he is the author of *Problèmes Concrets d'Analyse Fonctionnelle,* Gauthier-Villars, Paris, 1951). His doctorate thesis was titled: "Sur les équations intégro-différentielles définissant des fonctions de lignes," but soon after he diverted much of his attention to the calculus of probability. This orientation was suggested by the material of his lectures in the Ecole Polytechnique. At the same time he gave many papers on the philosophy of mathematics and in one of them had the idea of theorems that might be true but whose demonstration would never be achieved, needing an infinity of mental steps. That was about 10 years before Gödel's work. In the axiom-of-choice debate he was a supporter of the axiom. He believed in the "existence" of a set apart from the mathematician who deals with and defines it.

In the calculus of probability he discovered many important tools, the first of them is the characteristic function (c.f.), which is the Fourier transform of the distribution function [$\int \exp itx\, dF(x)$]. He substituted this characteristic function for that of Poincaré, which was the Laplace transform [$\int \exp tx\, dF(x)$], existing only for certain classes of distribution. He classified the characteristic functions, opening the vast field of indefinitely divisible laws of probability; developed the notions of stable, semistable, and quasistable laws; and could write expressions for these characteristic functions. He gave a very useful definition of the distance between two probability laws, and achieved many fundamental results in random topology, chiefly in the field of random series. He introduced the concentration function of a distribution and used it for studying many convergence problems. It was in working about random series that he had the intuition of the truth of the theorem which is known as the Lévy–Cramér or Cramér–Lévy theorem, whose proof was given by Cramér* in 1937: If the sum of two independent random variables is a normal variable, each one of the two variables is normal.

All that was the starting point of the arithmetic-of-probability laws, to which he devoted a book (*Théorie de l'Addition des Variables Aléatoires,* 2nd ed., Gauthier-Villars, Paris, 1954). The infinitely divisible laws led him to the study of stochastic processes (he is the author of *Processus Stochastiques et Mouvement Brownien,* 2nd ed., Gauthier-Villars, Paris, 1965), on which he wrote many papers. He was, with Wiener,* the initiator of what is now called the Wiener–Lévy process [i.e., a Gaussian random function $X(t)$ with $EX(t) = 0$ and cov $(X(t_1), X(t_2)) = \min(t_1, t_2)$], that is, the typical Brownian motion process. He discovered a probability law (quite analogous to the von Mises–Smirnov law) which concerns the area swept by a random vector whose endpoint has two coordinates which are two independent Wiener–Lévy processes. This result can be used for testing the discrepancy between a theoretical and an empirical distribution, and so is a typical statistical result. Lévy discovered the so-called arcsine law, which is the probability law of the period during which a Wiener–Lévy process is positive. An extension of that result to the case of a Wiener–Lévy's bridge is known as Gnedenko's theorem and leads again to a comparison of empirical and statistical distribution. The extension of this theorem to multidimensional distribution is still open. He also dealt with Markov chains, Markov processes, martingales, and theory of games.

He worked very hard until his last days. A very important part of his scientific production was written during his last 10 years. One year before his death he wrote his memoirs under the title *Quelques Aspects de la Pensée d'un Mathématicien* (Ed. Alfred Blanchard, Paris, 1971).

All his papers and memoirs are published by Gauthier-Villars, Bordas, and Dunod (*Oeuvres Complétes de Paul Lévy,* 6 vols.). The entire output is about 250 titles.

D. DUGUÉ

Lüroth, Jakob

Born: February 18, 1844, in Mannheim, Germany.
Died: September 14, 1910, in Munich, Germany.
Contributed to: mathematics, statistics, mathematical logic

Lüroth was the only child of a respectable family, his father being a member of Mannheim's municipal parliament. Lüroth's early interest was in astronomy, and already in 1862 he published calculations of the orbits of two minor planets and began studying astronomy under Argelander at Bonn University. However, his bad eyesight compelled him to switch over to mathematics. He studied mathematics in Heidelberg, where among his teachers was Hesse (and, in physics, Kirchhoff) and where he acquired his doctor's degree in 1865. He then attended lectures by Weierstrass in Berlin and by Clebsch in Giessen.

His further life was successful even if not eventful. He was a professor at the Technische Hochschule Karlsruhe from 1869, then at the Technische Hochschule Munich, and from 1883 until his death at the University of Freiburg, where he served as its rector in 1899/1900. He died during a pleasure trip which he undertook with his wife and daughter.

Lüroth was a versatile scientist working in mathematical logic, theory of invariants, and various branches of geometry and following up on Clebsch, Schröder, and Staudt. In addition, he contributed to applied mechanics, geodesy, and the theory of errors. For a long time Lüroth suffered from a heart condition, but continued to work hard. In all, this eminent mathematician published 70 papers and two books, one of the latter being devoted to elementary numerical calculations and the use of logarithmic tables with a large number of significant digits. He also paid much attention to popularizing science, and chaired the Freiburg Naturforschende Gesellschaft for many years. In 1909, in a letter to a friend, Lüroth wrote: ". . . by now, I am already 25 years in Freiburg and belong here to the eldest members of the professorial staff. I am satisfied with my scientific work, although perhaps I fell short of what was expected of me in my youth." [7]

Lüroth was familiar with the Gaussian theory of errors, and kept in touch with Jordan, the author of the "three-sigma" rule for rejecting outlying observations. He reviewed Helmert's* classical study in higher geodesy [6] and reasonably called him a leading German geodesist. However, he missed an opportunity to mention Helmert's earlier treatise on the method of least squares [2], so that possibly he did not read it.

Lüroth published three papers on the theory of errors. In [3] he generalized Peters' formula for evaluating the probable error to the case of several unknowns. Elsewhere [5] he made an unclear attempt at determining the distribution of $f(x)$, given that of x. Of real interest is his paper [4], where he obtains a distribution, equivalent to the t-distribution up to a scale transformation, as the posterior distribution of μ_1, say, given m observations $(x_1, \ldots, x_m)$, with x_i being normally distributed with expectation $\sum_{j=1}^{n} c_{ij}\mu_j$ and precision constant h $(= 1/\sqrt{2}\sigma)$. His result is based on a uniform prior for $\mu_1, \ldots, \mu_n$ and h. Published in an astronomical journal, it escaped statistical notice. See Pfanzagl and Sheynin [8] for more details.

An extensive biography can be found in Brill and Nöther [1].

REFERENCES

[1] Brill, A. and Nöther, M. (1911). Jakob Lüroth. *Jahresb. Deutsch. Math. Ver.* **20,** 279–299. (With a list of publications.)

[2] Helmert, F. R. (1872). *Die Ausgleichsrechnung nach der Methode der kleinsten Quadrate.* Leipzig.

[3] Lüroth, J. (1869). Bemerkung über die Bestimmung des wahrscheinlichen Fehlers. *Astron. Nachr.,* **73,** 187–190.

[4] Lüroth, J. (1876). Vergleichung von zwei Werthen des wahrscheinlichen Fehlers. *Astron. Nachr.,* **87,** 209–220.

[5] Lüroth, J. (1880). Ein Problem der Fehlertheorie. *Z. Vermessungswesen,* **10,** 432–438.

[6] Lüroth, J. (1883–1884). Review of Helmert, F. R. (1880, 1884). Die mathematischen und physikalischen Theorien der höheren Geodäsie. Leipzig. *Z. Phys. Math., Hist.-lit. Abt.,* **28,** 55–58; **31,** 139–144.

[7] Lüroth, J. (1909). Letter to a friend, kept at the Staatsbibliothek Berlin, *Preussischer Kulturbesitz, H* 1889 (13).

[8] Pfanzagl, J. and Sheynin, O. (1995). A forerunner of the *t*-distribution. *Biometrika,* to appear.

J. PFANZAGL
O. SHEYNIN

Rényi, Alfréd

Born: March 20, 1921, in Budapest, Hungary.
Died: February 1, 1970, in Budapest, Hungary.
Contributed to: probability, statistical theory, information theory, number theory, combinatorial theory.

CAREER

Alfréd Rényi came from an intellectual family. After finishing his secondary school, he was not admitted into the university immediately because of racial laws, but one year later, after winning a mathematical competition, he gained admittance and studied mathematics and physics at the University of Budapest until 1944. Professor L. Fejér was a great influence on his formative career years, but Rényi also profited very much from the young, but already productive, mathematicians P. Turán and T. Gallai.

In 1944, Rényi was interned in a labor camp but escaped and, in the guise of a uniformed soldier, was able to help other persecuted persons. After the liberation of Budapest, he went to Szeged and earned a doctoral degree with F. Riesz. Then he was employed for one year by a social security institution. After marriage to the able mathematician Catherine Schulhof (a very lively person), he traveled to Leningrad with her on scholarship. Under the guidance of Yu. V. Linnik,* he obtained the candidate degree of mathematical sciences in one year instead of the prescribed three.

Beginning in 1947, again in Budapest, he acted as assistant lecturer at the University of Budapest; at the same time (between 1948 and 1950) he regularly visited the University of Debrecen as a professor. In 1950, the Mathematical Institute of the Hungarian Academy of Sciences was established, and he soon became the director of the Institute and head of the Department of Probability. From that time on he was also chairman of the Department of Probability Theory at the University of Budapest. He performed both these duties until the end of his life.

Around 1950, the Hungarian School of Probability was formed; it included P. Medgyessy, Ilona Palásti, A. Prékopa, G. Székely, I. Vincze, Margaret Ziermann, and L. Takács, who already had been active in this field as a student of the distinguished Hungarian statistician Ch. Jordan.*

During the next 20 years Rényi played an important role in the Mathematical and Physical Department of the Hungarian Academy of Sciences (HAS) on the Committee for Scientific Qualification, and in the János Bolyai Mathematical Society. In the frame of this activity, he organized the first and second Hungarian Mathematical Congresses in 1950 and 1960, respectively. In almost every year conferences and colloquia in different fields of probability theory and its applications were held. He had a considerable role in improving teaching in the field of mathematics in secondary schools and in the university among others, organizing several kinds of mathematical competitions. He initiated and founded the *Publications of the Mathematical Institute of HAS* (now *Studia Scientiarum Mathematicarum Hungaricae*).

In recognition of his scientific activity he became corresponding member of the HAS in 1949 and ordinary member in 1956. He received the silver grade of the Kossuth Prizein 1949 and the golden grade in 1956. As a member of the International Statistical Institute he was elected vice president in 1965. He was a member of the editorial boards of ten (Hungarian and foreign) periodicals. As a visiting professor he lectured at Michigan State University (1961), the University of Michigan (1964), Stanford University (1966), Cambridge University (1968), and the University of North Carolina (1968).

Following the unexpected death of Catherine Rényi in August 1969, physicians diagnosed an inoperable lung carcinoma in Alfréd; he died on February 1, 1970.

SCIENTIFIC ACTIVITY

Rényi wrote his doctoral dissertation on Cauchy–Fourier series and his candidate thesis on the quasi-Goldbach problem, giving a remarkable development of the large-sieve method. Then he published a sequence of articles on the theory and applications of probability theory and mathematical statistics. He collaborated with L. Jánossy and J. Aczél on compound Poisson distributions, and with L. Pukánszky on measurable functions. He has also written on mixing sequences and on algebras of distribution. In 1953, his basic paper on order statistics appeared, establishing a method for the determination of limiting distribution laws; he determined the limiting law of the relative deviation between empirical and theoretical distributions, giving a new version of the Kolmogorov test. In 1954, his axiomatic foundation of the theory of probability based on conditional probability was published. Since 1945, he has published on analytic functions of a complex variable, on geometry, on algebra, and on Newton's method of approximation; he continued his work on varying topics for the remainder of his life.

Many joint papers with his friends P. Turán and P. Erdős appeared: papers on combinatorics, and particularly on random graphs; and also on the theory of functions and on the theory of numbers, in which topic Catherine Rényi was also sometimes his coauthor. Beginning in 1950, with the stimulus of problems posed by industrial and other experts, he solved practical problems and published on storage problems, breaking processes, energy needs of plants, rational dimensioning of compressors, chemical reactions, and replacement policy in stocks. In 1956 appeared his paper on information theory, where he dealt mainly with the concept of entropy in the several interpretations and applications of this concept. In a sequence of papers he considered entropy and statistical physics, considered entropy and mathematical statistics, and finally established a theory of search. He collaborated with H. Hájek* on a generalization of the Kolmogorov inequality, with R. Sulanke on geometrical probabilities, and with J. Neveu on inequalities in connection with probabilities of events. We cannot list all the topics he considered, as his papers number about 300, all achieved in a period of 25 years.

The most complete collection of Rényi's articles appears in *Selected Papers of Alfréd Rényi,* edited by P. Turán, in three volumes (Akadémiai Kiadó [Publishing

House of the HAS], Budapest, 1979), which include the English translation of several works written in Hungarian.

BOOKS

Rényi's first book, *Theory of Probability* (Tankőnyvkiadó [Publisher of Textbooks], Budapest, 1954), is a monumental synthesis of theory, applications, and practice, still not translated into any foreign language. Its theoretical part, in a revised and extended form, with an addendum on information theory, appeared in German as *Wahrscheinlichkeitstheorie mit einem Anhang über Informationstheorie,* which was translated, each time in revised form, into Hungarian, English, French, and Czech. He wrote a book on the foundation of the theory of probability based on his conditional-probability concept, which posthumously appeared in English. His other books are popularizations or essays. It can be a notable experience for a mathematician or an interested layman to read them. They include *Dialogues* (1967) (Socratic dialogues on mathematics and applications), *Die Sprache des Buches der Natur* (1968), *Ars Mathematica* (1970) (includes discussion of information theory), and *Letters on Probability* (1972) (a fictional correspondence between Fermat and Pascal).

István Vincze

Smirnov, Nikolai Vasil'yevich

Born: October 17, 1900, in Moscow, Russia.
Died: June 2, 1966, in Moscow, USSR.
Contributed to: theory of nonparametric statistics, theory of order statistics, nonparametric tests, goodness of fit, tables of statistics.

N. V. Smirnov's contributions to the development of the theory and applications of mathematical statistics (along with E. E. Slutskii's* and A. N. Kolmogorov's*) are undoubtedly the most significant achievements of the Russian statistical school.

Smirnov was born in Moscow into the family of a minor church office worker. Having completed his high-school education and after service in the army (initially in a medical battalion during World War I and later in the Red Army), he devoted much of his time to studying philosophy and philology.

In 1921 he enrolled in the Mathematical Branch of Moscow University, being convinced that fruitful results in the field of liberal arts can be achieved only after a careful study of natural sciences. After graduation from Moscow University in 1926, Smirnov was engaged in pedagogical activities, lecturing at Moscow University, Timoryazev Agricultural Academy, and Moscow City Pedagogical Institute. At that time he chose the area of his future scientific work to be probability theory and mathematical statistics. In 1938 Smirnov successfully defended his Doctor of Science dissertation "On approximation of the distribution of random variables." This work served as a foundation for the theory of nonparametric problems in mathematical statistics. It was in this field that Smirnov eventually gained world-wide renown.

Smirnov truly excelled in solving difficult computational problems of mathematical analysis and especially in evaluations of multiple integrals over complex regions. (These types of calculations arise naturally in evaluation of significance levels of tests and their powers.) The basic idea of the analytic methods he developed is based on an investigation of the asymptotic behavior of multiple integrals as the number of variables tends to infinity, which is a main feature of the asymptotics of numerous statistical problems.

The subsequent period of Smirnov's scientific activity (1938–1966) is connected with the Steklov Mathematical Institute of the USSR Academy of Sciences in Moscow, where he worked until the last day of his life (eventually—during the last year—as the head of the Branch of Mathematical Statistics). In the Institute he obtained his new fundamental results in nonparametric statistical theory and classified the limiting distributions of order statistics [6–10].

His works dealing with the distribution of various statistics of nonparametric tests, probabilities of large deviations, and limiting distributions of order statistics obtained worldwide renown and are cited in practically all advanced textbooks and monographs on mathematical statistics.

N. V. Smirnov devoted much effort to continuing the activities initiated by his friend E. E. Slutskii in connection with compilation of statistical tables required for probabilistic and statistical calculations [11–14]. He also coauthored with L. N.

Bol'shev* the well-known *Tables of Mathematical Statistics,* which at that time (1965) were the most comprehensive available in this field. He was also engaged in dissemination of the use of mathematical statistics in the natural sciences and engineering and in training highly qualified experts. His students—renowned scientists such as L. N. Bol'shev and D. M. Chibisov—occupied and continue to occupy important positions in a number of scientific centers in Russia and western countries. His textbooks (e.g. ref. [15]), slanted towards practical applications, were, at least until 1995, very popular in Russia and abroad.

Smirnov's talents are highly original, and it is hard to pinpoint who had a decisive influence on his career. It is difficult to trace how he arrived at his ingenious results. When asked about these matters, he often jokingly replied that it is all available in some exercise manual, or that he arrived at the formula of a limiting distribution in his dreams.

Smirnov was one of those scientists who devote their lives to the solution of a small number of very difficult problems and who do not create large schools or present brilliant lectures, so that their activities with a small group of close students often remain unnoticed. He was not a self-complacent person. His standards were very high and he commonly denigrated the importance of his own results. He shunned discussions and arguments, preferring real action over "grandiose plans." He did not initiate new substantial scientific adventures, although the above-mentioned textbooks ought to be considered an innovative public activity. He was amenable, tolerant, and soft-spoken in his personal relations.

N. V. Smirnov's achievements were recognized in the USSR—he was granted a State Prize in 1951 for his contributions to nonparametric statistics and was elected in 1960 a corresponding member of the USSR Academy of Sciences.

REFERENCES

[1] Smirnov, N. V. (1939). On deviations of the empirical distribution curves (in Russian). *Mat. Sb.,* **6**(48), 3–24.

[2] Smirnov, N. V. (1944). Approximation of distributions of random variables based on empirical data (in Russian). *Usp. Mat. Nauk,* **10,** 179–206.

[3] Smirnov, N. V. (1939). Estimation of a deviation between empirical distribution curves in two independent samples (in Russian). *Bull. Moscow State Univ.,* **2**(2), 3–14. (The basic paper on Smirnov's two-sample test.)

[4] Darling, D. A. (1957). The Kolmogorov–Smirnov Cramér–von Mises tests. *Ann. Math. Statist.,* **28,** 823–838.

[5] Hodges, J. L. (1957). The significance probability of the Smirnov two-sample test. *Ark. Math.,* **3,** 469–486.

[6] Smirnov, N. V. (1947). On a test of symmetry of a distribution of a random variable (in Russian). *Dokl. Akad. Nauk SSSR,* **56,** 13–16.

[7] Smirnov, N. V. (1949). Limit distributions for terms of variational series (in Russian). *Trudy Steklov. Mat. Inst.,* **25,** 5–79.

[8] Smirnov, N. V. (1953). Limit distributions for terms of variational series. *AMS Transl.* **67,** 1–67. (English translation of ref. [7].)

[9] Smirnov, N. V. (1961). Probabilities of large values of non-parametric one-sided goodness of fit tests (in Russian). *Trudy Steklov. Mat. Inst.,* **64,** 185–210.

[10] Smirnov, N. V. (1967). Remarks on the limiting distribution of the terms of a variational series. *Teor. Veroyat. Primen.,* **12,** 391–392. (English transl. In *Theor. Probab. Appl.*)

[11] Smirnov, N. V. (1948). Table for estimating the goodness of fit of empirical distributions. *Ann. Math. Statist.,* **19,** 279–287.

[12] Smirnov, N. V., ed. (1960). *Tables of the Normal Probability Integral, Normal Density and its Normalized Derivatives.* Isdatel'stvo Akad. Nauk SSSR, Moscow.

[13] Smirnov, N. V. and Bol'shev, L. N., eds. (1962). *Tables for Calculation of the Bivariate Normal Probability Distribution Function,* Izdatel'stvo Akad. Nauk SSSR, Moscow.

[14] Bol'shev, L. N. and Smirnov, N. V., eds. (1965). *Tables of Mathematical Statistics.* Fizmatgiz, Moscow.

[15] Smirnov, N. V. and Dunin-Barkovskii, I. V. (1965). *A Course in Probability Theory and Mathematical Statistics for Engineering Applications* (in Russian), 2nd enlarged ed. Nauka, Moscow.

S. A. AIVAZIAN (TRANSLATED AND EDITED BY S. KOTZ)

Wilks, Samuel Stanley

Samuel S. Wilks

Born: June 17, 1906, in Little Elm, Texas, USA.
Died: March 7, 1964, in Princeton, New Jersey, USA.
Contributed to: multivariate analysis, nonparametric methods, statistical education in the USA, Institute of Mathematical Statistics

Samuel Wilks was the eldest of three sons of Chance C. Wilks and his wife, Bertha May Gammon Wilks. His father farmed a 250-acre ranch outside Little Elm, Texas. His early education was in a one-room school-house. In high school, he boarded in nearby Denton. After graduating from high school, he attended North Texas Teachers' College, obtaining a B.A. degree in architecture in 1926. During the years 1926 to 1929 he was associated with the Department of Mathematics at the University of Texas, obtaining an M.A. degree in 1928. He studied statistics from 1929 to 1931 at the University of Iowa under the prominent American statistician (the "father of American mathematical statistics") H. L. Rietz. He obtained a Ph.D. degree in 1931 with a dissertation on the distributions of statistics in samples from a bivariate normal population with matched sampling of one variable. This was published in *Metron* (9, 87–126) in 1932.

Wilks married Virginia Orr in September, 1931. In the years 1931–1932, Wilks was working with Harold Hotelling* at Columbia University, New York on problems in multivariate analysis, and became acquainted with Walter Shewhart, the pioneer of statistical quality control. Wilks spent 1932–1933 in England, first at University College, London, collaborating with E. S. Pearson,* and later at Cambridge University, with J. Wishart. The Wilks' only child, a son, Stanley, was born in London in October, 1932. In 1933 Samuel was offered a post at Princeton University. From 1933 to 1936, he was an instructor in mathematics, rising to associate professor in 1938. His lectures in the late thirties led to his famous notes on *Statistical Inference*. After World War II began, he was working with the National Defense Research Committee, devoting more and more of his time to the war effort as the demand increased. In 1944 he was promoted to professor of mathematics at Princeton, and became director of the Mathematical Statistics Section in 1945, remaining in that post until his untimely death in 1964.

In this period, Wilks published his lecture notes, more or less informally, as *Statistical Inference* (1937—the "little orange book"), *Mathematical Statistics* (1943—gray paperbound), and *Elements of Statistical Analysis* (1948—the "little blue book"). *Mathematical Statistics* was finally published, hardbound, after considerable revisions (notably during a sabbatical leave at Cambridge University), by Wiley in 1962.

Wilks was prominent in organizing the Institute of Mathematical Statistics (IMS) in 1935, and was the founding editor of the *Annals of Mathematical Statistics* for over eleven years (1938–1949). By the end of his editorship, this journal had become the leading periodical devoted to mathematical statistics in the world. He was president of the IMS in 1940, and presented the Rietz lecture in 1959. He assisted in bringing together various American statistical societies to form the Committee of Presidents of Statistical Societies (COPSS). After Wilks' death, this committee established the prestigious Wilks Award. Wilks was a fellow of the Econometric Society and the Royal Statistical Society, and a member of the International Statistical Institute. He was president of the American Statistical Association in 1950 and a member of the Russell Sage Foundation's' board of trustees.

His activities in statistical education are reflected in five books. The last one—*Introductory Engineering Statistics* (with I. Guttman, published by Wiley) appeared posthumously in 1965.

Wilks established, in cooperation with W. A. Shewhart, the highly esteemed Wiley Publications in Statistics series, which, at the time of his death, contained about thirty titles in mathematical statistics and about the same number in applied statistics.

A man of exceptional energy and dedication, Wilkes was involved in numerous committees, trained many graduate students, and was active in application of statistical methods in industry. His main research, in the field of multivariate analysis, started with derivation of maximum-likelihood estimators and proceeded to development of likelihood criteria for testing various hypotheses on multivariate normal distributions (including the famous Wilks Λ-criterion). He introduced the concept of "generalized variance" and studied multivariate versions of intraclass correlation models—his name, along with those of T. W. Anderson, P. C. Mahalanobis,* and J. Wishart, is familiar to even occasional users of these techniques. Wilks pioneered research on tolerance intervals and order statistics. His further fields of versatile endeavor include design of experiments, statistical inference in geology, and the history of statistics.

His book *Mathematical Statistics* (referred to above), together with M. G. Kendall's *Advanced Theory of Statistics,* vol. 1 (1943) and H. Cramér's *Mathematical Methods in Statistics* (1946), forms a triad which ushered in a golden age of statistical science. The second version (1962, published by Wiley) serves, even today after 30 years, as an evergreen standard text and reference. (It was translated into Russian in 1966.)

Wilks' friendly personal interest in his students and colleagues, his sociability, and his sense of humor endeared him to his contemporaries. His untimely death was a blow to the development of statistical science, but his life was an inspiration for it.

REFERENCES

[1] Anderson, T. W. (1965). Samuel Stanley Wilks: 1906–1964. *Amer. Math. Statist.,* **34,** 1–26.

[2] Mosteller, F. (1964). Samuel S. Wilks: Statesman of statistics. *Amer. Statist.,* **18**(2), 11–17.

[3] Pearson, E. S. (1964). Obiturary: Samuel Stanley Wilks, 1906–1964. *J. R. Statist. Soc. A,* **127,** 597–599.

Wold, Herman Ole Andreas

Born: December, 25, 1908, in Skien, Norway.
Died: February, 16, 1992, in Uppsala, Sweden.
Contributed to: time-series analysis, econometrics, multivariate analysis.

Herman Ole Andreas Wold was born in southern Norway in 1908 as the sixth child of Edvard and Betsy Wold. Due to hard economic times in Norway, the family, except the two oldest children, moved to Lidköping in southwestern Sweden in 1912, and became Swedish citizens. Here Edvard Wold started a small fur and hide business, which still operates, today under the ownership of Torbjörn Wold, a nephew of Herman Wold. Herman Wold grew up and went to elementary school in Lidköping, but at that time there was no high school in town, and he therefore went to high school in Skara 40 kilometers (25 miles) away.

In 1927 Wold enrolled in the University of Stockholm, where he graduated in 1930 with a degree in mathematics, economics, and physics. Harald Cramér* was one of his teachers, and Wold became greatly interested in Cramér's work and in statistics. After some years' work in an insurance company, Wold enrolled as Cramér's graduate student, and in 1938 he presented his thesis "A study in the analysis of stationary time series." In the thesis he proved a theorem about the decomposition of a time series into two components, the *Wold decomposition,* which forms an essential element in the foundation of time-series analysis and forecasting.

Basically, Wold showed that any stationary time series can be separated into a deterministic component precisely predictable from its past, plus a random component which can be modeled as a weighted sum of "innovations." Wold's decomposition showed that the three classic time-series models—the model of hidden periodicities, the moving-average model, and the autoregressive model—could be seen as different cases of the same general model. This theorem is also referred to as the Cramér–Wold device (e.g., ESS, vol. 7, p. 601, and Supplement, p. 36) and had an immediate impact on the fields of statistics and economics.

Ragnar Frisch was the first opponent at the dissertation, and was very critical of the decomposition theorem. After a long and lively dissertation however, Wold passed with honors, and became docent of statistics and insurance mathematics at Stockholm University.

After a few more years in Stockholm, Wold became the first professor of statistics at Uppsala University in 1942, where he stayed until 1970. He then moved to Gothenburg as professor of statistics, staying until his retirement in 1975, when he moved back to Uppsala. He married Anna-Lisa Arrhenius in 1940, and they had three children: Svante, Maria, and Agnes. The three children all became scientists: Agnes an immunologist at Gothenburg University, Maria a data analyst at Pharmacia AB in Uppsala, and Svante a chemometrician at Umeå University. Svante Wold collaborated with Herman Wold on multivariate modeling and analysis of chemical systems and processes, and continues the development of the partial least-squares (PLS) methodology as applied in natural science and technology.

Wold liked to travel, and visited many statistics and econometrics departments all over the world, for both short and long periods, among them Columbia University in New York, the Wharton School in Philadelphia, the University of Wisconsin at Madison, and the University of Geneva, Switzerland.

After the dissertation, Wold's research was centered on demand analysis and econometrics; his book *Demand Analysis* (with his assistant Laws Jureén) published in 1952, became a classic in the field. He continued to work with the modeling of complicated econometric systems, and was very active in the international discussion about how best to formulate and estimate these models. In the 1960s he developed the so-called fixpoint method of estimation for econometric systems. This solved some problems for systems with many variables and collinear variables. This, in turn, led him to his final domain of interest, multivariate analysis using "soft" modeling and projection methods (principal-components analysis, and its extension, PLS projection to latent structures). Wold saw these methods as least-squares correspondents to multivariate (ML) maximum-likelihood models such as ML factor analysis and LISREL. The latter class of methods were developed by his pupil Jöreskog, who together with Wold edited two volumes on the PLS and ML approaches to multivariate analysis, namely *Systems under Indirect Observation,* North-Holland, Amsterdam, 1982.

Wold was a very practical man, and wanted estimation and modeling methods to work with a minimum of assumptions, for incomplete data, with many variables and collinear variables, etc.; and he developed PLS accordingly. This has made PLS of great use for the analysis of large collinear data sets in the social sciences, business economics, and market research. PLS, in its simplest two-block form, is also extensively employed in chemistry, chemical engineering, chemometrics, and parts of biology and medicine for applications such as multivariate calibration, composition-property and structure-property modeling, image analysis, environmental analysis, and the monitoring, modeling, and optimization of chemical and biological processes. The Swedish Chemical Society in 1995 constituted the Herman Wold Medal in memory of his contributions to chemistry and chemometrics, to be given to a distinguished chemist who creatively uses statistics in her/his research.

Wold believed that teaching was as important as research, and spent much time on developing undergraduate and graduate courses of statistics; he also wrote two undergraduate textbooks (in Swedish). Besides writing and editing the books mentioned above, Wold was the editor or coeditor of a few other books, and author or coauthor of around 40 scientific papers, mainly in statistics and econometrics.

Wold was a fellow of the Institute of Mathematical Statistics, the American Statistical Association, and the Swedish Academy of Sciences, an honorary fellow of the Royal Statistical Society, and an honorary member of the American Economic Association and of the American Academy of Arts and Sciences. He served as vice president of the International Statistical Institute in 1957–1961, and as president of the Econometric Society in 1966. He has honorary doctorates from the Technical University of Lissabon and from the University of Abo, Finland.

SVANTE WOLD

Wolfowitz, Jacob

Born: March 19, 1910, in Warsaw, Poland.
Died: July 16, 1981, in Tampa, Florida.
Contributed to: statistical inference, sequential analysis, inventory theory, queuing theory, information theory, decision theory.

Jacob Wolfowitz

Jacob Wolfowitz was born in Warsaw, Poland on March 19, 1910, and came to the United States with his family in 1920. He received the baccalaureate from the College of the City of New York in 1931. Positions were scarce in 1931, a year of severe economic depression, and he supported himself as a high-school teacher while studying for the doctorate in mathematics at New York University. He received the Ph.D. degree in 1942.

Wolfowitz had met Abraham Wald* in the autumn of 1938, when Wald came to Columbia University to study statistics with Harold Hotelling.* Wald and Wolfowitz quickly became close friends and collaborators, their first joint paper [18] appearing in 1939. During the period of United States involvement in World War II, they worked together on war-related research at the Statistics Research Group of Columbia University. In 1945 Wolfowitz became an associate professor at the University of North Carolina at Chapel Hill. In 1946 he joined the faculty of Columbia University, leaving in 1951 to join the Department of Mathematics at Cornell University. In 1970 he became a professor of Mathematics at the University of Illinois in Urbana. After retiring from the University of Illinois in 1978, he became Distinguished Professor of Mathematics at the University of South Florida in Tampa, a position he held until his death following a heart attack, on July 16, 1981. He had held visiting professorships at the University of California at Los Angeles, at the Universities of Paris and Heidelberg, and at the Technion–Israel Institute of Technology in Haifa.

Wolfowitz's research is remarkable for its combination of breadth and depth. He made important contributions to all of the major areas of mathematical statistics, and also to inventory theory, queuing theory, and information theory. Several of his papers make contributions to several different areas simultaneously.

Wolfowitz's earliest research interest was nonparametric inference. His first two published papers, written jointly with Wald, were on nonparametric inference: ref. [18] constructs a confidence band for an unknown continuous cumulative distribution function based on a random sample from the distribution; ref. [19] proposes and analyzes the celebrated two-sample test based on runs. Wolfowitz wrote several other papers on the theory and application of runs, ref. [27] containing an application to qualify control. The term "nonparametric" was originated by Wolfowitz in ref. [26]. His interest in nonparametric inference did not end with these early papers. In ref. [7], with Dvoretzky and Kiefer, he proved that the empirical cumulative distribution function is an asymptotically minimax estimator of the population cumulative distribution function for a variety of reasonable loss functions. In refs. [13] and [15] Kiefer and Wolfowitz extended these results to the problem of estimating joint cumulative distribution functions.

Wolfowitz's research on the minimum-distance method is an application of techniques developed in nonparametric inference to parametric inference. The method estimates the unknown parameters by those values of the parameters that minimize a distance between the empirical cumulative distribution function and the parametric family of cumulative distribution functions. This method gives consistent estimators in some very complicated problems. The papers in refs. [32–35] and the joint paper [9] with Kac and Kiefer cover the devlopment, analysis, and applications of the minimum-distance method.

Starting with ref. [28], which discusses the sequential estimation of a Bernoulli parameter, Wolfowitz made many important contributions to sequential analysis. In ref. [29] he developed a Cramér–Rao type of lower bound for the variance of an estimator based on sequential sampling, under certain regularity conditions. In ref. [21] he and Wald studied the structure of Bayes decision rules when sequential sampling is used. One of the papers [20] he was proudest of was written with Wald and proves the optimum character of the Wald sequential probability ratio test. In ref. [8] Dvoretzky, Wald, and Wolfowitz showed that randomization can be eliminated in sequential decision problems under certain conditions. In ref. [22] Wald and Wolfowitz showed that under mild conditions, in sequential decision problems, if randomization is used after each observation, we get the same class of risk functions as when randomization is used only once, to choose a nonrandomized decision rule at the start of the process. Dvoretzky, Kiefer, and Wolfowitz [4, 5] solved sequential decision problems when observation is continuous over time. In ref. [31] Wolfowitz showed that the optimal sequential estimator of a normal mean when the variance is known is essentially a fixed-sample-size estimator. In ref. [24] Weiss and Wolfowitz constructed an asymptotically efficient sequential equivalent of Student's t-test, and in ref. [23] these authors used an adaptive sequential scheme to construct optimal fixed-length nonparametric estimators of translation parameters.

Kiefer and Wolfowitz [10] modified the Robbins–Monro stochastic approximation procedure to estimate the point at which an unknown regression function achieves its maximum.

In a regression model, a particular choice of the values of the independent variables is called a *design*. An optimal design is a design which enables the user of the model to estimate given functions of the unknown regression coefficients as efficiently as possible. Kiefer and Wolfowitz [14, 16, 17] made important contributions to the theory underlying the construction of optimal designs.

The inventory problem is the problem of deciding how much inventory to hold during each of a sequence of time periods, when there are penalties for holding either too much or too little inventory and demand for the product is random. Dvoretzky, Kiefer, and Wolfowitz [2, 3, 6] made pioneering contributions to this subject and really started the subject known nowadays as "dynamic programming"; this is the theory of which sequence of nonsampling decisions is optimal when a decision must be made in each time period in a sequence of time periods. In ref. [6] the authors showed that under certain circumstances the well-known (s, S) policy is optimal: this policy is to order enough to make the total inventory equal to S as soon as the stock on hand goes below s.

Wolfowitz's research on maximum-likelihood estimators started with ref. [30] and led to the development, in collaboration with Weiss, of maximum-probability estimators. For large samples, these estimators have the highest probability of being close to the true unknown parameters: among a wide class of estimators they often coincide with maximum-likelihood estimators, but exist in cases where the latter do not. The monograph in ref. [25] describes most of the results in this area.

Kiefer and Wolfowitz [11, 12] made fundamental contributions to the theory of queues with many servers, by showing the existence of limiting distributions of waiting times and queue lengths as time approaches infinity.

Starting in 1957, Wolfowitz [36] devoted a rapidly increasing proportion of his time to what he called coding theorems of information theory, describing how rapidly information can be sent when random errors occur in the transmission and the probability of correct decipherment must be at least equal to a preassigned value. This problem can be considered as a generalization of statistical decision theory, in the following sense. In statistical decision theory, there is an unknown parameter with a given set of possible values, and based on observed random variables whose distribution depends on the parameter, we must guess the value of the parameter. In coding theory, we have the additional choice of the set of possible values of the parameter: Each value in the set we choose becomes one of the entries in our codebook, the codebook being simply a list of the words which we are allowed to transmit over the channel. We want to choose as many words as possible, but in such a way as to achieve the desired lower bound on correct decipherment. Wolfowitz proved both direct coding theorems, which state that the codebook can contain at least a certain number of words, and converse theorems which state that the codebook cannot contain more than a certain number of words. His work in this area represents deep generalizations of the theory which existed before he started his research. Most of his results are contained in a monograph [39].

In addition to the mathematical theory, Wolfowitz was interested in practical and philosophical issues. Reference [37] contains a criticism of a set of axioms used to support the Bayesian approach to statistical decision theory. Reference [38] contains an interesting criticism of the theory of testing hypotheses for not having practical application.

Wolfowitz was a renowned teacher and lecturer, unsurpassed in his ability to clarify the intuition underlying the most complicated results. He was selected as Rietz Lecturer and as Wald Lecturer by the Institute of Mathematical Statistics, and as Shannon Lecturer by the Institute of Electrical and Electronic Engineers. His list of other academic honors is a long one: an honorary doctorate from the Technion; election to the U.S. National Academy of Sciences and to the American Academy of Arts and Sciences; election as a fellow of the International Statistics Institute, the Econometric Society, the American Statistical Association, and the Institute of

Mathematical Statistics; a term as president of the Institute of Mathematical Statistics; visiting professorships at several universities; and selection as a Guggenheim Fellow.

Wolfowitz's reading was not confined to mathematical subjects. He read detective stories for relaxation and kept up with political and social conditions in all of the large nations of the world and many of the smaller ones. He was a man of strong opinions, with a particular detestation of tyranny. He took a leading part in organizing protests against Soviet repression of minorities and dissidents, and was able to aid several victims of such repression.

A fuller account of Wolfowitz's research can be found in ref. [1]. A complete list of his 120 publications is given in ref. [40].

REFERENCES

[1] Augustin, U., Kiefer, J., and Weiss, L. (1980). In *Jacob Wolfowitz: Selected Papers,* J. Kiefer, ed. Springer, New York, pp. ix–xxi.

[2] Dvoretzky, A., Kiefer, J., and Wolfowitz, J. (1952). *Econometrica,* **20,** 187–222.

[3] Dvoretzky, A., Kiefer, J., and Wolfowitz, J. (1952). *Econometrica,* **20,** 450–466.

[4] Dvoretzky, A., Kiefer, J., and Wolfowitz, J. (1953). *Ann. Math. Statist.,* **24,** 254–264.

[5] Dvoretzky, A., Kiefer, J., and Wolfowitz, J. (1953). *Ann. Math. Statist.,* **24,** 403–415.

[6] Dvoretzky, A., Kiefer, J., and Wolfowitz, J. (1953). *Econometrica,* **21,** 586–596.

[7] Dvoretzky, A., Kiefer, J., and Wolfowitz, J. (1956). *Ann. Math. Statist.,* **27,** 642–669.

[8] Dvoretzky, A., Wald, A., and Wolfowitz, J. (1951). *Ann. Math. Statist.,* **22,** 1–21.

[9] Kac, M., Kiefer, J., and Wolfowitz, J. (1955). *Ann. Math. Statist.,* **26,** 189–211.

[10] Kiefer, J. and Wolfowitz, J. (1952). *Ann. Math. Statist.,* **23,** 462–466.

[11] Kiefer, J. and Wolfowitz, J. (1955). *Trans. Amer. Math. Soc.,* **78,** 1–18.

[12] Kiefer, J. and Wolfowitz, J. (1956). *Ann. Math. Statist.,* **27,** 147–161.

[13] Kiefer, J. and Wolfowitz, J. (1958). Trans. Amer.Math. Soc., **87,** 173–186.

[14] Kiefer, J. and Wolfowitz, J. (1959). *Ann. Math. Statist.,* **30,** 271–294.

[15] Kiefer, J. and Wolfowitz, J. (1959). *Ann. Math. Statist.,* **30,** 463–489.

[16] Kiefer, J. and Wolfowitz, J. (1960). *Canad. J. Math.,* **12,** 363–366.

[17] Kiefer, J. and Wolfowitz, J. (1965). *Ann. Math. Statist.,* **36,** 1627–1655.

[18] Wald, A. and Wolfowitz, J. (1939). *Ann. Math. Statist.,* **10,** 105–118.

[19] Wald, A. and Wolfowitz, J. (1940). *Ann. Math. Statist.,* **11,** 147–162.

[20] Wald, A. and Wolfowitz, J. (1948). *Ann. Math. Statist.,* **19,** 326–339.

[21] Wald, A. and Wolfowitz, J. (1950). *Ann. Math. Statist.,* **21,** 82–99.

[22] Wald, A. and Wolfowitz, J. (1951). *Ann. Math.,* **53,** 581–586.

[23] Weiss, L. and Wolfowitz, J. (1972). *Z. Wahrsch. Verw. Geb.,* **24,** 203–209.

[24] Weiss, L. and Wolfowitz, J. (1972). *J. R. Statist. Soc. B,* **34,** 456–460.

[25] Weiss, L. and Wolfowitz, J. (1974). *Maximum Probability Estimators and Related Topics.* Springer, New York.

[26] Wolfowitz, J. (1942). *Ann. Math. Statist.,* **13,** 247–279.

[27] Wolfowitz, J. (1943). *Ann. Math. Statist.,* **14,** 280–288.
[28] Wolfowitz, J. (1946). *Ann. Math. Statist.,* **17,** 489–493.
[29] Wolfowitz, J. (1947). *Ann. Math. Statist.,* **18,** 215–230.
[30] Wolfowitz, J. (1949). *Ann. Math. Statist.,* **20,** 601–602.
[31] Wolfowitz, J. (1950). *Ann. Math. Statist.,* **21,** 218–230.
[32] Wolfowitz, J. (1952). *Skand. Aktuar.,* **35,** 132–151.
[33] Wolfowitz, J. (1953). *Ann. Inst. Statist. Math.,* **5,** 9–23.
[34] Wolfowitz, J. (1954). *Ann. Math. Statist.,* **25,** 203–217.
[35] Wolfowitz, J. (1957). *Ann. Math. Statist.,* **28,** 75–88.
[36] Wolfowitz, J. (1957). *Illinois J. Math.,* **1,** 591-606.
[37] Wolfowitz, J. (1962). *Econometrica,* **30,** 470–479.
[38] Wolfowitz, J. (1967). *New York Statistician,* **18,** 1–3.
[39] Wolfowitz, J. (1978). *Coding Theorems of Information Theory,* 3rd ed. Springer, New York.
[40] Wolfowitz, J. (1980). *Selected Papers.* Springer, New York.

L. WEISS

SECTION **4**

Probability Theory

Bernstein, Sergei Natanovich

Born: March 5 (n.s.), 1880, in Odessa (a port on the Black Sea in the Ukraine).
Died: October 26, 1968, in Moscow, USSR.
Contributed to: theory and application of differential equations, function approximation theory, probability.

Berstein received his mathematical education in Paris on completing high school in Odessa (where his father was a doctor and university lecturer) in 1898, and also studied at Göttingen. He defended a doctoral dissertation at the Sorbonne at age 24 in 1904, and another in Kharkov in 1913. From 1908 to 1933 he taught at Kharkov University, and from 1933 he worked at the Mathematical Institute of the USSR Academy of Sciences, also teaching at the University and Polytechnic Institute in Leningrad, and continued to work at the Mathematical Institute in Moscow from 1943. He was much honored, including foreign membership in the Paris Academy of Sciences, and membership in the USSR and Ukrainian Academies. In his general mathematical work he united the traditions of the St. Petersburg school, founded by P. L. Chebyshev,* with those of modern western European thinking. The scope of his probabilistic work was in general ahead of its time, and his writings, including his textbook [4], which first appeared in 1927, were largely responsible for determining the course of development of this subject area in the USSR.

Bernstein's early publications of a probabilistic nature have a heavily analytical character: a constructive proof, using the sequence of what came to be known as Bernstein polynomials and Bernoulli's law of large numbers, of Weierstrass's uniform approximation theorem, and a consideration of the accuracy of approximation of the binomial distribution by the normal distribution, a problem to which he was to return several times. These early writings include an interesting attempt (1917) at the axiomatization of probability theory (see ref. [5, pp. 10–60] and refs. [6] and [8]).

One of the most significant areas of Bernstein's creativity, however, was a reexamination in a new light of the main existing theorems of probability theory of his times. For example, if $X_1, X_2, \ldots$ are independent random variables with finite variance, adjusted to have zero mean, and $S_n = \sum_{i=1}^{n} X_i$, then Chebyshev's inequality may be written

$$P\{|S_n| < tB_n\} > 1 - t^{-2},$$

where B_n = var S_n, and Kolmogorov's inequality strengthens this result by replacing $|S_n|$ by $\sup_{1 \leq k \leq n} |S_k|$. Bernstein succeeded in raising the lower bound $1 - t^{-2}$ to $1 - 2e^{-t^2}$ in both, under additional assumptions on the X_k, this variant assuming the name Bernstein's inequality, even more significantly he showed the refinements to hold where the X_k are no longer necessarily independent but form what is now termed a martingale-difference sequence, so that $\{S_n\}$ is a martingale. Another direction taken by Bernstein within this area was to generalize the conditions of Liapunov for the applicability of the central limit theorem for sums of random variables. In 1922, Bernstein proved this under conditions which, when specialized to the same setting, are equivalent to those of Lindeberg, whose paper appeared in the same year; a fundamental and perhaps his best-known paper [3] extended this work to sums of "weakly dependent" random variables X_k and Markov chains; he later proved it for martingales $\{S_n\}$. (He speaks of "sums of dependent variables, having mutually almost null regression," since for a martingale difference sequence $E[X_{k+1}|X_k, X_{k-1}, \ldots, X_1] = 0$.) A group of papers, and Appendix 6 of [4], deal in essence with the weak convergence of a discrete-time stochastic process to one in continuous time whose probability distribution satisfies a diffusion equation, together with an examination of the boundary theory of such an equation, anticipating later extensive development of this subject. Little known also are his surprisingly advanced mathematical investigations in population genetics (e.g., ref. [2]), including a synthesis of Mendelian inheritance and Galtonian "laws" of inheritance. In addition, Bernstein took a keen interest in the methodology of the teaching of mathematics at secondary and tertiary levels.

In the years 1952–1964 he devoted much time to the editing and publication of the four-volume collection of his works [5], which contains commentaries by his students and experts in various fields.

REFERENCES

[1] Alexandrov, P. S., Akhiezer, N. L., Gnedenko, B. V., and Kolmogorov, A. N. (1969). *Uspekhi Mat. Nauk*, **24**, 211–218 (in Russian) (Obituary; emphasis is mathematical rather than probabilistic, but interesting commentary on Bernstein's population mathematics.)

[2] Bernstein, S. N. (1924). *Uch. Zap. Nauchno-Issled. Kafedr Ukr., Otd. Mat.*, **1**, 83–115 (in Russian). [Also in ref. [5, pp. 80–107]; in part (Chap. 1) in English translation (by Emma Lehner) in *Ann. Math. Statist.*, **13**, 53–61 (1942).]

[3] Bernstein, S. N. (1926). *Math. Ann.*, **97**, 1–59. (Published in 1944 in Russian translation, in *Uspekhi Mat. Nauk*, **10**, 65–114. Also in ref. [5], pp. 121–176.)

[4] Bernstein, S. N. (1946). *Teoriya Veroiatnostei*, 4th ed. Gostehizdat, Moscow-Leningrad. (1st ed.: 1927; 2nd and 3rd eds.: 1934. Portions on nonhomogeneous Markov chains—pp. 203–213, 465–484—reprinted in ref. [5, pp. 455–483].)

[5] Bernstein, S. N. (1964). *Sobranie Sochineniy, Vol. 4: Teoriya Veroiatnostei i Matematicheskaia Statistika* [1911–1946]. Izd. Nauka, Moscow. [This fourth volume of Bernstein's collected works contains Russian-language versions of most of his probabilistic work. Volume 1 (1952) and vol. 2 (1954) deal with constructive function theory, and vol. 3 (1960) with differential equations.]

[6] Kolmogorov, A. N. and Sarmanov, O. V. (1960). *Teor. Veroyatn. Primen.*, **5**, 215–221 (in Russian). (Division of Bernstein's writings into groups, with commentary; complete listing of his probabilistic writings up to ref. [5]; full-page portrait.)

[7] Kolmogorov, A. N., Linnik, Yu. V., Prokhorov, Yu. V., and Sarmanov, O. V. (1969). Sergei Natanovich Bernstein. *Teor. Veroyatn. Primen.*, **14**, 113–121 (in Russian). (Obituary, with good detail on biography and probabilistic work, and a photograph on p. 112.)

[8] Maistrov, L. E. (1974). *Probability Theory. A Historical Sketch*. Academic Press, New York, pp. 250–252. [Translated and edited from the Russian edition (1967) by S. Kotz.]

Bonferroni, Carlo Emilio

Born: January 28, 1892, in Bergamo, Italy.
Died: August 18, 1960, in Florence, Italy.
Contributed to: probability theory, actuarial mathematics, linear algebra, geometry.

Carlo Emilio Bonferroni (1892–1960).

Carlo Emilio Bonferroni attended the University of Torino, where he was exposed to some very outstanding mathematicians as instructors. He studied analysis with Giuseppe Peano and geometry with Corrado Segre. After receiving the degree he was appointed an assistant to Filadelfo Insolera, holder of the chair of mathematics of finance. Soon he became an instructor in courses in classical mechanics and geometry at the Polytechnic Institute of Torino.

His first professorial appointment was in 1923 to the chair of mathematics of finance at the University of Bari, where he stayed for ten years. For seven out of those years he was the president of the University of Bari. In 1933 he was appointed to the chair of general and financial mathematics of the University of Florence, and he remained in Florence for twenty-seven years until his death on August 18, 1960. During this period he was the dean of the Faculty of Economics and Commerce in 1945–1949. Simultaneously he also taught courses in statistical methods at the Bocconi University in Milan and analysis and geometry for the Faculty of Architecture in Florence. The Tuscan Academy "La Colombaria" honored him with the presidency of the Class of Physical, Natural, and Mathematical Sciences. He was also a member of the Italian Institute of Actuaries.

He was married to Mrs. Jolenda Bonferroni. He had a strong interest in music and was an excellent pianist and composer. Among his hobbies was also mountain climbing.

While teaching courses in different subjects he was an active writer and researcher. He wrote about 65 research papers and books. Among statisticians he is mostly known as the author of Bonferroni's inequalities [1]. However, he had several original ideas in other fields of mathematics. His work can be grouped in three broad categories: (1) analysis, geometry, and mechanics, (2) actuarial mathematics and economics, and (3) statistics and probability. In the first category there are 14 entries in the bibliography, including three textbooks (or lecture notes). In the second category there are 19 entries (including one textbook), and finally his list of publications contains 33 entries in the general area of statistics and probability.

In the first group we shall quote here two of his theorems. One concerns an extension of the so-called "Napoleon" theorem. Napoleon observed and asked Laplace to prove that "if one constructs three equilateral triangles, each on one of three sides of an arbitrary triangle, their centers form an equilateral triangle." Bonferroni extended this theorem to the following: "Consider three triangles with angles, α, β, γ ($\alpha + \beta + \gamma = \pi$) constructed internally or externally on three sides of an arbitrary triangle respectively. Their centroids form a triangle with angles α, β, γ."

The following theorem in linear algebra with application to analysis was also established by Bonferroni: "The determinant obtained by horizontal multiplication of two conjugate matrices does not exceed the product of an element of the main diagonal times its minor, and if this product is nonzero, the determinant is equal to it only when the lines that cross at that element have every other element nonzero." Hadamard's theorem and other results on Hermitian bilinear forms follow as simple consequences.

An article by Carlo Benedetti in *Metron* contains an extensive review of Bonferroni's books. His textbook *Elements of Mathematical Analysis* had its fourth edition (1957) shortly before the author died. This reviewer has very high praise for its style, elegance, and level of generality.

The book on *Foundations of Actuarial Mathematics* includes several of his earlier ideas on the subject.

In the area of statistics and probability most of Bonferroni's research is concentrated in the period 1917–1942. There are only three papers published in the 1950s, all concerning the same subject of medians in continuous distributions.

The last revision of the text *Elements of General Statistics* (1941–1942) contains many of his ideas on the subject and some original proofs of results known at the time. He introduces and discusses various indices and coefficients. His coefficient of parabolic correlation of order k is revisited. Contingency indices of Pearson* and Chuprov* get unified treatment, together with other indices which he proposes.

Bonferroni was also interested in measures of concentration. He proposed a concentration index, unaware of earlier work of Dalton [18]. Another concentration index proposed by him could be considered a variation of Gini's* concentration ratio. This could have created some friction with Gini, in addition to that arising from Gini's defense of his assistant, Cisbani, who had a dispute with Bonferroni.

SOURCES AND SELECTED BIBLIOGRAPHY

[1] Alt, F. B. (1982). Bonferroni inequalities and intervals. In *Encyclopedia of Statistical Sciences*, S. Kotz and N. L. Johnson, eds. Wiley, New York, vol. 1, p. 294.

[2] Benedetti, C. (1982). Carlo Emilio Bonferroni (1892–1960). *Metron,* **40**(3–4), 3–36.

[3] Bonferroni, C. E. (1919). *Principi di Geometria (ad Uso degli Studenti del Politecnico di Torino)*. Gnocchi, Turin, Italy.

[4] Bonferroni, C. E. (1920). *Corso di Meccanica Raxionale (ad Uso degli Studenti del Politecnico di Torino)*. Gili, Turin, Italy.

[5] Bonferroni, C. E. (1921). Probabilità assolute, relative, di precessione, *Gior. Mat. Finanz.*

[6] Bonferroni, C. E. (1927–1928). *Elementi di Statistica Generale*, 1st ed. Bocconi, Milan, Italy. [Last (expanded) edition 1940.]

[7] Bonferroni, C. E. (1928). On the measures of the variations of a collectivity. *J. Inst. Actuaries Lond.* **59**, 62–65.

[8] Bonferroni, C. E. (1936). Teoria statistica delle classi e calcolo delle probabilità. In *Volume in Honor of Riccardo Della Volta.* University of Florence, Italy.

[9] Bonferroni, C. E. (1939). Di una estensione del coefficiente di correlazione. *Giorn. Econ.* (Expanded version in vol. 1, Institute of Statistics, Bocconi, 1941.)

[10] Bonferroni, C. E. (1940). Le condizione d'equilibrio per operazionè finanziarie finite ed infinite, *Giorn. Istituto. Ital. Attuar.* **11**, 190–213.

[11] Bonferroni, C. E. (1940). Di un coefficiente del correlazione. *Cong. Unione Mat. Ital.* (Also in vol. 1, Institute of Statistics, Bucconi, 1941.)

[12] Bonferroni, C. E. (1940). Un indice quadratico di concentrazione. Ibid.,

[13] Bonferroni, C. E. (1941). Nuovi indici di connessione fra variabili statistiche. *Cong. Unione Mat. Ital.*,

[14] Bonferroni, C. E. (1942). *Fondamenti di Matematica Attuariale,* 2nd ed. Gili, Turin, Italy. (1st ed. 1935–1936.)

[15] Bonferroni, C. E. (1950). Di una disuguaglianza sui determinanti ed il teorema di Hadamard. *Boll. Unione Mat. Ital.*,

[16] Bonferroni, C. E. (1950). Una teorema sul triangolo ed id teorema di Napoleone, Ibid.

[17] Bonferroni, C. E. (1957). *Elemente di Analisi Matematica,* 6th ed. Cedam, Padua, Italy. (1st edn. 1933–1934.)

[18] Dalton, H. (1920). *The Inequalitites of Income,* Routledge, London; Dulton, New York.

PIOTR W. MIKULSKI

Cantelli, Francesco Paolo

F. P. Cantelli

Born: December 20, 1875, in Palermo, Italy.
Died: July 21, 1966, in Rome, Italy.
Contributed to: foundations of probability theory, stochastic convergence, laws of large numbers, actuarial and financial mathematics.

Francesco Paolo Cantelli is one of the main probabilists in Europe in the first decades of the current century. He attended the University at Palermo, where he took his degree in mathematics in 1899, with a thesis in celestial mechanics. In fact, his main scientific interest in the first years after taking his degree was in astronomy. In one of his more interesting studies in this field, he established, in a rigorous way and on the basis of the indications given by Dante in the *Divina Commedia*, the positions of heavenly bodies in the years 1300 and 1301. Thanks to this research, he could confirm 1301 as the year in which Dante's imaginary trip took place; see ref. [3].

From 1903 to 1923 Cantelli worked as an actuary at the Istituti di Previdenza della Cassa Depositi e Prestiti and performed some important research in the field of financial and actuarial mathematics. At the same time, he began his studies in probability theory. At the beginning, probability was, for Cantelli, basically an indispensable tool for his astronomical and actuarial researches. Soon, however, probability theory became his main interest, not only for its possible applications to other fields, but chiefly as an autonomous subject. Cantelli's point of view about the interpretation of probability substantially agrees with that of the main probabilists of the time, such as Lévy,* Fréchet, and Castelnuovo, and can be included under the heading of the empirical approach; cf. refs. [1], [9], [13], [17], and [18]. However, Cantelli was convinced that, in order to make probability acceptable to mathematicians as a branch of mathematical analysis, the abstract theory had to be clearly distinguished from the formulation of empirical postulates.

In 1915, at the University of Rome, he introduced, together with Castelnuovo, courses in probability and actuarial mathematics. They represented the germ of the School of Statistical and Actuarial Sciences, officially established in 1927. In 1935, that school merged with the Special School of Statistics, founded some years before by Corrado Gini,* and formed the first Faculty of Actuarial, Demographic and Statistical Sciences in Italy.

In his career, Cantelli obtained a number of acknowledgments for his scientific activity and held various national and international offices. In the 1920s, he was, among other things, the actuary of the pension board of what was then called the Society of Nations in Geneva and was later succeeded by Harald Cramér.* In 1930, he founded the scientific review *Giornale dell'Istituto Italiano degli Attuari*, of which he was editor up to 1958. In those years, the *Giornale* was one of the main scientific reviews in the world in the field of probability, statistics, and actuarial mathematics and published fundamental papers by outstanding probabilists and statisticians, among them Kolmogorov,* Neyman,* de Finetti,* Lévy, Cramér,* and, of course, Cantelli himself.

Cantelli's main contributions to probability theory are related to stochastic convergence. We note that in 1916–1917, when Cantelli developed these studies, the abstract theory of probability was not yet clearly formulated; cf. ref. [18]. Cantelli discussed and pinpointed various forms of convergence of a sequence of random events and variables; see refs. [6], [7], and [12]. He observed how the concept of convergence of a sequence of random variables could be viewed as a generalization of the notion of convergence for a numerical sequence. The generalization can assume various forms, and to each of them a different notion of stochastic convergence is associated; he was the first to clearly distinguish these various concepts. He used the term *convergence in the sense of probability theory* for what is now known as convergence in probability. He analyzed this kind of convergence in depth and applied it to derive, in ref. [7], the following form of the weak law of large numbers, valid for not necessarily independent random variables: if $\{X_n\}_{n\geq 1}$ is a sequence of random variables, with finite expectations $E[X_i] = m_i$ such that $\sum_{i=1}^{n} m_i/n \to m$, and if $n^{-2}\,\mathrm{Var}(\sum_{i=1}^{n} X_i) \to 0$, then $n^{-1}\sum_{i=1}^{n} X_i \to m$ is probability. It is important to observe that, apart from some results by Markov* in the first years of the century, the weak law of large numbers was known, at the time, only for independent random variables.

In 1917, Cantelli [8] introduced a new kind of probabilistic convergence, which he meaningfully called *uniform convergence in probability*. He defined a random sequence $\{X_n\}_{n\geq 1}$ to be uniformly convergent in probability to a random variable X if, for each $\epsilon > 0$,

$$P\left(\bigcap_{k=0}^{+\infty} \{|X_{n+k} - X| \leq \epsilon\}\right) \to 1$$

Of course, this is exactly what is now known as convergence with probability one (a.s. convergence).

Connected with this kind of stochastic convergence, he contributed also some important results on the "*uniform law of large numbers*," now called the strong law of large numbers. We recall that E. Borel [2] was the first probabilist to give, in 1909, a version of the strong law of large numbers. However, Borel's reasoning was not completely correct, and moreover he only took into account the classical case of the frequencies in a Bernoulli scheme with success probability $p = 0.5$. Cantelli, in ref. [8], substantially extended Borel's result, proving the following version of the strong law of large numbers: let $\{X_n\}_{n\geq 1}$ be a sequence of independent (not necessarily identically distributed) random variables with zero expectations and for which there exists $K > 0$ such that $E[X_n^4]$ for every $n \geq 1$. Then $n^{-1}\sum_{i=1}^{n} X_i$ converges to zero with probability one. The proof of this result, now known as Cantelli's strong law of large numbers, was based on some notable inequalitites of the kind due to Chebyshev* and on the well-known Borel–Cantelli lemma, whose proof was given by

Borel in ref. [2] for the case of independent events and was completed by Cantelli in ref. [8].

It is interesting here to note a heated dispute, mostly between Cantelli and Slutskii,* about the priority for the strong law of large numbers (apparently given by the latter in ref. [20] to Borel, whereas Cantelli claimed it was his own); cf. ref. [19].

These facts are also useful in revealing the vehemence and the strength of the author in defending his own positions and points of view. Harald Cramér (see ref. [21]), who knew Cantelli very well, wrote about him: "He was a very temperamental man. When he was excited, he could cry out his views with his powerful voice: a very energetic fellow."

Cantelli made important contributions also to the foundations of probability theory. In particular, Cantelli had a great interest in the formalization of an axiomatic theory of probability. We recall that in the early decades of the century, in Europe, research into abstract probability theory was undergoing substantial development. To this, Cantelli himself made a fundamental contribution, publishing, in 1932, his "*Teoria astratta della probabilita*" [9]. In this paper, Cantelli presents an abstract formulation of the concepts of probability theory; cf. refs. [16] and [17]. This work can be considered as the first showing a complete and rigorous correspondence between related notions of probability and measure theory. It was published one year before the famous work by A. N. Kolmogorov* [15], which is the basis of the modern axiomatic probability theory. Unlike Kolmogorov's formulation, in his paper Cantelli always assumes as sample space Ω the set [0, 1], endowed with Lebesgue measure. Given an arbitrary probability distribution, he defines, on [0, 1], a measurable function (by Cantelli called *variabile ponderata*) having as law the assigned one. In a certain sense, Cantelli thus anticipated Skorohod's ideas. By means of the abstract formulation of the theory, free from empirical postulates, Cantelli wished to remove the perplexities of the contemporary mathematicians in accepting probability as a branch of mathematical analysis.

A further contribution of Cantelli's to probability theory is a generalization of Kolmogorov's law of the iterated logarithm; see refs. [9] and [10].

A detailed survey of Cantelli's works in probability can be found in Benzi [1].

In mathematical statistics, Cantelli's most famous result is related to the a.s. uniform convergence of the empirical distribution function associated with a sequence of independent and identically distributed random variables. This result, now known as the Glivenko-Cantelli theorem, or the fundamental theorem of mathematical statistics, was proved by Glivenko, in ref. [14], in the case in which the common distribution of the random variables is continuous, and it was extended (and proved in a different, simpler way) by Cantelli for arbitrary distributions; see refs. [11] and [12].

Another important result in statistics is connected with the problem of curve fitting. Cantelli, in ref. [4], introduced a new method, called *metodo delle aree*, to choose, in a given family, the curve which better represents a given set of data $(x_1, y_1,), \ldots, (x_m, y_m)$. After selecting n points in the interval (min $(x_1, \ldots, x_m)$, max $(x_1, \ldots, x_m)$), if the curve has equation $y = \phi(x, c_1, \ldots, c_n)$, one determines the coefficients $c_1, \ldots, c_n$ which make equal all the corresponding areas under the fitted and the empirical curves.

Let us conclude with a few notes on Cantelli's contributions in financial and actuarial mathematics. One of these is related to the general analysis of capitalization laws. He studied them as functions of two variables: investment and disinvestment dates. In particular, he provided a new classification of these laws. In his fundamental work "*Genesi e costruzione delle tavole di mutualita*" [5], Cantelli formulated and solved the so-called problem of accumulated capital, studying the retrospective mathematical reserves of insurance firms.

REFERENCES

[1] Benzi, M. (1988). *Historia Math.* **15**, 53–72. (A review of Cantelli's contributions in probability theory).

[2] Borel, É. (1909). *Rend. Cir. Mat. Palermo*, **27**, 247–271. (Includes the first version of the strong law of large numbers, for Bernoulli trials.)

[3] Cantelli, F. P. (1900). *Atti dell'Accademia Pontaniana*. Napoli.

[4] Cantelli, F. P. (1905). *Sull'adattamento delle Curve ad una Serie di Misure e Osservazioni*. Tipografia Bodoni e Bolognesi, Roma.

[5] Cantelli, F. P. (1914). *Boll. Notizie Credito e Previdenza*, **3–4**, 247–303.

[6] Cantelli, F. P. (1916). *Rend. R. Accad. Lincei*, **25**, 39–45.

[7] Cantelli, F. P. (1916). *Rend. Circ. Mat. Palermo*, **41**, 191–201.

[8] Cantelli, F. P. (1917). *Rend. R. Accad. Lincei*, **26,** 39–45.

[9] Cantelli, F. P. (1932). *Giorn. Ist. Ital. Attuari*, **3**, 257–265.

[10] Cantelli, F. P. (1933). *Giorn. Ist. Ital. Attuari*, **3**, 327–350.

[11] Cantelli, F. P. (1933). *Giorn. Ist. Ital. Attuari*, **4**, 421–424.

[12] Cantelli, F. P. (1935). *Ann. Inst. H. Poincaré*, **5**, 1–50.

[13] Castelnuovo, G. (1919). *Calcolo delle Probabilitá*. Societá Editrice Dante Alighieri, Milano.

[14] Glivenko, V. I. (1933). *Giorn. Ist. Ital. Attuari*, **4**, 92–99. (Includes a proof of the a.s. uniform convergence of the empirical distribution function in the continuous case.)

[15] Kolmogorov, A. N. (1933). *Grundbegriffe der Wahrscheinlichkeitsrechnung*. Springer, Berlin.

[16] Ottaviani, G. (1939). *Giorn. Ist. Ital. Attuari*, **10**, 10–40. (Includes a discussion of the abstract probability theory proposed by Cantelli.)

[17] Ottaviani, G. (1966). *Giorn. Ist. Ital. Attuari*, **2**, 179–191. (Cantelli's obituary, with a survey of his life and scientific activity.)

[18] Regazzini, E. (1991). *Technical Report I.A.M.I. 91.8*. (A comprehensive treatment of the history of probability in the first half of the current century.)

[19] Seneta, E. (1992). *Historia Math.* **19**, 24–39. (A historical discussion about the strong law of large numbers including some questions about its priority.)

[20] Slutskii, E. (1925). *C. R. Acad. Sci.*, **187**, 370. (The paper starting the controversy about priority for the strong law of large numbers.)

[21] Wegman, E. J. (1986). *Statist. Sci.*, **4**, 528–535. (Includes, among other things, some recollections of H. Cramér on Cantelli.)

BIBLIOGRAPHY

Cantelli, F. P. (1958). *Alcune Memorie Matematiche: Onoranze a Francesco Paolo Cantelli.* Giuffrè, Milano (Selected papers of Cantelli's.)

Chetverikov, N. S. (1975). *Statisticheskie Issledovania. Teoriia I Praktika.* Nauka, Moscow. (Includes interesting documents about the controversy between Slutskii and Cantelli.)

EUGENIO MELILLI

Cauchy, Augustin-Louis

Born: August 21, 1789, in Paris, France.
Died: May 22, 1857, in Sceaux, France.
Contributed to: mathematics, mathematical physics, celestial mechanics, probability, mathematical statistics.

The contributions of this great mathematician to mathematical statistics occur in connection with the problem of estimation (using modern notation) of the $r \times 1$ vector $\boldsymbol{\beta} = \{\beta_i\}$ from a vector $\mathbf{Y}$ of n observations in the classical linear model: $\mathbf{Y} = X\boldsymbol{\beta} + \boldsymbol{\epsilon}$ where $\boldsymbol{\epsilon} = \{\epsilon_i\}$ is an error vector, and $X = \{x_{ij}\}$, a known fixed $n \times r$ matrix, $n \geq r$, of full column rank r. The nineteenth-century mathematicians (Gauss,* Laplace,* Bienaymé,* and Cauchy) regarded this problem as one of finding an $r \times n$ matrix $K = \{k_{ij}\}$ (or as they would put it, a system of "multipliers") such that

$$\boldsymbol{KX} = \boldsymbol{I}, \tag{1}$$

so that consequently $\boldsymbol{\beta}$ is estimated by (the linear estimate) $\overline{\boldsymbol{\beta}} = \{\overline{\beta}_i\}$, where $\overline{\boldsymbol{\beta}} = \boldsymbol{KY} = \boldsymbol{\beta} + \boldsymbol{K\epsilon}$, $\boldsymbol{K}$ being chosen under the constraint (1), in some optimal way. The least-squares choice of $\boldsymbol{K}$—that minimizing $\boldsymbol{\epsilon}^T\boldsymbol{\epsilon}$ and leading to $\boldsymbol{K} = (\mathbf{X}^T\mathbf{X})^{-1}\mathbf{X}^T$—had been justified by Gauss, on probabilistic grounds, in 1809 as the most probable (maximum likelihood) estimator if $\boldsymbol{\epsilon}$ is $N(\mathbf{0}, \sigma^2\boldsymbol{I})$, and in 1821 as the minimum-variance (best linear unbiased) estimator if $E\boldsymbol{\epsilon} = \mathbf{0}$, var $\boldsymbol{\epsilon} = \sigma^2\boldsymbol{I}$. Intermediate to these justifications of Gauss were approaches of Laplace in 1812, which assume that independent, identically distributed (i.i.d.) ϵ_i are described by a symmetric density confined to a finite interval, $[-g, g]$. Laplace shows (heuristically, and at least for $r = 1$) that the standardized random variable

$$\frac{\overline{\beta}_i - \beta_i}{\left(\sigma^2 \sum_{h=1}^n k_{ih}^2\right)^{1/2}} \equiv \frac{\sum_{h=1}^n k_{ih}\epsilon_h}{\left(\sigma^2 \sum_{h=1}^n k_{ih}^2\right)^{1/2}} \rightarrow N(0, 1)$$

as $n \rightarrow \infty$, so that, for large n, a symmetric fixed-probability-level interval about the origin for the error $\overline{\beta}_i - \beta_i$ is of the form $\pm\, z_0(\sigma^2\sum_{h=1}^n k_{ih}^2)^{1/2}$, z_0 constant, whose length is therefore minimized by choosing $\boldsymbol{K}$ to minimize under the constraint (1), the quantity $\sigma^2\sum_{h=1}^n k_{ih}^2$. Since this last is var $\overline{\beta}_i$, simultaneous minimization for all $i = 1, \ldots, r$ leads again to the least-squares choice of $\boldsymbol{K}$.

Cauchy had started to write on the choice of $\boldsymbol{K}$ in 1853, one of his intentions being to show that there are situations where the least-squares choice is not optimal. Bienaymé resented the apparent criticism of Laplace, and there ensued a heated controversy between them in the meetings of the Academy of Sciences, and on the pages of its journal, the *Comptes Rendus*, for that year [4]. Within it, Cauchy shows, using heuristic and involved reasoning, that if the i.i.d. ϵ_i, $i = 1, \ldots, n$, have characteristic function $\exp(-c|\theta|^\alpha)$, c, $\alpha > 0$, then the "most probable" estimate $\overline{\boldsymbol{\beta}}$, in a certain sense, comes about from choosing K to satisfy (1) and to simultaneously minimize $\sum_h |k_{ih}|^\alpha$ for each i. This marks the first time in the history of probability that the (symmetric) stable laws appear; Cauchy's "most probable" estimate $\overline{\boldsymbol{\beta}}$ has been rediscovered recently in econometric contexts, as that which minimizes the scale parameter of the distribution of each of the $(\overline{\beta}_i - \beta_i) \equiv \sum_{h=1}^n k_{ih}\epsilon_h$ when the common distribution of the ϵ_i is such a symmetric stable law.

Toward the close of the controversy, Cauchy considers the same asymptotic problem as Laplace (see above), i.e., the distribution of a linear function of residuals $\sum_{h=1}^n \lambda_h \epsilon_h$ as $n \to \infty$ under Laplace's distributional assumptions on the ϵ_i, but allowing the λ_i's to depend on n, using characteristic functions, inversion formulas, and careful estimates of the various integrals that occur. All these ideas were rather innovative for their time, but, restricted to a small amount of space, Cauchy's presentation was a very concentrated sketch; and since it occurred in the context of his dispute with Bienaymé, and within the framework of least squares, it is not surprising that this rigorous work of Cauchy on the central limit theorem was not understood, and was passed over at the time. In 1892, I. V. Sleshinsky [7] took up Cauchy's sketch, filled in the proofs, adding new steps as necessary, to produce by characteristic-function methods a first rigorously proved, if somewhat restricted, version of the central limit theorem.

A final contribution to mathematical statistics by Cauchy, alluding also to the nonoptimality of least squares, arises from noting that (for fixed n) if the ϵ_i again satisfy Laplace's distributional assumptions, then the maximum attainable error of estimate in $\overline{\beta}_i - \beta_i$ is $g\sum_{h=1}^n |k_{ih}|$, which is therefore minimized for each i by choosing $\boldsymbol{K}$ to satisfy (1) and to minimize $\sum_{h=1}^n |k_{ih}|$ for each i. Cauchy gives a correct solution for the appropriate $\boldsymbol{K}$ in the case $r = 1$, and proposes, without proof, a correct solution in general [6]. Note that this $\boldsymbol{K}$ is the same as that needed to resolve the case $\alpha = 1$ in the above-described stable-laws setting.

Cauchy made certain other contributions to the theory of the classical linear model from an *interpolational* (error-theoretic) rather than *probabilistic* (as hitherto described) point of view. An early contribution along these lines is an attempt, dating from 1814 although not published until 1824 and 1831 [1, 2], to extend by geometric reasoning to the case $r = 2$ Laplace's treatment of the case $r = 1$, with a view to obtaining the estimate of $\boldsymbol{\beta}$ from the criterion $\sum_{i=1}^n |\epsilon_i| = \min$, which procedure has been revived in modern times since the estimate is robust although nonlinear. In 1835, Cauchy formulated and ingeniously resolved a then new aspect of the interpolation view of the classical linear model, of how many β_i, $i = 1, 2, \ldots,$ to fit in succession until an "adequate" fit to the data obtains, proceeding in such a way that the estimates for $\beta_1, \ldots, \beta_r$ are unaffected in going to that for β_{r+1}. This idea was revived within the context of the controversy of 1853, and is thence directly linked [4], via Bienaymé, with the idea of successive orthogonalization of the columns of the design matrix X, and its connection with least-squares estimates, in the work of Chebyshev,* at least in a polynomial interpolation setting.

Although the term "Cauchy distribution" arises from the 1853 controversy, in which it occupies a prominent place, it is known to have occurred in earlier writings of Poisson* [8]. The equation $f(x) + f(y) = f(x + y)$, variants of which are encountered in the theory of Markov processes and of the stable laws, was shown by Cauchy to have general solution, under prior assumption of continuity of f, $f(x) = cx$; it is often called Cauchy's functional equation, although it was known earlier also (to d'Alembert, for example).

REFERENCES

[1] Cauchy, A. *Oeuvres Complètes d'Augustin Cauchy.* Gauthier-Villars, Paris. (Collected works; volumes in several series, with various dates of publication.)

[2] Freudenthal, H. (1971). In *Dictionary of Scientific Biography,* Vol. 3, C. C. Gillispie, ed. Scribner's, New York, pp. 131–148. (Cauchy's life and scientific work surveyed in their entirety; contains a brief sketch of his "error theory.")

[3] Gnedenko, B. V. and Sheynin, O. B. (1978). In *Matematika XIX Veka* [*Mathematics of the 19th Century*]. Nauka, Moscow, pp. 184–240. (Pages 205–207 of this Russian-language account sketch some of Cauchy's contributions.)

[4] Heyde, C. C. and Seneta, E. (1977). I. J. Bienaymé: Statistical Theory Anticipated. Springer-Verlag, New York. (Chapter 4 contains an extensive modern analysis of Cauchy's work and its consequences in reference to the classical linear model and the central limit theorem, with particular emphasis on 1853.)

[5] Seal, H. L. (1967). Biometrika, **54**, 1–24. (A well-known account endowing the historical development of least-squares theory with somewhat fictitious continuity in regard to the roles of Bienaymé and Cauchy.)

[6] Seneta, E. (1976). *Ann. Soc. Sci. Brux.*, **90**, 229–235.

[7] Sleshinsky [Sleschinsky, Sleszyński], I. V. (1892). *Zap. Mat. Otd. Novoross. Obshch. Estestvoispyt. (Odessa),* **14,** 201–264 (in Russian).

[8] Stigler, S. M. (1974). *Biometrika*, **61**, 375–380.

[9] Valson, C. A. (1868). *La Vie et les travaux de Baron Cauchy.* Gauthier-Villars, Paris. (Well-known hagiography.)

E. SENETA

Chebyshev (or Tshébichef), Pafnuty Lvovich

Born: May 26 (n. s.), 1821, in Okatovo (Kaluga region), Russia.
Died: December 8 (n. s.), 1894, in St. Petersburg, Russia.
Contributed to: number theory and analysis, the theory of mechanisms, approximation theory, probability theory.

Chebyshev's contributions to probability theory form a relatively small portion of his contributions to mathematics and practical mechanics, although in regard to mathematical statistics, this is supplemented by his interpolational work in the context of linear least squares. He was a leading exponent of the Russian tradition of treating the probability calculus as an integral part of mathematical training, and through the "Petersburg mathematical school," of which he was the central figure, his overall influence on mathematics within the Russian empire was enormous. His distinguished disciples within this framework included A. A. Markov* and A. M. Liapunov,* who, *inter alia*, extended his own remarkable probabilistic work.

In his early mathematical education at Moscoe University, where he enrolled in 1837, Chebyshev was strongly influenced by N. D. Brashman. Under this influence he produced his first two contributions to probability in 1846: his master's thesis and the article that is ref. [8], both of which seek to give an elementary but rigorous analytical discussion of some of the then principal aspects of probability theory. Reference [8] in particular is notable in that it contains an analytical deduction of the weak law of large numbers (WLLN) of Poisson*: that if X is the number of successes in n independent trials, where p_i, $i \geq 1$, is the probability of success in the ith, then

$$\Pr\left[\left|\frac{X}{n} - \bar{p}(n)\right| < \epsilon\right] \to 1$$

as $n \to \infty$ for any $\epsilon > 0$, where $\bar{p}(n) = \sum_{i=1}^{n} p_i/n$. This publication, however, passed unnoticed, and the law, with its "floating mean" $\bar{p}(n)$, remained an object of controversy among the French mathematicians who dominated probability thoery at the time, for years to come[5, Sec. 3.3]. The proof is also interesting insofar as it proceeds by obtaining upper bounds for the lower and upper-tail probabilities of the number of successes. (The modern approach, via the Bienaymé–Chebyshev inequality of which we shall speak shortly, was then unknown.) Indeed, both early works already display one of the features peculiar to Chebyshev's work: the estimation for finite n of the *deviation* from its limit of a quantity that approaches the limit as $n \to \infty$, which was evidently communicated to both Markov and Liapunov.

In 1847, Chebyshev began to teach at St. Petersburg University, eventually becoming full professor in 1860, in which year he took over the course in probability theory (on the retirement of V. Ya. Buniakovsky), which reawakened his interest in the subject area. He had been promoted to the highest academic rank of the St. Petersburg Academy of Sciences in the previous year. Subsequently, he wrote only two papers [11, 12] in probability theory, which, nevertheless, had great influence.

In ref. [11] he obtains the Bienaymé–Chebyshev inequality for the arithmetic mean of independently but not necessarily identically distributed random variables, each of which has only a finite number of sample points, and uses it to deduce the corresponding WLLN, with subsequent application to the cases of both Poisson (which he had treated in ref. [8]) and J. Bernoulli.* I. J. Bienaymé* had arrived at both the inequality and a WLLN by the simple reasoning still used in 1853 for general probability distributions, and may have arranged the juxtaposition of a reprinting of his own article next to a French printing of Chebyshev's in Liouville's journal. Indeed, the year 1858 [9] seemed to mark the beginning of a mutual correspondence and admiration between the two men, leading to the eventual election of each to a membership in the other's Academy of Science. Even though in 1874 Chebyshev gave Bienaymé credit in print for arriving at the inequality via "the method of moments," whose discovery he ascribed to Bienaymé, and this view was later reiterated by Markov,* it is a valid point that it was more clearly stated and proved by Chebyshev. In any case, through the subsequent writings of the strong Russian probability school, Chebyshev's paper has undeniably had the greater publicity, to the extent that the inequality has often borne Chebyshev's name alone.

In ref. [12], Chebyshev uses "the method of moments" for the first time as a tool in the proof of the central limit theorem for not necessarily identically distributed summands X_i, $I \geq 1$ (under the implicit assumption of independence, to which, in the manner of the times, he never referred explicitly). His assumptions and proof are incomplete, and have been the subject of much discussion in the Soviet historical literature. In 1898, Markov used the same method to overcome the inadequacies, and shortly after this the very general version, proved with the aid of characteristic functions, was obtained by Liapunov. Chebyshev's attempt at *rigorous* proof and Markov's follow-up have often been thought to be the first such, although in a more restricted setting, a rigorous proof had been largely given by A. L. Cauchy* in 1853, and completed by I. V. Sleshinsky [7], who recognized the gaps in Chebyshev's treatment, by characteristic-function methods. Chebyshev's paper is also notable for his pointing out the possibility of refining the central limit theorem by means of an asymptotic expansion in what are now known as the Chebyshev–Hermite polynomials.

In respect to mathematical statistics, Chebyshev's influence stems from the paper that is ref. [9], in which he is concerned with fitting a polynomial

$$y = \sum_{j=1}^{n} \beta_j x^{j-1}$$

to n pairs of observations (Y_i, x_i), $i = 1, \ldots, n$, and does so by producing from the q powers $1, x, \ldots, x^{q-1}$ a set of q polynomials $T_0(x) = 1, T_1(x), \ldots, T_{q-1}(x)$ which are *orthogonal* with respect to the points $x_1, \ldots, x_n$ in that

$$\sum_{i=1}^{n} T_s(x_i)T_t(x_i) = 0 \qquad (s \neq t)$$

(assuming equal weights). These orthogonal polynomials, at least in the case where x_i, $i = 1, \ldots, n$, are at equidistant intervals, have come to bear his name. He is aware that the coefficients produced by his procedure are those arising out of a linear least-squares fit, but is more concerned with the connection of the problem with his continued-fractions theory approach, another central theme of his *oeuvre*. The statistical significance of this work is finally made clear, through the catalytic effect of Bienaymé, in Chebyshev's paper given here as ref. [10], in which he recognizes the ease of modifying the expression for the residual sum of squares with increasing q by progressive orthogonalization, in the process of deciding where to stop the "expansion." In the setting of the general linear model $\mathbf{Y} = \mathbf{X}\boldsymbol{\beta} + \boldsymbol{\epsilon}$, in which the $n \times r$ design matrix $\mathbf{X} = \{x_{ij}\}$ has the special form $x_{ij} = x_i^{j-1}$ in the polynomial context, the interpolational problem of how many β_i, $i = 1, 2, \ldots$, to fit in succession until an "adequate" fit to the data obtains, proceeding in such a way that the estimates $\beta_1, \ldots, \beta_r$ are unaffected in going to that for β_{r+1}, had been revived by Cauchy in 1853 in the course of his controversy with Bienaymé. (However, Cauchy's own earlier solution does not have the numerous interpolational and statistical advantages of orthogonalization.)

Chebyshev had a deep belief in the mutual benefits of the interaction of theory and practice; a famous quotation [2] reads in part: "And if theory gains much when new applications or new developments of old methods occur, the gain is still greater when new methods are discovered; and here science finds a reliable guide in practice." When he retired from teaching in St. Petersburg University in 1882, he continued to maintain close contact with his disciples and young scientists; and he died greatly loved and esteemed among his colleagues. However, to the English-speaking world the significance of his probabilistic and statistical work, at least, was not immediately apparent [1], although in other areas he had an international reputation.

REFERENCES

[1] Anonymous (1895). *Nature (Lond),* **52,** 345. (An English-language obituary.)

[2] Chebyshev, P. L. (1944–1951). *Polnoe Sobranie Sochineniy,* 5 vols. Izd. AN SSSR, Moscow. (Russian-language collected works with commentaries; the quotation is in vol. 5, p. 150.)

[3] Chebyshev, P. L. (1955). *Izbrannie Trudy.* Izd. AN SSSR, Moscow. (Russian-language selected works, with commentaries.)

[4] Gnedenko, B. V. and Sheynin, O. B. (1978). In *Matematika XIX Veka.* [*Mathematics of the 19th Century*]. Nauka, Moscow, pp. 184–240. (Pages 216ff. Contain, with a portrait, a more balanced Russian-language view than usual of the question of Chebyshev's priority and significance in probability theory.)

[5] Heyde, C. C. and Seneta, E. (1977). *I. J. Bienaymé: Statistical Theory Anticipated.* Springer-Verlag, New York. (Contains a detailed account of the interaction between Bienaymé, Cauchy, and Chebyshev.)

[6] Maistrov, L. E. (1974). *Probability Theory: A Historical Sketch.* Academic Press, New

York. (Translated and edited from the Russian-language work of 1967 by S. Kotz. Strong bias to Russian contributions.)

[7] Sleshinsky (Sleschinsky, Sleszyński), I. V. (1892). *Zap. Mat. Otd. Novoross. Obshch. Estestvoispyt. (Odessa),* **44,** 201–264 (in Russian).

[8] Tchébichef, P. L. (1864). *Crelle's J. Reine Angew. Math.,* **33**, 259–267. (Also in refs. [2], [3], and [13].)

[9] Tchébichef, P. L. (1858). *Liouville's J. Math. Pures Appl.,* **3**, 289–323. (Translation by Bienaymé of a Russian article of 1855 with prefatory footnote by the translator. Also in refs. [2], [3], and [13].)

[10] Tchébichef, P. L. (1859). *Mem. Acad. Sci. St. Petersburg (8),* **1**(15), 1–24. (Also in refs. [2] and [3].)

[11] Tchébichef, P. L. (1867). *Liouville's J. Math. Pures. Appl. (2),* **12**, 177–184. (Published simultaneously in Russian in Mat. Sb. (2), 2, 1–9. Also contained in refs. [2], [3], and [13].)

[12] Tchébichef, P. L. (1890–1891). *Acta Math.,* **14**, 305–315. [Originally published in Supplement to Zap. Imp. Akad. Nauk (S.P.–B.), **55**(6) (1887). Also contained in refs. [2], [3], and [13].]

[13] Tchébichef, P. L. (n.d.). *Oeuvres,* 2 vols. A. Markov and N. Sonin, eds. Chelsea, New York.

[14] Youshkevitch, A. P. (1971). In *Dictionary of Scientific Biography,* vol. 3, C. C. Gillispie, ed. Scribner's, New York, pp. 222–232. (Chebyshev's life and scientific work surveyed in their entirety; contains a valuable list of secondary materials.)

E. Seneta

Feller, William

Born: July 7, 1906, in Zagreb, Croatia.
Died: January 14, 1970, in Princeton, New Jersey, USA.
Contributed to: probability theory and its applications.

William Feller attended the University of Zagreb from 1923 to 1925 when he obtained a M.Sc. Degree, and the University of Göttingen 1925–1926, where he received a Ph.D. in 1926. He remained in Göttingen until 1928, and then was at the University of Kiel as a *Privatdozent* from 1928 until 1933, when, refusing, to sign a Nazi oath, he was forced to leave. After one year in Copenhagen, he spent five years at the University of Stockholm, and then moved to Providence, Rhode Island, in 1939, as an associate professor at Brown University and as the first executive editor of *Mathematical Reviews*. In 1945 he went to Cornell University as a professor and in 1950 he finally moved to Princeton University as Eugene Higgins Professor of Mathematics, a position he held until his death. He was a fellow of the Royal Statistical Society (Great Britain) and a member of the National Academy of Sciences (USA), Danish Academy of Sciences, Yugoslav Academy of Sciences, and American Academy of Arts and Sciences (Boston). He was awarded a National Medal of Science in 1969 and was cited by President Nixon for "original and definitive contributions to pure and applied mathematics, for making probability available to users, and for pioneering work in establishing *Mathematical Reviews*."

His list of publications contains 104 entries and a two-volume text, *An Introduction to Probability Theory and Its Applications,* which has achieved a remarkable popular appeal and has defined the manner and style of most introductory books to date. In addition to his work in probability, whose main outlines are described below, his papers include contributions to calculus, geometry, functional analysis, and mathematical biology.

His work in probability theory may be grouped under four headings: central limits, renewal theory, Markov chains, and, most importantly, Markov processes.

The classical limit theorems of probability theory are about the asymptotic behavior of the sums $S_n = X_1 + \cdots + X_n$ of a sequence of independent random variables $X_1, X_2, \ldots$. By 1930, the central limit theorem was nearly in its final form, assuming that the means m_n and variances v_n^2 of the S_n are finite: some weak additional conditions were known to ensure that $(S_n - m_n)/v_n$ has a distribution close to the standard normal for large n. In 1935, Feller [1] and Paul Lévy,* independently, showed that m_n and v_n have little to do with central limit theorems and proceeded to give simple conditions that are necessary and sufficient. Feller returned to the problem later in refs. [2–4] to simplify the proofs, calculate explicit estimates, and study examples. The modern form of the central limit theorem is essentially due to Feller and Lévy.

Feller's development of renewal theory illustrates both his power as a mathematician and his interest in the applications of probability theory. His paper [5] presents what was known in 1940, furnishing proofs, unifying the subject, and bringing the problem to the attention of mathematicians. His paper [6] with Erdős and Pollard was the first to prove the renewal theorem in the modern sense, a result of extreme importance in theoretical applications. He was the first to introduce the theory of regenerative phenomena, and the first to recognize that the importance of renewal theory derives from its applicability to regenerative processes like Markov chains and semi-Markov processes. He never tired of the subject, and the current theory of discrete regeneration (in continuous time, but with independent identically distributed intervals between regenerations) was fashioned by him using his formalism of renewal equations with Blackwell's renewal theorem; see refs. [7–10].

Feller's greatest achievements were in the theory of Markov processes. In a remarkable paper in 1931, Kolmogorov* had shown that the transition function of a Markov process satisfies certain integrodifferential equations, the best-known instance being the Brownian motion and the heat equation. Consequently, there arose the question of existence and uniqueness of a Markov process corresponding to a given equation. Feller treated such problems in refs. [11] and [12], but the exact relations remained unclear. The problem was to study the parabolic equation $\partial u/\partial t = Lu$, where L is a second-order differential operator on an interval. His idea was to view a boundary condition as a restriction on the domain of L, so that the restricted operator became the infinitesimal generator of an appropriate semigroup. The papers [13–20] settled the main problems, extended them to abstract semigroups, and interpreted the boundary conditions in terms of sample paths. Next, he generalized the notion of differential operators appearing in diffusion theory. He found that every such operator (excepting some degenerate cases) can be written, almost uniquely, in the form $L = (d/d\mu)(d/ds)$, where s is a coordinate function (the scale parameter) and μ is an increasing function (the speed measure). His papers [21–27] reduced the study of the most general diffusion on an interval to that of a Brownian motion. Together with the later work of Dynkin, who explained Feller's results in terms of sample paths, these are rightly considered among the major achievements of modern probability theory.

Finally, Feller was a great expositor and applied probabilist. His book [28] has done more to spread the knowledge of probability than any other book. In the jacket cover of the third edition of vol. 1, G.-C. Rota put it justly:

> . . . one of the great events in mathematics in this century. Together with Weber's *Algebra* and Artin's *Geometric Algebra* this is the finest textbook in mathematics in this century. It is a delight to read and it will be immensely useful to scientists in all fields.

REFERENCES

Central Limit Theorems

[1] Feller, W. (1935). Über den zentralen Grenzwertsatz der Wahrscheinlichkeitsrechnung. *Math. Z.*, **40**, 521–559.

[2] Feller, W. (1943). Generalization of a probability limit theorem of Cramér. *Trans. Amer. Math. Soc.*, **54**, 361–372.

[3] Feller, W. (1945). On the normal approximation to the binomial distribution. *Ann. Math. Statist.*, **16**, 319–329.

[4] Feller, W. (1968). On the Berry-Esseen theorem. *Z. Wahrsch. Verw. Geb.*,**10**, 261–268.

Renewal Theory

[5] Feller, W. (1941). On the integral equation of renewal theory. *Ann. Math. Statist.*, **12**, 243–367.

[6] Feller, W., Erdős, P., and Pollard, H. (1948). A property of power series with positive coefficients. *Bull. Amer. Math. Soc.*, **55**, 201–204.

[7] Feller, W. (1948). Fluctuation theory of recurrent events. *Trans. Amer. Math. Soc.*, **67**, 98–119.

[8] Feller, W. and Orey, S. (1961). A renewal theorem. *J. Math. and Mech.*, **10**, 619–624..

[9] Feller, W. (1961). A simple proof of renewal theorems. *Commun. Pure Appl. Math.*, **14**, 285–293.

[10] Feller, W. (1964). On semi-Markov processes. *Proc. Nat. Acad. Sci. U.S.A.*, **51**, 653–659.

Markov Processes

[11] Feller, W. (1936). Zur Theorie der stochastischen Prozesse (Existenz und Eindeutigkeitssatze). *Math. Ann.*, 113–160.

[12] Feller, W. (1940). On the integro-differential equations of purely discontinuous Markoff processes. *Trans. Amer. Math. Soc.*, **48**, 488–515.

[13] Feller, W. (1951). Two singular diffusion problems. *Ann. of Math.*, **54**, 173–182.

[14] Feller, W. (1952). Diffusion processes in genetics. *Proc. 2nd Berkeley Symp. Math. Statist. Probab.*, pp. 227–246.

[15] Feller, W. (1952). Some recent trends in the mathematical theory of diffusion. *Proc. Int. Congress Math. 1950*, vol. II, pp. 322–339.

[16] Feller, W. (1952). The parabolic differential equations and the associated semigroups of transformations. *Ann. of Math.*, **55**, 468–519.

[17] Feller, W. (1952). On generalization of Marcel Riesz' potentials and the semigroups generated by them. *Commun. Sem. Math. Univ. Lund. Suppl.* (for Marcel Riesz), 73–81.

[18] Feller, W. (1953). Semi-groups of transformations in general weak topologies. *Ann. of Math.*, **57,** 287–308.

[19] Feller, W. (1953). On the generation of unbounded semi-groups of bounded linear operators. *Ann. of Math.,* **58**, 166–174.

[20] Feller, W. (1954). Diffusion processes in one dimension. *Trans. Amer. Math. Soc.*, **77**, 1–31.

[21] Feller, W. (1954). The general diffusion opereator and positivity preserving semi-groups in one dimension. *Ann. of Math.*, **61**, 417–436.

[22] Feller, W. (1955). On second order differential operators. *Ann. of Math.*, **61**, 90–105.

[23] Feller, W. (1955). On differential operators and boundary conditions. *Commun. Pure Appl. Math.*, **3**, 203–216.

[24] Feller, W. (1956). On generalized Sturm–Liouville operators. *Proc. Conf. Differential Equations*, University of Maryland, 1955, pp. 251–290.

[25] Feller, W. (1957). Generalized second order differential operators and their lateral conditions. *Illinois J. Math.*, **1**, 459–504.

[26] Feller, W. (1957). On the intrinsic form for second order differential operators. *Illinois J. Math.*, **2**, 1–18.

[27] Feller, W. (1959). Differential operators with the positive maximum property. *Illinois J. Math.*, **3**, 182–186.

General

[28] Feller, W. (1950–1971). *An Introduction to Probability Theory and Its Applications.* Wiley, New York [vol. 1: 1950, 1957 (2nd ed.), 1968 (3rd ed.); vol. 2: 1960, 1971 (2nd ed.)].

ERHAN ÇINLAR

Gnedenko, Boris Vladimirovich

Boris Vladimirovich Gnedenko.

Born: January 1, 1912, in Simbirsk (now Ul'yanovsk), Russia.
Died: December 30, 1995, in Moscow, Russia.
Contributed to: limit theorems of probability, infinite divisibility, reliability, mathematical statistics, mathematical education.

B. V. Gnedenko was born in Simbirsk, on the Volga river, which is also the birthplace of V. I. Lenin. His family moved to Kazan when he was three years old. He completed his high-school education at the age of fifteen, and was a student at the University of Saratov from 1927 to 1930. He then taught for four years at the Ivanovo Textile Institute. His first publications—on (textile) machinery failures—were written during this period (1930–1934).

Gnedenko was a graduate student, under Khinchin, at Moscow University from 1934 to 1937 and was a lecturer at that university from 1938 to 1945. In 1942 he was awarded a doctorate in physical–mathematical sciences, and given the title of professor in the same year. In 1945 he moved to the University of L'vov in Ukraine. He was elected as a corresponding member of the Ukrainian Academy of Sciences in the same year. He became an Academician of that organization in 1948, and was associated with its Institute of Mathematics in Kiev from 1950 to 1960.

He returned to Moscow University in 1960, and remained there until his death, serving, for many years, as head of the Department of Probability Theory.

During the period 1937–1941, Gnedenko did outstanding work in probability theory, including a complete characterization of infinitely divisible laws and their domains of attraction, and also work on stable laws. These studies led eventually to the famous monograph by Gnedenko and A. N. Kolmogorov,* *Limit Theorems for Sums of Independent Random Variables*, first published in 1949. His seminal paper, in (1943 [1] proved rigorously that there are only three types of limiting distributions for extreme (greatest or least) values in random samples, and characterized the three corresponding domains of attraction. An English translation of this paper, with commentary by R. L. Smith, is included in *Breakthroughs in Statistics*, vol. 1 (Springer-Verlag, 1991, pp. 185–225).

In 1949 there was published Gnedenko's textbook *A Course in Probability Theory*, which has become famous, going through six Russian editions, and being translated into Arabic, English, German, and Polish. Other books written by Gnedenko include:

1. *History of Mathematics in Russia*, written before World War II, but not published until 1946. (In the last years of his life, he was working on a second edition of this book.)

2. With Yu. K. Belyaev and A. D. Solov'yev, *Mathematical Methods of Reliability*. This was published in Russia, in 1965 and has since been translated into several other languages. It has been among the most influential treatises in the field.
3. With I. V. Kovalenko, *Introduction to Queueing Theory*, first published in 1967, with a second edition in 1987. It has been translated into English.
4. A monograph, *From the History of the Science of the Random*, published in 1981.
5. With I. A. Ushakov, *Probabilistic Reliability Engineering* (Wiley, New York, 1995).

Gnedenko's list of publications contains over 200 items. He was a member of the International Statistical Institute (elected in 1959), a Fellow of the Institute of Mathematical Statistics (1966), and an honorary Fellow of the Royal Statistical Society (1964). He was awarded the State Prize of the USSR in 1978, and also the Chebyshev and Lomonozov Prizes.

He was active in the International Statistical Institute in the sixties and seventies, establishing close contact with scientists from (at that time) "Socialist" countries, including Bulgaria, Cuba, and German Democratic Republic, and Poland.

He was—in the words of his colleagues Belyaev, Kolmogorov, and Solov'yev—"a fervent propagandist of progressive methods of training who deals with the philosophical problems of the natural sciences from a Marxist point of view" [*Uspekhi Mat. Nauk*, **27**(4), 175 (1972)]. This assessment was omitted in a similar article ten years later [*Uspekhi Mat. Nauk*, **37**(6), 246 (1982)] and was replaced by a more intimate statement: "The home of Boris Vladimirovich and Natalia Konstantinova, his life-long companion" (who passed away several years before her husband), "is always open to visitors."

In spite of his very important scientific contributions and other activities, Gnedenko was never elected a full member of the Academy of Sciences of the USSR (as was Nemchinov*), although—as already mentioned—he became a full member of the Ukrainian Academy of Sciences at the age of 36. There appears to have been little or no interaction, scientific or personal, between him and the well-known Soviet statistician A. Ya. Boyarskii, even though they were working in the same university (as heads of different departments) for more than 25 years.

Gnedenko's official biography does not note membership of the Communist Party of the Soviet Union, but—having been imprisoned for six months in 1938—he hardly ever deviated in his writings from the current rigid, but unstable, partly line, including or omitting the name "Stalin" as circumstances required.

With the demise of A. N. Kolmogorov in October 1987, Gnedenko became the elder statesman of Soviet probability theory. During his lifetime he trained over 100 doctoral students (from the USSR and Eastern Bloc countries), of whom at least 30 became full professors, and seven obtained the title of Academician.

REFERENCE

[1] Gnedenko, B. V. (1943). Sur la distribution limite du terme maximum d'une série aléatoire, *Ann. of Math.*, **44**, 423–453.

POSTSCRIPT

(The following information was supplied by Professor Sergei A. Aivazian. It relates to the later years of the life of Professor Gnedenko.)

In 1960, Gnedenko returned to Moscow. In 1965, A. N. Kolmogorov invited him to take charge of the Department of Probability Theory at Moscow State University. This department had developed, under Kolmogorov's leadership for some thirty years, the famous "Kolmogorov probabilistic–statistical school" in the USSR. The consequent "second period" of Gnedenko's activities was characterized by a shift in the emphasis of his scientific interests towards applications in the fields of reliability theory and quality control, with special attention to problems of statistical education. He wrote a number of excellent textbooks and methodological manuals on the theory of probability and mathematical statistics, and was very active in the pedagogical process in the University.

On the occasion of his eightieth birthday, an interview with Gnedenko was published in *Probability Theory and its Applications* [**37**(4), 724–746 (1992)]. In this interview he was asked: "What are your goals, as one carrying the torch to continue Kolmogorov's activities?" He replied:

> My first goal is not to allow the outstanding Kolmogorov School to vanish. Times are very difficult—the country is in a deep economic and political crisis. The country seems to have no interest in science, and some people proclaim "let the scientists go abroad and wait there until things improve." They forget that, by doing this, they would create an outflow of the most productive middle generation, which should transmit its knowledge to the younger generation. If this were done systematically, the School would vanish within 10 to 15 years, and it would be impossible to re-establish it. The youth must be in a constant state of scientific excitement; young people must be aware of important unsolved problems, and be educated in a spirit of desire to engage in these problems, and consider them the central problems of their lives. . . . My second goal, which Andrei Nikolaevich [Kolmogorov] asked me to fulfill, is to do my utmost to ensure that mathematical statistics would flourish in our Department, both as mathematics and as a practical discipline, highly important for clarification of the laws of nature and human society.

It was to the solution of these problems that B. V. Gnedenko devoted the final segment of his life.

S. A. AIVAZIAN

Hannan, Edward James

Born: January 29, 1921, in Melbourne, Australia.
Died: January 7, 1994, in Canberra, Australia.
Contributed to: stochastic processes, inference on stationary random processes, time-series analysis, group representations in applied probability, linear systems, statistical education.

Edward James Hannan.

EARLY LIFE

Ted Hannan was one of two twins born in 1921 to James Thomas and Margaret Josephine (née McEwan) Hannan. He attended Xavier College at Kew, Melbourne, where he excelled in his school studies. After taking his leaving certificate (a statewide school diploma) at age 15, he became a clerk in the Commonwealth Bank of Australia (CB). In 1941 he enlisted in the 2nd Australian Imperial Force and saw active service as an infantry lieutenant in New Guinea. On repatriation in 1946, he enrolled under the Commonwealth Reconstruction and Training Scheme for a Bachelor of Commerce degree at the University of Melbourne. For this, he studied ten subjects in economics, two in mathematics, and one in statistics. Having graduated in December, 1948, he married Irene Trott and took up a job in March 1949 at the CB in Sydney (later the Reserve Bank of Australia). He worked there mainly as a statistician, building a model of the Australian economy.

ACADEMIC CAREER

In 1953, Hannan was selected to spend a year in the Department of Economics at the Australian National University (ANU). There, Professor P. A. P. Moran of the Department of Statistics recognized his exceptional talent in mathematics and arranged for his transfer to Statistics. Hannan never left Canberra: after earning his Ph.D. in 1956, he remained as a research fellow in Moran's department until 1958. In 1959, he was appointed professor and head of the Department of Statistics at the Canberra University College, which was soon to become the ANU's undergraduate School of General Studies (SGS). In 1971, to concentrate on his research, he moved back as the second professor in Moran's department at the ANU's postgraduate Institute of Advanced Studies (IAS). On Moran's retirement in 1982, Hannan succeeded him as department head until 1985.

When the Applied Probability Trust (APT), a foundation designed to foster mathematical and probabilistic research, was created by Joe Gani in 1964, Hannan, who had helped with its formation, became one of the four trustees. In June 1964 the APT's *Journal of Applied Probability* was launched. Hannan served as one of the editors of this and its companion *Advances in Applied Probability* (1969) until his death. On his retirement in 1986, the ANU appointed Hannan emeritus professor, with facilities in his old department at the SGS (now called The Faculties). He continued to work there until the very day of his death. During his academic career, Hannan trained 17 graduate students; many of them later became researchers or professors of statistics in Australia, Britain, China, Europe, Japan, and The USA. Between 1953 and 1993, he wrote over 130 papers and four influential books [1–4]. A detailed analysis and complete bibliography of his work appear in Robinson [5]; further biographical details are available in *Who's Who in Australia 1994* [6] and the comprehensive obituary by Gani [7].

Although Hannan never held a permanent academic position other than at the ANU, his involvement with statistics and statisticians throughout the world was considerable. He spent his sabbatical leaves overseas, lecturing at universities and addressing numerous conferences all over the world. At various times he held visiting appointments at American universities such as Brown, Johns Hopkins, MIT, North Carolina, Princeton, and Yale. He also spent periods at the IBM Almaden Research Center in San Jose, the Institute of Applied Mathematics, the Chinese Academy of Science in Beijing, the London School of Economics, and the Technical University of Vienna. These numerous professional contacts and his active editorial role allowed him to take part in the advances of the American and European schools of statistics, and to exert a strong influence on the development of time-series analysis throughout the world.

STATISTICAL CONTRIBUTIONS

Hannan made many important contributions, mainly to inference on stationary random processes and to time-series analysis. His strength lay in the mastery of mathematical methods for the analysis of problems in these areas, many of them arising from realistic models in hydrology, climatology, geophysics, and econometrics. Between 1955 and 1958 his research was concerned with testing for zero autocorrelation. He recognized the inadequacy of finite-sample theory and in ref. [8] began to make comparisons based on the Pitman efficiency. With Watson [9], he examined the results of misspecifying the error spectral density, and in ref. [10] he suggested an asymptotic approximation for Studentizing the sample means of a stationary series. Later, he modified Fisher's test for a jump in the spectral distribution function by considering a nonparametric spectrum under the alternative hypothesis. In 1960 his first book *Time Series Analysis* [1] appeared: it was rapidly translated into Russian and Japanese.

A paper which Hannan always considered to be important was that with Hamon [11]. A frequency-domain weighted least-squares estimate was proposed for combining narrow-band regressions, weighting inversely with smoothed estimates of the nonparametric error spectrum. He later showed that this was asymptotically as efficient as the generalized least-squares estimate in the presence of nonparametric error autocorrelation; this was an early example of adaptive estimation. Over a decade later, adaptiveness to regression error heteroscedasticity of unknown form was demonstrated. Hannan, in Ref. [12] and with Groves [13] and Terrell [14], used the same approach in more elaborate models in oceanography and econometrics. Having first introduced the idea of omitting frequencies to cope with the errors-in-variables problems in 1963, Hannan justified this view with Robinson [15].

Hannan's interest in seasonality began in the early 1960s. Two early papers using operators to estimate seasonality were followed in ref. [16] by the modeling of the seasonal component by a cosinusoid whose coefficients were stationary processes, so that its spectrum had smooth peaks. In a similar model in 1967, the roots of the coefficients lay on the unit circle; this pioneering idea was to open up a fertile field of research. His second book *Group Representations and Applied Probability* [2] was published in 1965. It was followed two years later by ref. [17], his paper presenting a deep description of filters. Both works demonstrated his command of mathematical concepts. In 1970, his third book *Multiple Time Series* [3] appeared. This is his single most important contribution to the field: it included his research over the previous decade, mostly in the multivariate context. The book, later translated into Russian, was considered difficult to read, containing as it did many unsolved research problems among its exercises.

The following decade was devoted to solving several complex problems in the frequency domain. With Thomson [18], Hannan established a central limit theorem for the discrete Fourier transform of vector time series, and later proposed improved estimates of coherence and group delay. In ref. [19], he studied cosinusoidal regression with unknown real-valued frequency, and with Cameron [20] he analyzed noise-corrupted measurements recorded at different locations. He considered the properties of periodograms with Zhao-Guo Chen and Hong-Zhi An [21]. The study of linear ARMA and transfer-function models depends on the identification problem; this is difficult because of lags in the innovations, and possible input as well as output variables. Hannan [22] gave conditions for identification of stationary vector ARMA and ARMAX models of known order. Their properties were clarified through the concept of the McMillan degree in a series of papers, among them ref. [23], with Dunsmuir and Deistler [24], and with Kavalieris [25].

A practical issue in the use of ARMA and ARMAX models is the order of a scalar autoregression. If $\hat{\sigma}_p^2$ is the estimate of the variance of the innovations, p is usually chosen to minimize $\log \hat{\sigma}_p^2 + pC_n$. With Quinn [26], Hannan showed that by setting $C_n \sim 2(\log \log n)/n$, which decreases faster than many previously proposed values of C_n, a consistent estimate of p is obtained. He also proved the consistency of deterministic procedures for obtaining the order of scalar ARMA models, and in ref. [27] he estimated the McMillan degree. Reference [28] provides an excellent overview of the subject. Although Hannan was primarily interested in the mathematical aspects of his research, he had a sharp eye for potential applications. In ref. [29], he used an economic asymptotically efficient method in the frequency domain for the estimation of ARMA models. He extended his study to ARMAX models with Nicholls [30], and developed a recursive ARMA estimation method with Rissanen [31]. This he later modified with Kavalieris [32].

Hannan used a wide range of mathematical techniques, among them Fourier analysis and martingale theory. He pioneered the application of martingale theory to time series in the early 1970s, as in ref. [33], where he improved on earlier asymptotic results. Basic results on asymptotic properties of sample autocovariances were derived with Heyde [34] and with Hong-Zhi An and Zhao-Guo Chen [35]. Hannan's fourth book *The Statistical Theory of Linear Systems* [4], with Deistler, published in 1988, gathered together his work on linear time-series models. After retiring in 1986, he continued to work on the application of stochastic complexity to spectral bandwith with Rissanen [36], to nonparametric density estimation with Hall [37], and on parametric signal filtering using Laguerre polynomials with Wahlberg [38]. He was still collaborating actively with Kavalieris on ARMA models, and with Huang and Quinn on a book on frequency estimation, just before his death.

RECOGNITION

After the publication of his book [1] in 1960, Hannan received international recognition, and was invited to lecture in Britain, China, Europe, Japan, and the USA. He was elected to fellowship of the Australian Academy of Science in 1970, and the Academy of Social Sciences in Australia in 1980. He was also a fellow of the Econometric Society (1967), an honorary fellow of the Royal Statistical Society, and an elected member of the International Statistical Institute. Hannan served on the editorial boards of many journals other than the APT's among them the *Annals of Statistics, Econometrica*, the *International Economic Review*, the *Journal of Forecasting*, the *Journal of Multivariate Analysis*, and the *Journal of Time Series Analysis.*

In 1979, the Australian Academy of Science awarded him its Lyle Medal for distinction in research in Mathematics or Physics, in recognition of his fundamental contributions to the theory of statistical inference. Seven years later, the Statistical Society of Australia presented him with the Pitman Medal, its highest research award for making Australia into a centre of excellence for time series analysis. Also in 1986, for his 65th birthday, he was honored with a Festschrift entitled *Essays in Time Series and Allied Processes*. After his death in 1994, the Australian Academy of Science created the Hannan Medal, an award for distinction in the mathematical sciences.

CONCLUDING REMARKS

Ted Hannan lived a full life and was very much a family man: he and Irene had four children and six grandchildren. He loved literature and was a great admirer of the poet W. B. Yeats; but he also enjoyed Australian Rules football and would barrack hilariously for his favorite team. Statistics, and particularly time-series analysis, owe much to his imaginative insights and his rigorous mathematical methods. He contributed prolifically to the subject, and was the founder of one of Australia's major departments of statistics at The Faculties, ANU.

REFERENCES

[1] Hannan, E. J. (1960). *Time Series Analysis*. Methuen, London (Also published in Russian in 1964 with an appendix by Yu. B. Rozanov, and in Japanese.)

[2] Hannan, E. J. (1965). *Group Representations and Applied Probability*. Methuen, London.

[3] Hannan, E. J. (1970). *Multiple Time Series*. Wiley, New York.

[4] Hannan, E. J. and Deistler, M. (1988). *The Statistical Theory of Linear Systems*. Wiley, New York.

[5] Robinson, P. M. (1994). Memorial article: Edward J. Hannan, 1921–1994. *J. Time Series Anal.*, **15**, 563–576.

[6] *Who's Who in Australia 1994* (1994). Information Australia Group, Melbourne.

[7] Gani, J. (1994). Edward James Hannan 1921–1994. *Hist. Rec. Austral. Sci.*, **10**(2), 173–185.

[8] Hannan, E. J. (1958). The asymptotic powers of certain tests of goodness of fit for time series. *J. R. Statist. Soc.* B, **20**, 143–151.

[9] Hannan, E. J. and Watson, G. S. (1956). Serial correlation in regression analysis. *Biometrika*, **43**, 436–448.

[10] Hannan, E. J. (1957). The variance of the mean of a stationary process. *J. R. Statist. Soc. B*, **19**, 282–285.

[11] Hamon, B. V. and Hannan, E. J. (1963). Estimating relations between time series. *J. Geophys. Res.*, **68**, 6033–6041.

[12] Hannan, E. J. (1967). The estimation of a lagged regression relation. *Biometrika*,**54**, 315–324.

[13] Groves, G. W. and Hannan, E. J. (1968). Time series regression of sea level on weather. *Rev. Geophys.***6**, 129–174.

[14] Hannan, E. J. and Terrell, R. D. (1968). Testing for serial correlation after least squares regression. *Econometrica*, **36**, 133–150.

[15] Hannan, E. J. and Robinson, P. M. (1973). Lagged regression with unknown lags. *J. R. Statist. Soc.* B, **35**, 252–267.

[16] Hannan, E. J. (1964). The estimation of a changing seasonal pattern. *J. Amer. Statist. Ass*, **59**, 1063–1077.

[17] Hannan, E. J. (1967). The concept of a filter. *Proc. Cambridge Phil. Soc.*, **63**, 221–227.

[18] Hannan, E. J. and Thomson, P. J. (1971). Spectral inference over narrow bands. *J. Appl. Probab.*, **8**, 157–169.

[19] Hannan, E. J. (1973). The estimation of frequency. *J. Appl. Probab.*, **10**, 510–519.

[20] Cameron, M. and Hannan, E. J. (1979). Transient signals. *Biometrika*, **66**, 243–258.

[21] An, H.-Z., Chen Z.-G., and Hannan, E. J. (1983). The maximum of the periodogram. *J. Multivariate Anal.*, **13**, 383–400.

[22] Hannan, E. J. (1971). The identification problem for multiple equation systems with moving average errors. *Econometrica*, **39**, 751–665.

[23] Hannan, E. J. (1979). Statistical theory of linear systems. In *Developments in Statistics*, vol. 2, P. Krishnaiah, ed. Academic Press, New York, pp. 83–121.

[24] Deistler, M., Dunsmuir, W., and Hannan, E. J. (1980). Estimation of vector ARMAX models. *J. Multivariate Anal.*, **10**, 275–295.

[25] Hannan, E. J. and Kavalieris, L. (1984). Multivariate linear time series models. *Adv. Appl. Probab.*, **16**, 492–561.

[26] Hannan, E. J. and Quinn, B. G. (1979). The determination of the order of an autoregression. *J. R. Statist. Soc., B*, **41**, 190–195.

[27] Hannan, E. J. (1981). Estimating the dimension of a linear system. *J. Multivariate Anal.*, **11**, 459–473.

[28] Hannan, E. J. (1987). Rational transfer function approximation. *Statist. Sci.*, **2**, 135–161.

[29] Hannan, E. J. (1969). The estimation of mixed moving average autoregressive systems. *Biometrika*, **56**, 579–593.

[30] Hannan, E. J. and Nicholls, D. F. (1972). The estimation of mixed regression, autoregression, moving average and distributed lag models. *Econometrica*, **40**, 529–547.

[31] Hannan, E. J. and Rissanen, J. (1982). Recursive estimation of mixed autoregressive-moving average order. *Biometrika*, **69**, 81–94.

[32] Hannan, E. J. and Kavalieris, L. (1984). A method for autoregressive-moving average estimation. *Biometrika*, **72**, 273–280.

[33] Hannan, E. J. (1973). The asymptotic theory of linear time series models. *J. Appl. Probab.*, **10**, 130–145.

[34] Hannan, E. J. and Heyde, C. C. (1972). On limit theorems for quadratic functions of discrete time series. *Ann. Math. Statist.* **43**, 2058–2066.

[35] An, H.-Z., Chen, Z.-G., and Hannan, E. J. (1982). Autocorrelation, autoregression and autoregressive approximation. *Ann. Statist.*, **10**, 926–936.

[36] Hannan, E. J. and Rissanen, J. (1988). The width of a spectral window. *J. Appl. Probab.*, **25A**, 301–307.

[37] Hall, P. G. and Hannan, E. J. (1988). On stochastic complexity and non-parametric density estimation. *Biometrika*, **75**, 705–714.

[38] Hannan, E. J. and Wahlberg, B. (1993). Parametric signal modelling using Laguerre filters. *Ann. Appl. Probab.*, **3**, 467–496.

J. GANI

Jordan, Karoly (Charles)

Born: December 16, 1871, in Budapest, Hungary.
Died: December 24, 1959, in Budapest, Hungary.
Contributed to: probability theory, calculus of finite differences, geometric probability, demography, elliptic functions, interpolation meteorology.

K. Jordan, together with A. Rényi,* was a founder of the flourishing school of Hungarian probability theory.

Stemming from the family of a well-to-do leather-factory owner, Jordan received his secondary education in the city of his birth, matriculating in 1889. He then studied in Paris (École Préparatoire Monge) and in Zürich Polytechnic, where he was awarded his diploma in chemical engineering in 1893. After spending a year at Owen's College, Victoria University in Manchester, he moved in 1894 to the University of Geneva, where he obtained his Docteur és Sciences Physique in 1895 and was granted the title of private docent for his work in physical chemistry. Returning to Hungary in 1899, he studied mathematics, astronomy, and geophysics at the Pázmány University in Budapest. During the years 1906–1913 he served as the director of the Budapest Institute of Seismology. During World War I he taught mathematical subjects at a military academy. During the years 1920–1950 he was associated with the University of Technical and Economical Sciences in Budapest, becoming a full professor in 1933. He was elected a corresponding member of the Hungarian Academy of Sciences in 1947, and was awarded prizes in 1928 and later in 1956 for his outstanding achievements in mathematics. He also was a fellow of the Royal Statistical Society and a member, or an honorary member, of numerous statistical, mathematical, and meteorological societies.

His first wife, Marie Blumauer—whom he married in Geneva in 1895—lost her life in the birth of their third child in 1899. He remarried to Marthe Lavaleé in 1900. Three more children were born of this marriage. She passed away in July of 1959; and less than six months later, K. Jordan died on Christmas Eve at the age of 88.

During his 30–year tenure at the University of Technical and Economical Sciences in Budapest, he was active in teaching and research and was known for his human warmth and devotion to students—many of whom later on played a prominent role in development of probability and statistics in Hungary and abroad. He was a man of extraordinary integrity, and did not hesitate to condemn injustice, even in the face of adversity. His book on *Mathematical Statistics,* published as early as 1927 almost simultaneously in Hungarian and in French, represents one of the earliest treatises on this topic, providing an up-to-date account of the field, including the author's own results.

His classical treatise on *Calculus of Finite Differences,* originally published in English in Budapest in 1939, was twice reprinted in the USA, and even today, it serves as the basic text on this subject.

In the Introduction to the U. S. edition, H. C. Carver, the founder of the *Annals of Mathematical Statistics,* notes that finite calculus is an important tool in mathematical statistics, and characterizes the text as follows:

> The author has made a most thorough study of the literature that has appeared during the last two centuries on the calculus of finite differences and has not hesitated in resurrecting forgotten journal contributions and giving them the emphasis that his long experience indicated they deserve in this day of mathematical statistics.

His magnum opus, *Chapters on the Classical Calculus of Probability*, appeared in Hungarian in 1956 and was posthumously translated into English by P. Medgyessy—a distinguished contemporary Hungarian probabilist—and was published in 1972 (as a commemoration of K. Jordan's hundredth birthday). The book summarizes the results of 50 years of research. It is a scholarly volume, and a gold mine of information for both theoretical probabilists and applied statisticians. Jordan's personal library of some 5000 volumes, including 1000 rare books, was destroyed during the Hungarian revolution in October of 1956. While in the hospital—recovering from a mild heart attack which resulted from destruction of his home and valuable possessions—the 85–year-old Jordan was working on printing errors that crept into his newly published book. A brief quotation from his chapter on *Probabilistic Theorems* (p. 193) characterizes the lucidity of the author's style and his ability to emphasize and clarify profound fundamental concepts:

> Players of roulette note the results diligently in order to choose the moment in which—according to their belief—their chances of winning are greater. It would be difficult to convince them that this is useless; their chance always remains unaltered. Their view, that the player has, over the bank, the advantage of being able to choose the right moment for beginning the game, — is erroneous.
>
> The chief reason for the false belief of the players is the belief in *equalization*. According to this, e.g. in case of a great number of observations red occurs just as many times as black. This belief arises from the misunderstanding of the empirical postulate according to which every event occurs in a number which is approximately proportional to its probability; if the probabilities are equal then they occur approximately equally often. Every one is aware of this, even if he has never dealt with the Calculus of Probability. The error lies in the interpretation of the word "approximately"; the difference between the number of reds and blacks, i.e. the *arithmetic deviation*, does not approach zero (it increases infinitely), but the difference of their logarithms or, in other words, the number of reds divided by that of blacks, i.e. the *geometric deviation* approaches unity; but this does not help the players.

A complete bibliography of K. Jordan's works (90 items) appears in the March issue of Vol. 32 (1961) of the *Annals of the Mathematical Statistics,* with an invited obituary by his ex-student, L. Takacs, from which this entry is adapted. The bibliography is reprinted with minor changes in the beginning of the English version of the *Chapters of the Classical Calculus of Probability*.

Kolmogorov, Andrei Nikolayevich

Born: April 25, 1903, in Tombov, Russia (about 400 miles southeast of Moscow).
Died: October 20, 1987, in Moscow, USSR.
Contributed to: history, philosophy and foundations of mathematics, set theory, measure theory, probability theory, stochastic processes, functional analysis, information theory, statistics, approximation theory, mathematical linguistics, crystallography, theory of turbulent flow, classical dynamics.

A. N. Kolmogorov, one of the most influential mathematicians of the twentieth century, was born in Tombov, where his mother, Mariya Yakovlevna Kolmogorova, stopped on a journey to the Crimea. She died in childbirth, and his aunt, Vera Yakovlevna Kolmogorova, took care of him in his childhood. His father was a learned agronomist, the son of a priest, and on his mother's side he was of gentry origin. In the fall of 1920, Kolmogorov enrolled in Moscow University. Initially, his main interest was in Russian history of the fifteenth and sixteenth centuries. In 1922, he became a student of V. V. Stepanov (an expert in trigonometric series) and the famous Russian mathematician, N. N. Luzin. In June of 1922, at the age of 19, he constructed an example of a Fourier–Lebesgue series everywhere divergent, and then a Fourier–Lebesgue series divergent at each point. These were novel and unexpected results.

We now briefly survey Kolmogorov's contributions to probability theory and mathematical statistics. In 1924, he started a long and fruitful collaboration with A. Ya. Khinchin. They, jointly, obtained necessary and sufficient conditions for convergence of series whose terms are mutually independent random variables. In 1928, Kolmogorov obtained necessary and sufficient conditions for the law of large numbers, and in 1929 he proved the law of the iterated logarithm for sums of independent random variables under very general conditions. (Earlier this law had been proved by Khinchin, first for Bernoulli and then for Poisson variables.) In the same year was obtained the remarkable necessary and sufficient condition for the strong law of large numbers—namely, the existence of a finite mathematical expectation for the summands. In 1933, his classical and famous monograph on *Foundations of the Calculus of Probability* was published in German; it was translated into Russian in 1936 and into English in 1950 and 1956. A second Russian edition appeared in 1974. In 1931, in his paper, "*Analytical Methods of Probability Theory*," the foundations of the modern theory of Markov processes were laid and relations between probability theory and the theory of differential equations were revealed. In late 1939, Kolmogorov developed statistical methods in the theory of turbulent fluid flow, and in 1941 he worked on interpolation and extrapolation of stationary time series. This, together with (but independently of) Wiener's* work, developed later into the Kolmogorov–Wiener theory of stationary processes and still later into the theory of branching processes. His monograph, with B. V. Gnedenko,* on *Limit Theorems for Sums of Independent Random Variables*, appeared in 1949. His interest in information theory in two articles in 1965 and 1969, published in the Russian journal *Problems of Information Transmission*, and in a survey paper in *Uspekhi Mat. Nauk*, **3**(4), 27–36, (1983). [*See also IEEE Trans. Inf. Theory*, **IT-14**, 662–664, (1968).]

His contributions to statistics include, among others, works on the least-squares method, which influenced Yu. V. Linnik's* work on this topic; on empirical distributions (the classical Kolmogorov statistic

$$D_n = \sqrt{n} \sup_t |F_n(t) - F(t)|$$

and the Kolmogorov–Smirnov test); on unbiased estimators; on statistical forecasting; on tables of random numbers; on Wald's identity; on measures of dependence (generalizations of correlation coefficients); on methods of medians in the theory of errors; on Mendel's theory; on the lognormal distribution; and on the estimation of parameters of a process related to a model of the earth's rotation. His pedagogical work began in 1922 and flourished in his capacity as the director of the Scientific Research Institute of Mathematics at Moscow University. His famous seminars in probability theory at Moscow University inspired a great many young Soviet mathematicians and statisticians; and a great majority of the contemporary leading Soviet scientists in this area, as well as a number of foreigners (including Parthasarathy, Vere-Jones, and Martin-Löf), were among his students. He was an Academician of the USSR from 1939 and was an honorary member of over 20 foreign scientific organizations, including the Royal Statistical Society, the International Statistical Institute, the U.S. Academy of Sciences, and the Paris Academy of Sciences. His professional activity was centered on Moscow State University (MSU), with which he was associated for over 65 years (1920–1987), being a faculty member for over 60 years (from 1925). Among his activities at MSU was the establishment of the Statistical Laboratory in 1961.

BIBLIOGRAPHY

Bogolyubov, N. N., Gnedenko, B. V., and Sobolev, S. L. (1987). A. N. Kolmogorov: On his 80th birthday. In *A. N. Kolmogorov's Selected Works*, vol. 3. Nauka, Moscow.

Anon. (1987). Kolmogorov, A. N. 1903–1987. *IMS Bull.*, **16**, 324–325. (Reprinted with permission of *The Times*, (London), October 26, 1987.)

Liapunov, Alexander Mikhailovich

Born: June 7 (N. S.), 1857, in Yaroslavl, Russia.
Died: November 3, 1918, in Odessa, USSR.
Contributed to: equilibrium-stability theory, probability theory.

Liapunov entered high school in 1870, having received his elementary education at home, initially from his father, who had been until 1855 an astronomer at Kazan University. He entered Petersburg University in 1876 at the flowering of the Petersburg mathematical "school" founded by Chebyshev,* who exerted a profound influence on the direction of Liapunov's academic development. This period saw the beginning of his association with A. A. Markov.* In 1885 he took up an academic appointment at Kharkov University, where he was heavily engaged in the teaching of mechanics, and was active in the affairs of the Kharkov Mathematical Society, as were his students V. A. Steklov and N. N. Saltykov. He left Kharkov for St. Petersburg in 1902, on his election to its Academy of Sciences, to work within its framework, but counted his years in Kharkov as the happiest of his life. Honors bestowed on him included foreign membership of the Academy of the Lincei in Rome. His life ended in suicide, following the death of his wife from tuberculosis in Odessa, where they had gone for her health, and where he had a brother.

Liapunov's interest in probability was stimulated by his attendance at Chebyshev's lectures in 1879–1880 on his not completely rigorous proof of the central limit theorem for independent (but not identically distributed) summands. Chebyshev's proof was modified in 1898 by Markov, and this modification, together with the fact that in his last years in Kharkov Liapunov taught probability theory, seems to be responsible for his papers on the central limit problem in 1900–1901. Although his few published writings on probabilistic topics, completely described in ref. [1], are confined to this problem and to a single year, they culminate in that very general version of the central limit theorem known as Liapunov's theorem [4], which has had a profound effect on probability theory. The first rigorous proof of a central limit theorem by extensive use of characteristic functions (as is the case with Liapunov's proof) should be ascribed to Cauchy* and I. V. Sleshinsky (see ref. [2]), but Liapunov's approach and conclusion still have a number of remarkably novel features.

Let X_i, $i \geq 1$, be a sequence of independent random variables such that $\mu_i \equiv EX_i$ and $E|X_i - \mu_i|^{2+\delta}$ (for some $\delta > 0$ independent of i) are both well defined for $i \geq 1$. Then, writing $B_n^2 = \text{var } S_n$, where $S_n = \sum_{i=1}^n X_i$, Liapunov's theorem states that

$$L_n^2 \equiv \frac{\sum_{i=1}^n E|X_i - \mu_i|^{2+\delta}}{B_n^{2+\delta} \to 0} \Rightarrow F_n(x) \to \Phi(x)$$

uniformly for all x, as $n \to \infty$, where $F_n(x) = P[(S_n - ES_n)/B_n \leq x]$, $\Phi(x) = (2\pi)^{-1/2}\int_{-\infty}^{x} \exp(-y^2/2)\, dy$. The proof, in terms of discrete random variables, culminates in showing that $\Omega_n = \sup_x |F_n(x) - \Phi(x)|$ is bounded by a quantity that approaches zero as $n \to \infty$. In the case $\delta = 1$ he obtains $\Omega_n \leq \text{const} \cdot |L_n \log L_n|$; it follows that for identically distributed X_i, $\Omega_n \leq \text{const} \cdot \log n/\sqrt{n}$. Thus he not only satisfied Chebyshev's criterion (of rigorous proof by estimation of the error of approximation), but also initiated the topic of *rate of convergence* to a limit distribution. Liapunov's bounds have subsequently been extended and improved by many workers. In the course of the proof he obtains the inequality for $l > m > n \geq 0$ for the absolute moments of a random variable X:

$$(E|X|^m)^{l-n} \leq (E|X|^n)^{l-m}(E|X|^l)^{m-n},$$

which in the case $n = 0$ is usually known as Liapunov's inequality, even though it was known to Bienaymé* [2].

Finally worth mentioning is the manuscript [7], pertaining to the classical linear model of full column rank r where the residuals ϵ_i, $i = 1, \ldots, N$, are independently distributed with var $\epsilon_i = k_i \mu^2$ (k_i assumed known), in which it is shown that $\hat{\mu}^2 = \Sigma_i(\hat{\epsilon}_i^2/k_i)/(N - r)$ is consistent for μ^2 as $N \to \infty$, where ϵ_i, $i = 1, \ldots, N$, are estimated residuals from a weighted least-squares fit, provided that $E(\epsilon_i^4)/[E(\epsilon_i^2)]^2$ is uniformly bounded for all $i \geq 1$. This "ratio" condition resembles in nature his conditions in his central-limit work; and he uses Markov's inequality in the deduction.

REFERENCES

[1] Gnedenko, B. V. (1959). *Istor. Mat. Issled.*, **12**, 135–160 (in Russian).

[2] Heyde, C. C. and Seneta, E. (1977). *I. J. Bienaymé: Statistical Theory Anticipated.* Springer-Verlag, New York.

[3] Krylov, A. N. (1919). *Ross. Akad. Nauk*, **13**, 6th Ser., 389–394 (in Russian). (Obituary.)

[4] Liapunov, A. M. (1901). Nouvelle forme du théorème sur la limite des probabilités. *Mém. Acad. Imp. Sci. St. Pétersbourg*, **12**(5), 1–24. [Russian translation in refs. [5] (pp. 219–250) and [6] (pp. 157–176).]

[5] Liapunov, A. M. (1948). *Izbrannie Trudy*. AN SSSR, Moscow. (Selected works, in Russian, with extensive commentaries and bibliographical material. Contains a biographical sketch and survey of Liapunov's nonprobabilistic work, based on refs. [3] and [8], by Smirnov, with two photographs.)

[6] Liapunov, A. M. (1954). *Sobranie Sochineniy,* vol. 1. AN SSSR, Moscow. (The first of four volumes of collected works in Russian, published 1954–1965; also contains Liapunov's three earlier memoirs on the central limit problem.)

[7] Liapunov, A. M. (1975). On the formula of Gauss for estimating the precision of observations (in Russian). *Istor. Mat. Issled.*, **20**, 319–322. (Commentary on this undated and unpublished manuscript discovered by O. B. Sheynin is given by him on pp. 323–328.)

[8] Steklov, V. A. (1919). *Izv. Ross. Akad. Nauk*, **13**, 6th Ser., 367–388 (in Russian). (Obituary.)

FURTHER READING

See the following work, as well as the references just given, for more information on the life and work of Liapunov.

Adams, W. J. (1974). *The Life and Times of the Central Limit Theorem.* Kaedmon, New York. (Pages 84–98 contain a biographical sketch and a survey of the approach and results of Liapunov's four memoirs on the central limit problem, following ref. [1]. There are also two photographs of Liapunov.)

E. SENETA

Linnik, Yurii Vladimirovich

Born: January 8, 1915, in Belaya Tserkov', Ukraine.
Died: June 30, 1972, in Leningrad, USSR.
Contributed to: analytical and ergodic number theory, limit theorems of probability theory, arithmetic of probability distributions, estimation and hypothesis testing, applications of statistical methods.

Subsequent to the birth of Yurii Vladimirovich Linnik, his father Vladimir Pavlovich Linnik became a famous designer of precision optical instruments and a full member of the Russian Academy of Sciences; his mother Mariya Abramovna Yakerina taught mathematics and astronomy in high school.

In 1932 Linnik entered the University of Leningrad, wanting to become a theoretical physicist, but three years later he understood that his real interests were in mathematics and he switched his major.

As a student he began profound research in the arithmetic of quadratic forms that he continued in graduate school. For his dissertation "On representation of large integers by positive ternary quadratic forms" he received in 1940 a Doctor of Science degree, much higher than that usually awarded upon successful completion of graduate studies, in recognition of outstanding results obtained in the dissertation.

From 1940 to his death in 1972 (with a break in 1941–1943 when he was a platoon commander during World War II) Linnik worked at the Leningrad Division of the Steklov Mathematical Institute (known by its Russian acronym LOMI) of the Russian Academy of Sciences, where he organized and headed the Laboratory of Statistical Methods. From 1944 on he was also a professor of mathematics at the University of Leningrad.

Linnik began his research in probability in the late 1940s after he had already obtained first-class results in number theory. His expertise in the analytical methods of number theory turned out to be very useful in his first papers on the rate of convergence in the central limit theory for independent symmetric random variables [1], where a strong nonuniform estimate with the best possible constant was obtained, and on the central limit theorem for nonhomogeneous Markov chains [2].

A series of Linnik's papers in the early 1960s, later summed up in the monograph [3] with Ibragimov, dealt with the probability of large deviations for the sums of independent identically distributed random variables. Here he introduced new methods that made possible treating random variables beyond the Cramér condition.

Linnik demonstrated his analytic power in his research in the arithmetic of probability distributions, a chapter of the probability theory of functions dealing with components (with respect to the convolution) of distribution functions. The first classical result here is due to Cramér,* who proved in 1936 that the normal distribution has only normal components. A little later Raikov proved that the Poisson distribution has only Poisson components. Linnik proved [4] that components of the composition of the normal and Poisson distributions are of the same form, a real achievement given that the characteristic functions of the normal and Poisson distributions behave absolutely differently. This and many other profound results on the arithmetic of probability distributions were summed up in the monograph [5] with Ostrovskii.

Linnik's first publications in statistics go back to the early 1950s. In an analytical masterpiece [6] he studied the phenomenon of the identical distribution of two linear forms in independent identically distributed random variables. This had been studied by Marcinkiewicz in 1938 under the condition that the random variables have finite moments of all orders. It turned out that without this condition the problem becomes much more difficult, requiring new methods that later were used in other analytical problems of probability and statistics.

A series of Linnik's papers deals with the problem of reconstructing the distribution of a population from that of certain statistics. These and other results formed what were later termed "characterization problems." The mongraph [7] with Kagan and Rao contains a systematic study of them.

Difficult mathematical problems of statistical origin always challenged Linnik. He was successful in proving (with Romanovskii and Sudakov) that in the classical Behrens–Fisher problem for two normal samples there exists a nonrandomized similar test of any size within a natural subalgebra of sufficient statistics if the sample sizes are of different parity; the other half of the problem is still open. He was proud when he noticed that considering the ratio of the variances as a complex number makes possible the use of the theory of analytic functions in analyzing regularity properties of similar tests in the Behrens–Fisher and other problems with nuisance parameters. In the monograph [8] a systematic study is carried out of the analytical approach to statistical problems with nuisance parameters.

Linnik's interest in applications of statistical methods to real-life problems was genuine and strong. His well-known monograph [9] on the method of least squares arose from a problem in geodesy.

He was one of the pioneers in applications of mathematical statistical methods to the analysis of polished-surface quality. It was Linnik's idea [10, 11] to consider a surface profilogram as a realization of a random process, an approach that made possible the study of relations between different characteristics of surface quality.

Linnik published ten monographs and over two hundred research papers. His contributions to mathematics were recognized by the highest awards of Russia (the state prize in 1947, the Lenin prize in 1970, and election to full membership in the Academy of Sciences in 1964.) He was a member of the Swedish Academy of Sciences, Doctor Honoris Causa of University of Paris, and a fellow of the International Statistical Institute.

One of the three volumes of *Selected Papers* by Linnik published by the Russian Academy of Sciences contains fifteen of his papers in probability and statistics [12].

Linnik's interests outside mathematics embraced literature, especially poetry and memoirs, and history. Fluent in seven languages, he wrote witty verses in Russian, German, and French and was an expert in military history. A famous Russian joke, "The (Communist) party line is straight, since it consists only of inflection points," which became a part of folklore, originated with Linnik.

However, mathematics was Linnik's real passion. Everybody who worked with him for any length of time remembers days, weeks, and even months with no rest to himself or to his collaborator till the research was finished. If Linnik was impressed by somebody's result, he became its passionate popularizer, and the author could rely on Linnik's support.

Linnik's life coincided with difficult years for Russia (and the Soviet Union) as a country and for Russian mathematics in particular. Linnik was one of those who should be credited for the fact that the status of probability/statistics in Russia was better than in some other fields.

REFERENCES

[1] Linnik, Yu. V. (1947) (in Russian). *Izv. Akad. Nauk SSSR, Ser. Mat.,* **11**, 111–138. (Reproduced in [12].)

[2] Linnik, Yu. V. (1949) (in Russian). *Izv. Akad. Nauk SSSR Ser. Mat.,* **13**, 65–94. (Reproduced in [12].)

[3] Ibragimov, I. A. and Linnik, Yu. V. (1965) (in Russian). *Independent and Stationary Dependent Random Variables.*Nauka, Moscow (English translation: American Mathematical Society, Providence, RI, 19)

[4] Linnik, Yu. V. (1957) (in Russian). On decomposition of the composition of Gaussian and Poisson distributions. *Teor. Veroyatn. i Primen.*, **2,** 34–59. (Reproduced in [12].)

[5] Linnik, Yu. V. and Ostrovskii, I. V. (1972) (in Russian). *Decomposition of Random Variables and Vectors*. Nauka, Moscow. (English translation: American Mathematical Society, Providence, RI, 1977).

[6] Linnik, Yu. V. (1953) (in Russian). *Ukrainskii Mat. Zh.,* **5**, 207–243 (Part I), 247–290 (Part II).

[7] Kagan, A. M., Linnik, Yu. V., and Rao, C. R. (1972). *Characterization Problems of Mathematical Statistics* (in Russian). Nauka, Moscow. (English translation: Wiley, New York, 1973.)

[8] Linnik, Yu. V. (1968). *Statistical Problems with Nuisance Parameters* (in Russian). Nauka, Moscow. (English translation: American Mathematical Society, Providence, RI, 1968.)

[9] Linnik, Yu. V. (1962). *Method of Least Squares and of the Theory of Processing of Observations* (in Russian), 2nd ed. Fizmatgiz, Moscow. (English translation of the 1st Russian ed.: Pergamon Press, Oxford/London/New York/Paris, 1961.)

[10] Linnik, Yu. V. and Khusu, A. P. (1954) (in Russian). *Ingenernyi Sb. Akad. Nauk SSSR,* **20**, 154–159.

[11] Linnik, Yu. V. and Khusu, A. P. (1954). *Quality of Surfaces under Processing* (in Russian). Moscow and Leningrad, pp. 223–229.

[12] Linnik, Yu. V. (1981). *Selected Papers. Probability Theory* (in Russian). Nauka, Moscow.

SERGEI A. AIVAZIAN
ABRAM M. KAGAN

Markov, Andrei Andreevich

Born: June 2 (o.s.), 1856, in Ryazan, Russia.
Died: July 20, 1922, in Petrograd, USSR (now St. Petersburg, Russia)
Contributed to: number theory, analysis, calculus of finite differences, probability theory, statistics.

Markov entered the fifth Petersburg gymnasium in 1866, and was a poor student in all but mathematics. He revealed even at this educational level a rather rebellious nature, which was to manifest itself in numerous clashes with the czarist regime and academic colleagues.

Entering Petersburg University in 1874, he attended classes given by P. L. Chebyshev,* A. N. Korkin, and E. I. Zolotarev. On completion of his university studies at the physicomathematical faculty in 1878, he received a gold medal and was retained by the university to prepare for a career as an academic. His master's and doctoral theses were in areas other than probability, but with the departure of Chebyshev from the university in 1883 Markov taught the course in this subject, which he continued to do yearly.

Markov was first elected to the St. Petersburg Academy of Science in 1886 at the proposal of Chebyshev, attaining full membership in 1896. Retiring from the university in 1905, he continued to teach probability theory there, and seeking to find practical applications for this, his then prime interest, he participated from the beginning in deliberations on the running of the retirement fund of the Ministry of Justice. He also took a keen interest in the teaching of mathematics in high schools, coming into conflict yet again with his old academic adversary, P. A. Nekrasov, an earlier error of whose had stimulated Markov to give thought to the weak law of large numbers (WLLN) for dependent random variables, and led indirectly to the probabilistic concept of Markov chains.

Markov was a member of the St. Petersburg mathematical "school" founded by Chebyshev and, with Liapunov,* became the most eminent of Chebyshev's disciples in probability. Their papers placed probability theory on the level of an exact mathematical science. Markov's doctoral dissertation, "On some applications of algebraic continued fractions," results from which were published in 1884 in several journals, already had implicit connections with probability theory in that he proved and generalized certain inequalities of Chebyshev [11] pertaining to the theory of moments, and in turn based on notions of Bienaymé.* The method of moments had been the tool used by Chebyshev in his development of the proof of a central limit theorem for not necessarily identically (but independently) distributed summands. Unlike Liapunov's later approach to this problem, by characteristic functions, Markov's continued to be focused on the method of moments; it is often said that he was Chebyshev's closest student and the best expositor of his teacher's ideas.

The stream of Markov's work in probability was initially, in fact, motivated by Chebyshev's [12] treatment of the central limit problem. In letters [2] to A. V. Vasiliev he notes that the significance of Chebyshev's result is obscured by the complexity and insufficient rigor of proofs, a situation he desires to correct. Chebyshev had asserted the central limit theorem held for random variables $U_1, U_2, \ldots$, each described by a density, if (i) $EU_i = 0$, each i; and (ii) $|E(U_1^k)| \leq C$ for each i and positive integer $k \geq 2$ where C is a constant independent of i and k. In the letters Markov states that a further condition needs to be added to make the theorem correct, and suggests this can be taken to be: (iii) $(\sum_{i=1}^n \text{Var } U_i)/n$ is uniformly bounded from zero in n. In the very last part of his actual correction [3] of Chebyshev's theorem, he replaces (iii) by the stronger (iiia): EU_n^2 is bounded from zero as $n \to \infty$. Liapunov's theorem published in 1901 differs not only in its approach (by characteristic functions) but in its much greater level of generality. This evidently led Markov to wonder whether the method of moments might not be suitably adapted to give the same result, and he finally achieved this in 1913 in the third edition of *The Calculus of Probabilities* [5]. The central idea is that of truncation of random variables, much used in modern probability theory: For a random variable U_k' put $U_{k'} = U_k$ if $|U_k| < N$ and $U_k' = 0$ if $|U_k| \geq N$. The random variables U_k', $k = 1, 2, \ldots$, being confined to a finite interval, are easily tractable. The effects of the truncation can be eliminated by judicious manipulation as $N \to \infty$.

Markov's outstanding constribution to probability theory was the introduction of the concept of a Markov chain as one model for the study of dependent random variables, which is at the heart of a vast amount of research in the theory of stochastic processes. Markov chains first appear in his writings in 1906 [4], in a paper which is primarily concerned with extending the WLLN to sums of dependent random variables. The motivation for this was twofold: First, to show that Chebyshev's approach to this problem, via the celebrated Bienaymé–Chebyshev moment inequality but only for independent random variables, could be taken further. The second motivation is less well known but the more relevant. P. A. Nekrasov in 1902 had noticed that "pairwise independence" (i.e., random variables uncorrelated pairwise) would yield the WLLN by Chebyshev's reasoning, but claimed erroneously, on mysterious grounds, that this was not only sufficient but necessary for the WLLN to hold [7, 9]. Markov's deductions (to the contrary) are based on the inequality in the form

$$\Pr\{|S_n - ES_n|/n \geq \epsilon\} \leq \frac{\text{Var } S_n}{(n\epsilon)^2} = \frac{1}{\epsilon^2}\left\{\frac{\sum_{i=1}^n \text{Var } X_i}{n^2} + \frac{2}{n^2}\sum_{i<j} \text{Cov}(X_i, X_j)\right\},$$

which motivated further work on the WLLN by Chuprov* and Slutskii* in a similar vein. Although Markov's interest in sequences of random variables forming a Markov chain was largely confined to investigating the WLLN and the central limit theorem, the paper [4] nevertheless also contains an elegant treatment of the crucial averaging effect of a finite row-stochastic matrix **P** (applied to a vector **w** to produce a vector **z**: **z** = **Pw**) at the heart of the ergodicity theory of finite inhomogeneous Markov chains [10], which Markov also uses to establish ergodicity for a finite homogeneous chain.

Markov's statistical work in part consisted of modelling the alternation of vowels and consonants in several Russian literary works by a two-state Markov chain, and his work in dispersion theory, which significantly influenced Chuprov [7]. An account of most aspects of his work may be found in his textbook [5], which was outstanding in its time, and served to influence figures such as S. N. Bernstein,* V. I. Romanovsky, and Jerzy Neyman*—and through him the evolution of mathematical statistics. Markov was also interested, as regards statistics, again through the influence of Chebyshev, in the classical linear model which he treated in his textbook in editions beginning in 1900, allowing different variances for the independently distributed residuals. The misnamed Gauss–Markov theorem appears to emanate from this source through the agency of Neyman [8]; there is little originality in Markov's treatment of the topic. The Markov inequality, $\Pr\{X \geq \epsilon\} \leq (EX)/\epsilon$ for a random variable $X \geq 0$ and $\epsilon > 0$, first occurs in the third edition (1913) of the textbook.

REFERENCES

[1] Maistrov, L. E. (1974). *Probability Theory. A Historical Sketch*. Academic Press, New York. (Translated and edited from the Russian-language work of 1967 by S. Kotz. Contains an extensive account of the work of Chebyshev, Markov, and Liapunov.)

[2] Markov, A. A.)1898). *Izv. Fiz.-Matem. Obsch. Kazan Univ*., (2 ser.) **8**, 110–128. (In Russian. Also in ref. [6], pp. 231–251].)

[3] Markov, A. A. (1898). *Izv. Akad. Nauk S.-P.B.* (5) **9**, 435–446. (Also in ref. [6, pp. 253–269].)

[4] Markov, A. A. (1906). *Izv. Fiz.-Matem. Obsch. Kazan Univ*. (2) **15**, 135–156. (In Russian. Also in ref. [6, pp. 339–361].)

[5] Markov, A. A. (1924). *Ischislenie Veroiatnostei*, 4th, posthumous ed. Gosizdat, Moscow. (Markov's textbook: *The Calculus of Probabilities*.)

[6] Markov, A. A. (1951). *Izbrannie Trudy*. AN SSSR, Leningrad. [Selected works on number theory and the theory of probability. Pages 319–338 contain Markov's proof, taken from the 3rd ed. of ref. [5], of Liapunov's theorem. There is a biography (with two photographs) by Markov's son, and commentaries by Yu. V. Linnik, N. A. Sapogov, and V. N. Timogeev, as well as a complete bibliography.]

[7] Ondar, Kh. O., ed. (1981). *Correspondence between Markov and Chuprov*. Springer-Verlag, New York. (An English translation by Charles and Margaret Stein, with an Introduction by Jerzy Neyman, of: *O Teorii Veroiatnostei i Matematicheskoi Statistike*. Nauka, Moscow.)

[8] Seal, H. L. (1967). *Biometrika*, **54**, 1–24.

[9] Seneta, E. (1979). *Austral. J. Statist.*, **21**, 209–220. (Section 5 outlines less well-known aspects of work on the WLLN and central limit problem in prerevolutionary Russia.)

[10] Seneta, E. (1981). *Non-negative Matrices and Markov Chains*, 2nd ed. Springer-Verlag, New York, pp. 80–91.

[11] Tchébichef, P. L. (1874). *Liouville's J. Math. Pures Appl.*(2) **19**, 157–160. [Also in his *Oeuvres*. A. Markov and N. Sonin, eds., Chelsea House, New York, 2 vols.]

[12] Tchébichef, P. L. (1887). Supplement to *Zapiski Imp. Akad. Nauk* (S.P.-B.), **55**, No. 6. [Also in *Acta Math.*, **14** (1890–1891), 305–315, and in his *Oeuvres*—see ref. [11].]

[13] Uspensky, J. V. (1937). *Introduction to Mathematical Probability*. McGraw-Hill, New York. (Contains an English-language discussion of less well-known contributions by Markov.)

E. SENETA

Ramsey, Frank Plumpton

Born: February 22, 1903, in Cambridge, England.
Died: January 19, 1930, in Cambridge, England.
Contributed to: foundations of probability, foundations of mathematics.

The following notes draw largely on a much fuller account by Newman [4]. All otherwise unascribed quotations are from that article.

Frank Plumpton Ramsey came from an academic background. "His father was a mathematician, Fellow and later President of Magdalene College . . . and his brother Michael became Archbishop of Canterbury. He was educated at Winchester and at Trinity College Cambridge, and was a Scholar of both those ancient foundations."

A major influence in Ramsey's life was his friendship with the Austrian philosopher Ludwig Wittgenstein. While still an undergraduate, Ramsey was the main translator of the German text of Wittgenstein's Tractatus [6] into English. The author was pleased with the translation, and the two became close friends, following a visit by Ramsey to Austria, in September 1923, to discuss the meanings of certain passages in the book. "The eccentric philosopher and the brilliant undergraduate hit it off immediately."

However, a letter from Ramsey to Wittgenstein, late in 1923, contains the following passage, indicating that there were some difficulties in Ramsey's life: "But I am awfully idle; and most of my energy has been absorbed since January by an unhappy passion for a married woman, which produced such psychological disorder, that I nearly resorted to psychoanalysis, and should probably have gone at Christmas to live in Vienna for nine months and be analyzed, had not I suddenly got better a fortnight ago, since when I have been happy and done a fair amount of work."

Nevertheless, in 1924 Ramsey did "spend six months in Vienna in psychoanalysis (rarer then than now)." This seemed to have some effect. In the fall of 1924, "he became a Fellow of King's College and University Lecturer in Mathematics and soon afterwards married Lettice Baker, who had been a student in the Moral Sciences Tripos."

Ramsey worked in both the foundations of mathematics and the foundations of probability. Here we are primarily concerned with the latter, although this work was essentially couched in terms of abstract mathematics, despite the use of such apparently "applied" terms as "risk," "utility," "valuations," etc.

Ramsey's work on the foundations of probability is set out in Chap. VII ("Truth and Probability") of ref. [5], published posthumously. Zabell [7] provides a penetrating analysis of this chapter, examining in particular its relation to earlier work by Keynes [3]. Following a summary of Keynes' theory of probability, Ramsey developed an approach based on concepts of "mathematical expectations, probabilities and valuations." "Essentially, given any two" of these three concepts, "the remaining one follows more or less naturally." In contrast to earlier writers, Ramsey

> effectively bootstrapped both the valuations *and* the probabilities from mathematical expectation, at the small cost of (a) a very general assumption about preferences; (b) an assumed existence of a certain kind of event; and (c) a further principle, original

with him, that no agent's subjective probabilities should be inconsistent. To be inconsistent means that: "He could have a book made against him by a cunning better and would then stand to lose in any event": this no-win situation is now usually called a Dutch book.

Ramsey did not prove the existence of valuations and probabilities constructed from his system of axioms. Newman [4] outlined a proof based on Davidson and Suppes [2].

Zabell [7] gives a detailed account of the concepts introduced by Ramsey—noting, in particular, an "*ethically neutral proposition:* the philosophical equivalent of tossing a coin." Zabell also notes that Ramsey avoids use of the "principle of indifference" —the choice of "equally likely" outcomes—which was regarded as a desirable accomplishment.

Ultimately, Ramsey's ideas did not have a notable influence on ideas about ways of formalizing the intuitive concept of probability. This was not a rapid process. Broadbent [1], in the course of an article mainly devoted to Ramsey's work in mathematical logic, dismissed his studies on the foundations of probability as "interesting, if not altogether convincing." This may well be a not unreasonable, though somewhat perfunctory, assessment of Ramsey's contributions, insofar as he did not always write as lucidly as one might wish. Perhaps because of this, his ideas did not receive a great deal of attention, even among the limited circle of cognoscenti, until some years after his death.

Ramsey died in 1930, at the early age of 26, from the effects of jaundice. The philosopher Wittgenstein, whose association with Ramsey we have already noted, "was at Ramsey's bedside in the hospital until a few hours before he died."

REFERENCES

[1] Broadbent, T. A. A. (1975). Ramsey, Frank Plumpton. In *Dictionary of Scientific Biography.* Scribner's, New York, vol. 11, pp. 285–286.

[2] Davidson, D. and Suppes, P. (1956). A finitistic axiomatization of subjective probability and utility. *Econometrica*, **24**, 264–275.

[3] Keynes, J. M. (1921). *A Treatise on Probability.* Macmillan, London.

[4] Newman, P. (1987). Ramsey, Frank Plumpton (1903–1930). In *The New Palgrave: A Dictionary of Economics in 4 volumes*, (J. Eatwell, M. Milgate, and P. Newman, eds.) Macmillan, Stockton Press, pp. 41–45.

[5] Ramsey, F. P. (1931). *The Foundations of Mathematics and Other Essays*, R. B. Braithwaite, ed. (London, Preface by G. E. Moore., Routledge and Kegan Paul, A later version was published in 1978, edited by D. H. Melkov, L. Mirsky, T. J. Smiley, and J. R. N. Stone.)

[6] Wittgenstein, L. (1922). *Tractatus Logico-Philosophicus.* Kegan Paul, Trench and Trubner, London. (Introduction by Bertrand Russell.)

[7] Zabell, S. A. (1991). Ramsey, truth and probability. *Theoria*, **57**, 211–238.

Wiener, Norbert

Born: November 26, 1894, in Columbia, Missouri.
Died: March 19, 1964, in Stockholm, Sweden.
Contributed to: cybernetics, stochastic processes, mathematical physics, communication theory.

Norbert Wiener was born in 1894 in the United States. His father, descended from a family of rabbinical scholars, had migrated from Russia and, without a university education, became a professor of Slavic languages at Harvard. Under his father's influence Norbert Wiener became a child prodigy, entering Tufts College in Boston at 11 and graduating with a Ph.D. from Harvard at 18. His early interest was in natural science, particularly biology, rather than mathematics, but failure in the laboratory led his father to suggest philosophy, and his Ph.D. thesis was on mathematical logic.

In 1913 he went to Cambridge, England, and was influenced by Bertrand Russell and G. H. Hardy; Wiener claimed the latter was the "master in my mathematical training." Russell pointed out to Wiener that a mathematical logician might well learn something of mathematics. He published his first mathematical work soon after, and his first substantial paper [10] in 1914. After a period as an instructor in philosophy at Harvard and in mathematics at Maine, he became an instructor in mathematics at Massachusetts Institute of Technology.

Motivated partly by Einstein's work on Brownian motion, he developed the idea of representing each Brownian path as a point in a function space on which a probability measure is defined. He showed that almost all paths were continuous but not differentiable. His ideas were presented fully [11] in 1923. They have had an influence on modern probability of the most profound kind, as can be seen from the following. If $X(t)$ denotes a one-dimensional Wiener process, then the increment $X(t+\delta) - X(t)$ is independent of $X(s)$, $s \leq t$, and has a distribution independent of t. As a consequence of this (and the continuity of sample paths) $X(t)$ is Gaussian. If Y_k is a sequence of sums of independent identically distributed (i.i.d.) random variables with zero mean and finite variance, then the sequence has the same probabilistic structure as $X_k = X(T_1 + T_2 + \cdots + T_k)$ for suitable random variables, T_k, that are functions of the $X_j, j \leq k$, but i.i.d. Thus much of the limit theory of probability can be studied via $X(t)$. Moreover $X(t)$ itself represents chaotic behavior, as the previous comment about differentiability shows. Thus $dX(t)$ can intuitively be thought of as the driving term (or input) to a stochastic differential equation (or system) out of which comes a solution with the relatively organized behavior expected of physical phenomena. In this sense, the phenomena might be regarded as having been explained.

In 1926 he began his work on generalized harmonic analysis, which was fully developed in ref. [13]. Beginning from the notion of a measurable function $f(t)$, he introduced the autocorrelations

$$\gamma(t) = \lim_{T\to\infty} \frac{1}{2T} \int_{-T}^{T} f(s+t) f(s)\, ds,$$

assumed to exist. From these he constructed what would now be called the spectral distribution function. Moreover he showed how the mean square $\gamma(0)$ could be represented as a linear superposition of contributions from every oscillatory frequency, each contribution being the squared modulus of the contribution of that frequency to $f(t)$ itself. Wiener's characteristically constructive method is here demonstrated and can be contrasted with the axiomatic approach commencing from a probability space and a Hilbert space of square-integrable functions, over that space, generated by the action of a unitary group of translations on an individual function. (See ref. [1, pp. 636–637] for some history of this.) Out of his work on generalized harmonic analysis grew his work on Tauberian theory [12], one aspect of which is his celebrated theorem that if $f(\omega)$, $\omega \in [-\pi, \pi]$, has an absolutely convergent Fourier series and $f(\omega)$ is never zero, then $\{f(\omega)\}^{-1}$ also has an absolutely convergent Fourier series. The Wiener–Lévy theorem [4, p. 280] is the natural generalization of this. His work from 1926 to 1930 was gathered together in his book [14].

In 1933, with Paley named as coauthor, he produced ref. [7], Paley being dead at the time of writing. Apart from the (so-called) Paley–Wiener theory (see ref. [4, pp. 173–177] for a discussion), this contained also mathematical results relating to the Wiener–Hopf equation (see below) necessary for Wiener's later work on linear prediction. For the latter Wiener sought to discover a weight function $K(s)$, that, for a stationary process $f(t)$ with finite variance, minimized

$$\lim_{T\to\infty} \frac{1}{2T}\int_{-T}^{T}\left|f(t+a)-\int_0^\infty f(t-s)\,dK(s)\right|^2 dt,$$

i.e., that minimized the mean squared error in predicting $f(t+a)$ from $f(s)$, $s \leq t$. He reduced the solution of this to that of the Wiener–Hopf equation

$$\int_0^\infty \gamma(t-s)\,dK(s) = \gamma(t), \qquad t \leq 0,$$

which had previously arisen in connection with the distribution of stellar atmosphere temperature. Wiener solved this by the methods mentioned in connection with ref. [7]. This work was published in ref. [16], publication having been delayed due to restrictions because of a supposed need for military secrecy. At much the same time Kolmogoroff* [5, 6] had been working on the same problem. (See ref. [15, p. 59], for a discussion by Wiener of the question of priority.) Though perhaps less constructive, Kolmogoroff's work, which began from the Wold decomposition [21] of a discrete-time stationary process into a purely nondeterministic part and a perfectly predictable part, was in some ways more general. Wiener commenced from a direct representation of $f(t+1)$ in terms of $f(s)$, $s \leq t$ (i.e., an autoregressive representation), whereas Kolmogoroff commenced from the representation of $f(t)$ in terms of the prediction error or innovation sequence (i.e., moving-average representation). Kolmogoroff made use of fundamental results due to Szegö [9], that, for example, express the variance of the innovations as the geometric mean of the spectral density.

Wiener became well known to the general scientific public for his basically philosophical work on cybernetics [17]. This term was introduced by Wiener in 1948 [15]. This was concerned with the analogy between man as a self-regulating system, receiving sensory data and pursuing certain objectives, and mechanical or electrical servomechanisms. He made some attempt with Siegel [8] to use his differential space as a basis for a theory of quantum systems. With Masani [19, 20] he extended prediction theory to the multivariate case. In ref. [18] he sought to describe a general class of random processes obtained from a Brownian motion as a sum of homogeneous multilinear functionals of that process, of various degrees. This has led to much further research. (See ref. [2].)

The breadth of Wiener's work means that his name occurs throughout mathematics and probability. An example over and above those already mentioned arises in connection with a test for recurrence in a random walk. This in turn relates to potential theory and to diffusion (See ref. [3, p. 257].)

REFERENCES

[1] Doob, J. L. (1953). *Stochastic Processes.* Wiley, New York.

[2] Hida, T. (1970). *Stationary Stochastic Processes.* Princeton University Press, Princeton, N. J.

[3] Itô, K. and McKean, H. P. (1965). *Diffusion Processes and their Sample Paths.* Springer-Verlag, Berlin.

[4] Katznelson, Y. (1968). *An Introduction to Harmonic Analysis.* Wiley, New York.

[5] Kolmogoroff, A. N. (1939). Sur l'interpolation et extrapolation des suites stationnaires. *C. R. Acad. Sci. Paris*, **208**, 2043–2045.

[6] Kolmogorof, A. N. (1941). Interpolation and extrapolation of stationary random sequences, *Izv. Akad. Nauk SSSR Ser. Math*, **5**, 3–14.

[7] Paley, R. E. A. C. and Wiener, N. (1934). *Fourier Transforms in the Complex Domain.* Amer. Math. Soc., Providence, R. I.

[8] Siegel, A. and Wiener, N. (1955). The differential space of quantum systems. *Nuovo Cimento (10)*, **2**, 982–1003.

[9] Szegö, G. (1920). Beiträge zur theorie der Toeplitzchen formen. *Math. Z.*, **6**, 167–202.

[10] Wiener, N. (1914). A simplification of the logic of relations. *Proc. Cambridge, Phil. Soc.*, **27**, 387–390.

[11] Wiener, N. (1923). Differential space. *J. Math. Phys.*, **2**, 131–174.

[12] Wiener, N. (1923). Tauberian theorems. *Ann. Math.*, **33**, 1–100.

[13] Wiener, N. (1930). Generalized harmonic analysis. *Acta Math.*, **55**, 117–258.

[14] Wiener, N. (1933). *The Fourier Integral and Certain of its Applications.* New York.

[15] Wiener, N. (1948). *Cybernetics, or Control and Communication in the Animal and the Machine.* Hermann et Cie, Paris and Technology Press (MIT Press), Cambridge, Mass.

[16] Wiener, N. (1949). *Extrapolation, Interpolation and Smoothing of Stationary Time Series.* MIT Press, Cambridge, Mass., and Wiley, New York.

[17] Wiener, N. (1950). *The Human Use of Human Beings.* Houghton Mifflin, Boston.

[19] Wiener, N. (1958). *Nonlinear Problems in Random Theory*. MIT Press, Cambridge, Mass., and Wiley, New York.

[19] Wiener, N. and Masani, P. (1957). The prediction theory of multivariate stochastic processes. I. The regularity condition. *Acta Math.*, **98**, 111–150.

[20] Wiener, N. and Masani, P. (1958). The prediction theory of multivariate stochastic processes, II. The linear predictor. *Acta Math.*, **99**, 93–137.

[21] Wold, H. (1938). *A Study in the Analysis of Stationary Time Series*. Almqvist and Wiksell, Uppsala, Sweden.

BIBLIOGRAPHICAL NOTES

Wiener wrote a most interesting two-volume autobiography that serves to give some impression of the greatness of his mind. The volumes are:

Wiener, N. (1953). *Ex-prodigy: My Childhood and Youth.* Simon and Schuster, New York.

Wiener, N. (1956). *I Am a Mathematician. The Later Life of a Prodigy.* Doubleday, Garden City, N. Y.

In 1966 the American Mathematical Society devoted a special edition of its *Bulletin* (vol. 72, No. 1, Part II) to Wiener. This contains articles by distinguished scholars about Wiener and his work, including a very interesting article about the man himself by N. Levinson and a complete bibliography of his writing.

E. J. HANNAN

Yanson (or Jahnson), Yulii Eduardovich

Born: November 5, 1835 (o.s.), in Kiev, Russian Empire.
Died: January 31, 1892 (o.s.), in St. Petersburg, Russian Empire.
Contributed to: official statistics, economics, demography.

Y. E. Yanson's initial tertiary training was in the historico-philological faculty at Kiev University. In 1861, he was appointed to an academic position in agricultural statistics and political economy and, after several such posts, first taught at St. Petersburg University in 1865. Yanson's role was within the historical development of Russian statistical presentation of important socioeconomic issues, particularly problems of agricultural economics. Sometimes considered a father of the discipline of statistics in the Russian Empire, he is best known for two works: (1) *Comparative Statistics of Russian and Western-European States*, a two-volume work, the first volume of which appeared in 1878; (2) *The Theory of Statistics*, which appeared in five editions between 1885 and 1913 [5]. Of the latter work, Yanson's contemporary A. I. Chuprov, the father of A. A. Chuprov* and himself a leading figure in the same areas as Yanson, said [1]: "For statistical methodology we have nothing superior to Yanson's book; in regard to the description of statistical establishments and the applications of statistics, it seems there is little comparable to be found in Western-European literature." This book was used as a text for the course of statistics at St. Petersburg University (where Yanson was full professor from 1873), and was studied diligently by Lenin prior to his examination in 1891 as external student by a commission of which Yanson was a member [4]. Lenin was to refer later to Yanson's statistical data in his writings and may be regarded as having gained his statistical technology [3] from this book.

Yanson was elected to the International Statistical Institute in 1885 and became corresponding member of the Russian Academy of Science in 1892. He was active in social reform in the manner of the liberal intelligentsia of his milieu, and practically, apart from peasant economics, in the careful planning of censuses and epidemiological investigations. A photograph may be found in ref. [4]. There is a good obituary by A. I. Chuprov [1]; and the encyclopedia entry [2] gives extensive information.

REFERENCES

[1] Chuprov, A. I. (1893). Yulii Eduardovich Yanson. Obituary—1893 (in Russian). In *Rechi i Stati,* vol. 1. Sabashnikov, Moscow (1909), pp. 518–525.

[2] Anon. (1904). Yanson, Yu. E. In *Entsyklopedicheskii Slovar'*. Brokhaus and Efron, St. Petersburg, vol. XLI^{A}, 681–684.

[3] Il'in, V. (1908). *Razvitie Kapitalizma v Rossii,* 2nd ed. "Pallada," St. Petersburg. [In English as: Lenin, V. I. (1964). *The Development of Capitalism in Russia.* Progress Publishers, Moscow. V. Il'in was a pseudonym used by Vladimir Il'ich Ulianov, later to become known as Lenin.]

[4] Sipovska, I. V. and Suslov, I. P., eds. (1972). *Istoriia Prepodavaniia i Razvitiia Statistiki v Peterburgskom-Leningradskom Universitete (1819–1971)*. Leningrad University, Leningrad, pp. 22–36.

[5] Yanson, Yu. E. (1887). *Teoriia Statistiki*, 2nd ed. Schröder, St. Petersburg (5th ed., 1913).

E. SENETA

Yastremskii, Boris Sergeyevich

Born: May 9, 1877, in Dergach, near Kharkov, Ukraine.
Died: November 28, 1962, in Moscow, USSR.
Contributed to: time series, applied statistics.

Son of the well-known Russian revolutionary S. V. Yastremskii, B. S. Yastremskii was the leader of the dogmatic, strictly Marxist–materialistic approach to statistical sciences in the USSR and had a substantial influence on the development of statistics in that country.

He published a total of 97 papers on both the theoretical and practical problems of statistics.

He started his career in 1913 by criticizing the theory of stability of statistical series, which was developed by W. Lexis,* and continued by criticizing the "idealistic treatment" of the law of large numbers and the "law of averages." He also wrote extensively against Pearson's system of distributions and the concept of spurious correlation. He coauthored (with A. Ya. Boyarskii and others) two "Marxist" textbooks on statistics in 1931 and 1936. A summary of his statistical ideas is contained in his last book *Mathematical Statistics,* published in Moscow in 1956. Further details are given in an article [1] commemorating his 90th birthday.

REFERENCE

[1] Boyarskii, A. Ya. And Kil'disher, G. (1967). *Vestnik Statist.,* **5,** 35–40.

SECTION **5**

Government and Economic Statistics

Bowley, Arthur Lyon

Born: November 6, 1869, in Bristol, England.
Died: January 21, 1957, in Haslemere, Cumbria, England.
Contributed to: application of sampling techniques in social surveys, index-number construction, wage and income studies, mathematical economics, statistical education, popularization of statistics.

Arthur Lyon Bowley.

Sir Arthur Lyon Bowley was one of the most distinguished British statisticians in the first half of the twentieth century. His parents were Rev. J. W. L. Bowley, vicar of St. Philip and St. James in Bristol, and his wife, Maria Johnson. He attended Christ's Hospital School (1879–1888) and Trinity College, Cambridge (1888–1892), reading mathematics (bracketed as Tenth Wrangler in 1891). He gained an M.A. in 1895 and an Sc.D. in 1913, and was made an honorary fellow of Trinity in 1938.

Under the influence of Alfred Marshall and others in Cambridge active at that time in developing social sciences, he applied his mathematical abilities to problems in economic statistics. During this period, there appeared (in 1893) his earliest publication—*A Short Account of England's Foreign Trade in the Nineteenth Century* (3rd ed., 1922, Allen and Urwin, London)—for which he was awarded the Cobden prize.

After leaving Cambridge, he taught mathematics at St. John's School, Leatherhead, Surrey from 1893 to 1899. During that period, his interest in statistics developed rapidly. His first paper in the *Journal of the Royal Statistical Society (JRSS)*, on "Changes in average wages in the United Kingdom between 1860 and 1891," was published in 1895—the year in which the London School of Economics (LSE) was established. On Marshall's recommendation, Bowley was invited to serve as a lecturer. His Wednesday evening courses in statistics (at 5:45 or 6:00 P.M.) at London School of Economics continued from October, 1895 for some 38 years with only occasional interruption. For 18 years Bowley combined his lectureship at London School of Economics with teaching mathematics and economics at University College, Reading (later Reading University) as a lecturer from 1900 to 1907, and professor from 1907–1913. It was only in 1915 that he was given the title of professor at London School of Economics, having been a reader since 1908. In 1919 he was elected to a newly -established Chair of Statistics—a post he held until his retirement in 1936. He still continued many of his activities, both in London School of Economics and elsewhere, notably acting as director of the Institute of Statistics in Oxford University (1940–1944) during World War Two. He was created a C.B.E. (Companion of the Order or the British Empire) in 1937, and knighted in 1950.

In 1904, Bowley married Julie, daughter of Thomas Williams, a land agent of Spalding, Lincolnshire. They had three daughters, one of whom, Marian, became professor of political economy at University College, London. Bowley was a reserved and shy person, but had a dry sense of humor. He did not readily make friends. However, Joan Box records in her account of the life of her father, R. A. Fisher* [*Life of a Scientist* (1978), Wiley, New York] that in the 1920s Fisher and Bowley were close neighbors in Harpenden, Hertfordshire and frequently enjoyed evenings of bridge together. This did not prevent Bowley from criticizing Fisher at a meeting of the Royal Statistical Society in 1935.

Bowley's famous textbook *Elements of Statistics*, first published in 1910, went through seven editions, the seventh appearing in 1937. His stated goal was to put "in the simplest possible way those formulae and ideas which appear to be most useful in the fields of economic and social investigation and of showing their relationship to the treatment followed in the text." Bowley's influence as a teacher, especially at the postgraduate level, was considerable. [One of his many students was Maria (Mary) Smith-Faulkner (1878–1968), a fiery communist revolutionary who emigrated to the Soviet Union, and was responsible for the development of Soviet statistics in accord with rigid Marxist ideology. She also translated works of D. Ricardo and W. Petty* into Russian.]

In the field of economics, his most influential work seems to be on British incomes. He authored a number of studies on the definition and measurement of national income, which occupied his attention intermittently for more than twenty years before the first official estimates were issued, under the guidance of J.M. Keynes, during World War Two. Bowley's main publication during this period is *Wages and Income in the United Kingdom since 1860*, published in 1937 by Cambridge University Press.

His other pioneering activity in this field was in the London and Cambridge Economic Service, over the years 1923 to 1953. He was highly respected among British official statisticians, exercising his influence through teaching and research on the one hand, and extensive international contacts on the other.

The International Statistical Institute (ISI) owes much to him as a member from 1903, as treasurer (1929–1936 and 1947–1949), and as author of several reports sponsored by the Institute. In 1949 he was honored by election as honorary president. He traveled extensively, influencing statisticians and others in many parts of the world. Among these trips was a visit to India in 1934 to review "Indian economic conditions and the present statistical organization of India." Publication of *A Scheme for an Economic Census of India* (with Professor Sir Dennis Robertson) resulted from this journey.

Bowley's most original contributions were in the application of sampling techniques to economic and social surveys. He was forming his ideas on these topics as early as the 1890s, and propagated them through the ISI, though he was at first opposed by many leading official statisticians. He was attracted by the possibilities of the so-called "representative method," and developed an appropriate mathematical formulation of sampling precision and optimal ways of interpreting the results of such sample surveys to laymen. These were later presented in several publications, including *The New Survey of London Life and Labour* (vol. 3, pp. 29–96 and 216–253, in 1932, and vol. 6, pp. 29–117, in 1934). His pioneering work contributed subsequently to the development of market research. This work was closely associated with Bowley's studies on index numbers, reported in *JRSS* (1926), *Economic Journal* (1897, 1928), *Econometrica* (1938), and *London and Cambridge Economic Service* [Memo 5 (1924) and Memo 28 (1929)]. [See also Jazairi (1983).]

Bowley published relatively little in either mathematical statistics or mathematical economics. These subjects developed rapidly during the latter parts of his life, and he quite naturally felt less intense interest in them. Mathematical techniques themselves were incidental to his main purposes, and he used them essentially as a matter of convenience. He was a practitioner in applied statistics *par excellence*, though he cannot be regarded as a pathbreaker (in the sense of R. A. Fisher, K. Pearson,* and other British statisticians) in either statistical methodology or mathematical economics.

Bowley was closely associated with the Royal Economic Society (elected a fellow in 1893 and a member of Council in 1901, and publishing 15 papers in the *Economic Journal*). He was elected a fellow of the Royal Statistical Society in 1894, became a member of Council in 1899, was vice-president in 1907–1909 and 1912–1914, and was president in 1938–1940. He was awarded the Society's Guy Medals in Silver in 1895, and in Gold in 1935. His other scientific honors include a D. Litt. from Oxford University (1944), a D.Sc. from Manchester University (1926), and fellowship of the British Academy (1922).

SELECTED SOURCES

Allen, R. G. D. and George, R. F. (1957). Obituary of Professor Sir Arthur Lyon Bowley. *J. R. Statist. Soc. A*, **120**, 236–241. (Contains an extensive bibliography.)

Jazairi, N. (1983). Index numbers. In *Encyclopedia of Statistical Sciences*, **4**, 54–62. Wiley: New York.

Roberts, F. C., compiler. (1979). Obituaries from *The Times* of London, 1951–1960. Newspaper Archive Department, Reading, England.

Anon. (1973). *Who Was Who in America, Vol. 5 (1969–1973)*. Marquis Who's Who, Chicago.

Williams, E. T. and Palmer, H. T., eds. (1971). *The Dictionary of National Biography*. Oxford University Press, Oxford, England.

Engel, (Christian Lorenz) Ernst

Born: March 26, 1821, in Dresden, Kingdom of Saxony (Germany).
Died: December 8, 1896, in Oberlössnitz, near Dresden, Germany.
Contributed to: advanced education in statistics, census techniques, economic statistics, effective organization of official statistics, statistics of household income and consumption.

Christian Lorenz Ernst Engel.

Engel, one of the most influential nineteenth-century statisticians in Germany, was the third son of George Bernhard Engel, a wine merchant and vintner, and his wife Christiane Rosina Möbius. In 1848 he married Johanna Friederike Amalie von Holleufer; they had two sons and a daughter.

Upon completion of his studies as a mining engineer at the Mining Academy in Freiberg, Saxony, from 1841 to 1845 he visited industrial regions in Germany, France, Belgium, and England. In Paris he drew his first inspirations for social economics from Frédéric Le Play (École des Mines), and in Brussels he met Adolphe Quetelet,* whose work and thinking impressed him strongly. However, later on Engel explicitly disagreed with Quetelet's deterministic view of social phenomena as a result of hidden laws, and spoke of statistical regularities that cannot constrain the free will of individuals.

After his return to Saxony he was appointed a member, and later on the chairman, of a commission investigating industrial and working conditions in the kingdom. His publication of 1848 on the Saxon glassworks demonstrates his broad interests [1]. At that time Saxony was one of the most developed regions in Germany, which was a loose federation of independent states. In 1850 Engel, before reaching 30 years of age, successfully organized the General German Industrial Exhibition in Leipzig.

From 1850 to 1858 Engel was head of the newly established Royal Saxon Statistical Bureau in Dresden. During those years he organized the Saxon official statistics and took special care to publish the results of the Bureau's work efficiently, giving not only figures, but also explaining their economic and social importance [15–17]. He submitted his contributions to the Bureau's journal on statistical theory and its applications to the *Staatswirtschaftliche Fakultät* of Tübingen University and received the doctorate of economics.

In 1858 Engel resigned from that office because he did not get the support he wanted from the government for further reorganization of official statistics. He then engaged in studies of financial problems of the acquisition of land for building. He had already been concerned with savings banks and insurance in connection with his mathematical and statistical research. In two memoranda (1856, 1858) he dealt with mortgage insurance, and then he put his concepts into practice by founding and leading the first Saxon mortgage insurance company.

He left this post in April 1860, when he was appointed director of the Royal Prussian Statistical Bureau. In Berlin Engel paved the way for more efficient work in the organization of statistical surveys. Among other improvements he called on voluntary enumerators and installed qualified enumerating commissions for the 1861 Population Census [3]. Later on he introduced the individual census form for self-enumeration. Engel also reformed the Bureau's publications by issuing a statistical yearbook [18], a series of statistical reports [19], and a journal [20].

Engel's rich inspirations and ideas and his skill in organizing are very remarkable and worthy of being remembered as well as his contributions to theoretical statistics in connection with his economic and statistical research work. His fundamental studies on the individual consumption and household expenditures were published in various monographs [2, 6, 13, 14]. They resulted in the well-known *Engel's law*, which deals with the relationship of expenditures for consumption in households to the income available. The basis of his investigations was family budget surveys [12]. He demonstrated that total results of social statistics can be obtained from these individual data [14]. In this context he introduced the calculation with equivalent units of consumption. Later on he extended his research to the demand for housing and recognized the necessity for government and local communities to help solve the problems of the housing crisis [9]. In other respects he was a liberal-minded person who also served as National-Liberal representative in the Prussian Parliament from 1867–1870 and who inspired and supported social self-help by cooperatives, savings banks, and insurance companies. He therefore assisted the foundation of the *Verein für Socialpolitik* in 1872, a group of learned economists favoring these institutions.

Economic statistics also owe much of their clearness and systematic representation to Engel's effort. The first industrial census in (the *Kaiserreich* of) Germany in 1875 was conducted according to his preliminary studies [8, 10].

In a book on the era of "steam technology" [11] Engel aimed at estimating capital growth in Germany and other states. He had also a strong interest in the development of international statistics and played an active role in the international conferences of his time. For the 1863 congress in Berlin he provided a preliminary report [4], and later he pulled his weight for the realization of the resolutions on an international level [5].

Besides his official tasks and writing many pioneering publications, Engel also acted as a teacher. Due to his initiative, a seminar was attached to the Royal Statistical Bureau for training staff members and other civil servants in statistics.

Engel conceived statistics in the modern sense, as a science on its own, as a structural theory of human societies which he called demology (*Demologie*). He was convinced that this science serves as means to recognize and analyze problems arising from the formation of societies. A comprehensive publication, to be called *Demos*, on the measurement of people's, families', and individual welfare, was planned but not completed. A first volume that was published dealt with investigations of the cost value of a human being [13].

In July 1882 Engel gave up all his official positions, partly for health reasons, partly because of political difficulties: in 1881 he had published a pseudonymous attack, which was detected, on Bismarck's agricultural protectionism. He withdrew to private life on his estate near Dresden. Still, he continued creative work on a scientific formulation of statistics. Thus he deserves to be remembered as a personality representing the learned statistician and economist with charisma and ideas, even allowing for his conceptual shortcomings regarding the application of probability theory to statistics. His teaching was described by his disciples as lively, inspiring, and stimulating. Without his vigorous and creative work modern statistics in no way would be the same as it is. He pointed the way into the future of statistics as a science and as an essential tool of applied research.

SELECTED WORKS BY ERNST ENGEL

[1] Engel, E. (1848). *Einige Betrachtungen über die Glasfabrikation in Sachsen.* Bamberg, Dresden.

[2] Engel, E. (1857). Die vorherrschenden Gewerbezweige in den Gerichtsämtern mit Beziehung auf die Productions- und Consumtionsverhältnisse des Königreiches Sachsen. *Z. Statist. Bureaus Kön. Sächs. Ministeriums Innern*, **3**, 153–182.

[3] Engel, E. (1861). *Die Methoden der Volkszählung.* Verlagsbuchhandlung des Königlich Preußischen Statistischen Bureaus, Berlin.

[4] Engel, E. (1863). *Der Internationale Statistische Kongreß in Berlin.* Verlagsbuchhandlung des Königlich Preußischen Statistischen Bureaus, Berlin.

[5] Engel, E. (1864). *Die Beschlüsse des Internationalen Statistischen Kongresses in seiner fünften Sitzungsperiode.* Verlagsbuchhandlung des Königlich Preußischen Statistischen Bureaus, Berlin.

[6] Engel, E. (1866, 2nd ed. 1872). *Der Preis der Arbeit.* Verlagsbuchhandlung des Königlich Preußischen Statistischen Bureaus, Berlin.

[7] Engel, E. (1871). Systeme der Demologie. In Das Statistische Seminar und das Studium der Statistik überhaupt, *Z. Kön. Preußi. Statist. Bureaus*, **11**(III/IV), 198–210.

[8] Engel, E. (1872). *Die Reform der Gewerbestatistik im Deutschen Reich und in den Übrigen Staaten von Europa und Nordamerika.* Verlagsbuchhandlung des Königlich Preussischen Statistischen Bureaus, Berlin.

[9] Engel, E. (1873). *Die Moderne Wohnungsnot.* Duncker und Humblot, Leipzig.

[10] Engel, E. (1878). *Die Industrielle Enquête und die Gewerbezählung im Deutschen Reich und im Preußischen Staat am Ende des Jahres 1875.* Simion, Berlin.

[11] Engel, E. (1880, 2nd. Ed. 1881). *Das Zeitalter des Dampfes in Technisch-statistischer Bedeutung.* Verlagsbuchhandlung des Königlich Preußischen Statistischen Bureaus, Berlin.

[12] Engel, E. (1882). *Das Rechnungsbuch der Hausfrau und seine Bedeutung im Wirtschaftsleben der Nation.* Volkswirtschaftliche Zeitfragen, vol. 24, Berlin.

[13] Engel, E. (1883). *Der Wert des Menschen, Teil I: Kostenwert des Menschen.* Volkswirtschaftliche Zeitfragen, vols. 37 and 38, Berlin.

[14] Engel, E. (1895). *Die Lebenskosten Belgischer Arbeiterfamilien Früher und Jetzt.* C. Heinrich, Dresden.

[15] Engel, E., ed. (1853). *Jahrbuch für Statistik und Staatswirthschaft des Königreichs Sachsen,* 1: *Land und Leute, Wohnplätze und Materielle Hülfsquellen.* Dresden.

[16] Engel, E., ed. (1855–1857). *Zeitschrift des Statistischen Bureaus des Königlich Sächsischen Ministeriums des Innern,* **1–3.** Dresden.

[17] Engel, E., ed. (1851–1855). *Statistische Mittheilungen aus dem Königreich Sachsen* **1–4.** Dresden.

[18] Engel, E., ed. (1863–1876). *Jahrbuch für die Amtliche Statistik des Preuβischen Staates* Berlin.

[19] Engel, E., ed. (1860/61–1881). *Zeitschrift des Königlich Preuβischen Statistischen Bureaus* **1–21.** Berlin.

[20] Engel, E., ed. (1860–1882). *Preuβische Statistik.* Circa 70 issues.

SELECTED REFERENCES AND OBITUARIES

Blenck, E. (1896). Zum Gedächtnis an Ernst Engel. *Z. Kön. Preuβ. Statist. Bureaus*, **36**, 231–238.

Burckhardt, F. (1961). Engel, Ernst. In *Handwörterbuch der Sozialwissenschaften*, vol. 3, pp. 222–223.

Feig, J. (1907). Gedenkworte für heimgegangene deutsche Statistiker—Ernst Engel. *Allgemeines Statist. Arch.*, **7**, 349–359.

Földes, B. (1920). Ernst Engel. *Allgemeines Statist. Arch.*, **11**, 229–245.

Hacking, I. (1987). Prussian Numbers 1860–1882. In *The Probabilistic Revolution*, L. Kruger, ed. MIT Press. Cambridge, Mass., pp. 377–393.

Hacking, I. (1990). *The Taming of Chance.* Cambridge University Press, pp. 33–34, 128–131.

Houthakker, H. S. (1968). Engel, Ernst. In *International Encyclopedia of Social Sciences,* vol. 5, pp. 63–64.

Meier, E. (1959). Engel, Christian Lorenz Ernst. In *Neue Deutsche Biographie,* vol. 4, pp. 500–501.

A more detailed list of Engel's publications is given in:

Anon. (1981). *Gesamtverzeichnis des Deutschsprachigen Schrifttums (GV) 1700–1910.* K.G. Saur, München, New York, London, Paris.

HEINRICH STRECKER
ROLF WIEGERT

Farr, William

Born: November 30, 1807, in Kenley, Shropshire, England.
Died: April 14, 1883, in London, England.
Contributed to: vital statistics, epidemiology, demography, official statistics.

William Farr entered the British General Registry Office in 1837, shortly after its foundation. He started as a compiler of abstracts, and served in the office, moving through its ranks and making major contributions to its development, until his retirement in 1880. He developed the British system of death registration into an instrument for measuring the sanitary conditions of the time, and was a prime mover in making mortality statistics internationally comparable.

He was born to parents in destitute circumstances, but was adopted, in infancy, by Joseph Pryce, the squire of Dorrington, near Shrewsbury. He was educated at home in his early years, and assisted Pryce in managing his affairs. In the years 1826–1828 he studied medicine with a Dr. Webster of Shrewsbury.

Upon the death of Pryce in 1828 he received an inheritance of £500, and in 1829 he went to Paris to study medicine for two more years. Returning to London, he studied medicine for two further years, at University College and Hospital, qualifying as a Licentiate of the Apothecaries' Society in 1832. He married a farmer's daughter (a Miss Langford) in 1833 and started medical practice.

In 1837 he wrote an article on "Vital Statistics" for McCulloch's *Account of the British Empire*, which laid the foundation for a new type of science to which Farr devoted the rest of his life. He lost his wife, through consumption, also in 1837. A medical writer, James Clarke (1812–1875), selected Farr to assist him in revising his book on this subject, and it was through Clarke's influence that he obtained a post at the General Registrar's Office in 1838. Farr was assistant commissioner for the British censuses of 1851 and 1861, and commissioner for that of 1871. He wrote the greater part of the reports for these three censuses. His medical knowledge, skills in calculation and tabulation, and rich, yet lucid literary style made him highly successful in this work. Here is an example:

> It would formerly have been considered a rash prediction in a matter so uncertain as human life to pretend to assert that 9000 of the children born in 1841 would be alive in 1921; such an announcement would have been received with as much incredulity as Halley's prediction of the return of a comet, after the lapse of 77 years. What knew Halley of the vast realms of aether in which that comet disappeared? Upon what grounds did he dare to expect its re-appearance from the distant regions of the heavens? Halley believed in the constancy of the laws of nature; hence he ventured from an observation of parts of the comet's course to calculate the time in which the whole would be described; and it will shortly be proved that the experience of a century has verified quite as remarkable predictions of the duration of human generations. [*Fifth Annual Report of the Registrar of Births, Deaths, and Marriages* (1843), p. 21, quoted in ref. (6, p. 58)].

Life tables for insurance and issues of general statistics were of special interest for Farr. His colleagues remarked that "he was somewhat crotchety as to the modes of expression." He married his second wife, M. E. Whittal in 1842. They had eight children, of whom only a son and four daughters survived him. Mrs. Farr died in 1876.

Farr was elected a fellow of the Statistical Society in 1839, and served as a member of Council—almost continuously—from 1840 to 1882, as treasurer 1855–1867, vice-president 1867–1870, and president 1871–1873. During the period 1841 to 1877 he read sixteen papers before the Society. The titles and dates are listed below:

Mortality of Lunatics	(1841)
Influence of Scarcities and of High Prices of Wheat on Mortality of England	(1846)
Statistics of Civil Service of England, with Observations on the Constitution of Funds to provide for Orphans and Widows	(1849)
Influence of Elevation on Fatality of Cholera	(1852)
Income and Property Tax	(1853)
Pay of Ministers of the Crown	(1857)
Health of the British Army, and Effects of Recent Sanitary Measures on its Mortality and Sickness	(1861)
Reports of the Official Delegates from England at the International Statistical Congress, Berlin, 1863	(1863)
Address as President of Section F of the British Association	(1864)
Infant Mortality, and alleged Inaccuracies of the Census	(1865)
Mortality of Children in the principal States of Europe	(1866)
Inaugural Address as President, 1871	(1871)
Opening Address as President, 1872	(1872)
Valuation of Railways	(1873)
Valuation of Railways, Telegraphs, Water Companies, Canals, and other Commercial Concerns, with Prospective, Deferred, Increasing, Decreasing, or Terminating Profits	(1876)
On some Doctrines of Population	(1877)

In his presidential address in 1871, he appealed to the universities "to accord to statistics all the importance it deserves."

Farr was responsible for more than forty volumes of annual reports of births, marriages, and deaths, and the great English Life Tables, published in 1843, 1853, and 1867. He also contributed numerous papers to *The Lancet* from 1835 onwards, and a paper on life tables in the *Philosophical Transactions of the Royal Society* in 1859. He was the principal manager of the Health Section of the British Association for the Advancement of Social Sciences; reports and proceedings of this Association and also of the British Medical Association contain many papers by Farr. His chronological history of medicine up to 1453 appeared in the British Medical Almanack of 1936. In the next year the Almanack contained Farr's list of many medical and mortality statistics up to 1836.

Farr received an honorary M.D. degree from New York University in 1847, received an honorary D.C.L. degree from Oxford University in 1857, was elected a fellow of the Royal Society in 1855, and was honored as a Companion of the Order of the Bath in 1880. He also was awarded the Gold Medal of the British Medical Association.

His last paper, on estimating the value of stocks having a deferred dividend, read at a Statistical Society meeting in 1876, was a landmark in a new field. In 1879, the Registrar-General retired, and Farr was expecting to succeed him. However, he was passed over, and resigned his post in 1880. Soon after his retirement, "paralysis of the brain" was diagnosed, and after more than two years of ill health, he died of bronchitis in April, 1883.

Farr was characterized as a somewhat forgetful, absent-minded scholar, unselfish and highly devoted to his work, with a worldwide reputation for his statistical expertise. According to Hilts [7], his "somewhat stocky figure, bald head, large nose and deep eyes were familiar features among English civil servants during the middle years of the nineteenth century." An acquaintance during his period in Paris remarked "Mr. Farr, while of a simple disposition, is endowed with a vastness of ideas and a philosophic mind." As an accomplished linguist, he had an enviable capacity to put aside, temporarily, an absorbing but demanding study and relax, indulging his love of art and literature.

A selection of Farr's statistical works, under the title *Vital Statistics*, with a portrait of the author, was published by the Statistical Society in 1885.

REFERENCES

[1] Anon. (1949). Farr, William (1807–1883). In *Dictionary of National Biography.* Oxford University Press, Oxford, U.K., vol. six, pp. 1090–1091.

[2] Anon. (1883). Resolution on the Death of Dr. William Farr, and a letter from E. Jarvis, ex-President of the American Statistical Association. *J. Statist. Soc.*, **46**, 350–351.

[3] Anon. (1883). Obituary. *Times* (London), April 23, pp. 16, 18.

[4] Eyler, J. (1979). *Victorian Social Medicine: The Ideas and Methods of William Farr.* Cheere, Baltimore, Md.

[5] Giffen, R. (1883). Inaugural address. *J. Statist. Soc.*, **46**, 593–599.

[6] Hacking, I. (1990). *The Taming of Chance*, Cambridge University Press, Cambridge, U.K.

[7] Hilts, V. L. (1970). William Farr (1807–1883) and the human unit. *Victorian Stud.,* **14**, 143–150.

Fisher, Irving

Born: February 27, 1867, in Saugerties, New York, USA.
Died: April 29, 1947, in New York City.
Contributed to: index-number theory and practice; general equilibrium theory; monetary theory; macroeconomics.

Irving Fisher—according to Schumpeter [5] "one of the ten greatest economists of the 19th and 20th centuries"—was the son of a Congregational minister. He attended schools in Rhode Island, Connecticut, and Missouri. Following his father's death, his family moved to New Haven, Connecticut, where he enrolled in Yale College. After graduating first in his class, he continued at Yale with graduate studies, specializing mainly in mathematics. His doctoral thesis (1892) at Yale was an exposition of Walrasian general equilibrium theory (though he wrote in the introduction that he was unaware of Walras' work while preparing his dissertation). He remained at Yale for the rest of his life, except for the years 1898–1901, during which he was on leave recuperating from tuberculosis.

He was a teacher of mathematics from 1892 to 1895 and professor of economics from 1895 to 1935, though he was most active in university affairs up to 1920. In 1926 he merged his own firm—the Index Visible Company—with Remington Rand, and served on the board of directors of the latter company until his death in 1947. The merger made him a wealthy man, but he subsequently lost close to ten million dollars.

Except for a few theoretical economists, Fisher was not appreciated in the university where he spent over 50 years. He married in 1893; his wife, Margaret, died in 1940. They had two daughters and one son, I. N. Fisher—his biographer [3, 4].

Fisher's best-known contribution to statistics is Fisher's ideal index number. His ability in theoretical work was allied to a deep concern with the quality of observational data. His index [1, 2] is the geometric mean of the Laspeyres and Paasche formulas; it was also developed independently, and almost simultaneously, by Correa Walsh [6]. Several theoretical studies based on Fisher's ideal index have been made, among the earliest being by the Russian statistician Konüs.*

Fisher is regarded by many as the greatest expert of all time on index numbers and their use worldwide. He was also among the early users of correlation, regression, and other statistical tools.

The bibliography compiled by his son lists some 2000 titles authored by Fisher, and some 400 signed by his associates or written about him. For about 13 years (1923–1936) lists of index numbers were issued weekly from the Index Number Institute, housed in his New Haven home. His principal source of fame among noneconomists resides, perhaps, in his remarkably simple equation (often misused)

$$i = r + \pi$$

connecting nominal interest (i), real interest (r), and inflation (π).

The Nature of Capital and Income (New York, Macmillan, 1906, 1927) is one of his most brilliant pieces of work. It introduced the distinction (now common) between stocks and flows.

Irving Fisher was president of the American Statistical Association (1932), a founder of the International Econometric Society (1930), and president of the American Economic Association (1918). He had few personal students at Yale, and did not establish a "Fisherian" school. However, very few contributed more to the advancement of his chosen subject.

REFERENCES

[1] Fisher, I. (1921). The best form of index number. *Amer. Statist. Ass. Quart.*, **17**, 533–537.

[2] Fisher, I. (1922). *The Making of Index Numbers.* Houghton Mifflin, Boston.

[3] Fisher, I. N. (1956). *My Father, Irving Fisher.* Comet Press, New York.

[4] Fisher, I. N. (1961). *A Bibliography of Writings of Irving Fisher.* Yale University Library, New Haven, Conn.

[5] Schumpeter, J. A. (1948). Irving Fisher, 1861–1947. In *Ten Great Economists from Marx to Keynes.* Oxford University Press, Oxford.

[6] Walsh, C. M. (1921). *The Problem of Estimation.* King and Sons, London.

Franscini, Stefano

Born: October 23, 1796 in Bodio, Ticino canton, Switzerland
Died: July 19, 1857 in Berne, Switzerland
Contributed to: government statistics

Stefano Franscini, often called the "father of Swiss statistics," was a native of the transalpine, and sole Italian-speaking canton, Ticino, of Switzerland. The main aims of his life's work were the establishment of Ticino as a flourishing canton, regarded as an equal by the other cantons of Switzerland, and the recognition of numerical statistical data as an essential element in effective government.

Franscini left his modest home to train for the priesthood in the seminary of the Archbishop of Milan. However, he soon abandoned this course of action, and taught in a primary school in Milan. During the next few years he worked intensively on several projects, including a grammar textbook that went into eighteen editions, and self-taught studies in a number of disciplines.

His energies were finally focused as the result of a train journey in 1821, with Carlo Cataneo, to Zurich, in the course of which he was impressed by the relatively advanced state of economic development of the area through which they traveled, as compared with that of Ticino. He developed an ambition that Ticino, also, might be found capable of such development.

To put this into effect, he addressed himself with ardor to historical, political, economic and, above all, statistical studies. Returning to Ticino, where he opened a new type of school based on "mutual education," his growing certainty of the importance of statistics for achievement of his aims led to production, in 1828 of the book *Statistica delle Svizzera* for which he is best-known. Indeed, it was so well received at the time, that he became involved in discussions on a new Swiss constitution. In the new government arising from this constitution, Franscini became chancellor before being elected a Counsellor of State for Ticino. He represented Ticino in several hearings at the Federal Diet, and was elected a member of the Federal Council in 1848. He was appointed head of the Department of the Interior. In this position, he was able to introduce measures for improving the gathering and interpreting of Federal statistics, and encouraging cooperation among the various cantons. However, his colleagues on the Council, while appreciating the potential value of his proposals, were loath to provide sufficient money for their full implementation, so that it was possible to make only limited headway on them.

Franscini was very disheartened by these delays. From his autobiography, written in 1852, it appears that he no longer enjoyed his administrative and political activities. In addition to other disappointments, proposal for a chair in Statistics at the Federal Polytechnic to which he had devoted much of his prestige and work, was refused. Furthermore, he experienced family troubles, and he planned to retire at the end of 1857. However, he died on July 19 of that year.

Three years after his death, his friend, and successor as head of the Department of the Interior, G.P. Pioda, succeeded in completing Franscini's lifetime dream—the creation of the Swiss Bureau of Statistics.

Explanations that have been offered for Franscini's inability to turn this dream into reality are based on his growing isolation and loss of influence in his later years due to his hardness of hearing, his lack of knowledge of German, and his preoccupation with the future of his native canton, Ticino.

In 1990 a building for scientific meetings—the Centre Stefano Franscini in Ascona, Switzerland—was opened and named in his honor.

Franscini wrote many books. A few are noted below. Gfeller (1898), in a research study, gave an extensive account of his work.

BIBLIOGRAPHY

Franscini, S. (1827). *Statistica Delle Svizzera*, Lugano, Switzerland: Giuseppe Ruggia, Co. [New edition, 1991, edited by Raffaello Ceschi, Locarno, Switzerland: Dadò.]

Franscini, S. (1835–37). *Svizzere Italiana* (German edition, 1835; Italian edition, 1837) [New edition, in four volumes, edited by Birgilio Gilardoni, published Bellinzona: Casagrande 1987–89.]

Gfeller, E. (1898). *Stefano Franscini, ein Förderer der schweizerischen Statistik,* Berne.

Gini, Corrado

Born: May 23, 1884, in Motta di Livenza, Italy (about 50 kilometers northeast of Venice).
Died: March 13, 1965, in Rome, Italy.
Contributed to: foundations of statistics, methodology of statistics, demography, biometry, sociology, economics.

Corrado Gini is the most influential statistician of the first half of the twentieth century in Italy. A few days after his death, Vittorio Castellano [3] gave an enlightening and faithful portrait of him both as a scientist and as a human being. Castellano notes that his work continued almost until the hour of his death, which he met

> In the same quick and conclusive manner that was characteristic of his human encounters. Gini was born into an old family of landed gentry, in the fertile plain between the rivers Piave and Livenza and this country ancestry gave him something solid and withdrawn. He could be sociable when he wanted, and a brilliant talker, but it cost him an effort which he made only on rare occasions, when the inner promptings which always pushed him towards some precise aim either relaxed their grip on him or else demanded the effort in the ultimate interest of his ends and purposes. He combined within himself the seemingly opposite: tradition and new beginnings, the humanities and a technical bent, and that strange mixture of shyness and aggressiveness which is the hallmark of an age of evolution and crisis.

As far as his educational background is concerned, Castellano duly notes "that it is hard to say whether it was an accident or a sure sense of his vocation which guided the early choice of his readings and studies." In point of fact, Gini went to the university at Bologna, where he took his degree in law in 1905, and for this degree he defended a thesis which in 1908 was published as a book *Il Sesso dal Punto di Vista Statistico*. This book was awarded the Royal Prize for Social Sciences from the Accademia Nazionale dei Lincei. At the time, law was the only faculty in Italy which included a mandatory course in statistics. Gini, in addition to all courses of law, attended courses in mathematics and biology.

Gini's educational background appears to be somewhat fragmentary, which is definitely not the case with his research, which he started soon after graduation. In fact, from the very beginning he concentrated all his scientific efforts in the field of the social sciences and statistics, which, to his way of thinking, were quite complementary disciplines. He viewed statistics as the indispensable tool for monitoring the advancement of scientific investigation into mass phenomena, from the early perception of their characteristics to the formulation of a law governing them. As far as Gini's contributions to social sciences are concerned, let us recall that he worked out a theoretical basis of a coherent system of positive sociology according to which society is thought of as an organism which has some fundamental properties in common with biological organisms singled out by the biochemists. He crowned the work of the biochemists by introducing the concepts of evolutional and involutional equilibrium and of self-regulating development. In this context, population becomes a social body with an organization of life's structures, and demography, economics, history become different branches in the study of population. Gini provided an interpretation of some of modern society's economic mechanisms as mechanisms tending to preserve and re-establish equilibrium. In particular, he analyzed inter- and intranational migrations, developed a methodology to evaluate the income and wealth of nations and introduced measures of income and wealth inequalities, and singled out the so-called *Gini identity* for price index numbers.

Gini's work in statistical methodology is rarely the outcome of an abstract stance. As a rule, it resulted from his tackling concrete problems he invariably faced with his in-depth knowledge of fields as diverse as biology, demography, sociology, and economics. Yet his insights into statistical methodology are so far-reaching and challenging as to offer a comprehensive and stimulating corpus. Detailed reviews of Gini's work can be found in Gini [18, 23], Pietra [26], Castellano [3], Boldrini [2], Herzel and Leti [25], Benedetti [1], Forcina [7], and Giorgi [24]. Since he and his followers paid greater attention to complete statistical populations than to samples, Gini's main original achievements concern the definition and the analysis of statistical indices, which bring out distinctive features of sets of statistical data, with respect to well-defined purposes. We split Gini's contributions to this area into three main ideal sections: theory of mean values, theory of variability, theory of association between statistical variates. A final paragraph will be devoted to his critical analysis of the theories of statistical inference.

As far as the first group of contributions is concerned, Gini [19] introduced a helpful classification of mean values on the basis of some peculiar aspects of their analytical representation. At the same time, he provided general formulas for each of the classes singled out. Moreover, he dealt with the problem of extending the concept of mean value to qualitative variates (*mutabile* in Italian jargon), starting from suitable definitions of a deviate of a realization (*modalità* in Italian) of the qualitative characteristic under examination from a point of location. A complete treatment of Gini's work in the field is included in his ponderous book *Le Medie* [22]. The extension of the notion of mean value from quantitative to qualitative characteristics is, in Gini's way of thinking, closely connected with his unitary standpoint on variability of statistical characteristics. In *Variabilità e Mutabilità* [10] he started a systematic study of the variability of both quantitative and qualitative variates and, apropos of the former, he introduced into statistical literature the concept of mean difference, already considered by a few renowned astronomers (Jordan, von Andrae, Helmert*). Gini's mean difference provides a basic measure of the mutual diversity of the realizations of any statistical variate, against measures of the dispersion of a variate around a given point of location.

In connection with variability, Gini paid great attention to the concentration of positive transferable characteristics (e.g. wealth of a country). Roughly speaking, a characteristic of this type is more or less concentrated according as the proportion of the total exhibited by a given proportion of individuals is larger or smaller. In refs. [8] and [10], he made precise the concept of concentration by resorting to concentration curves and by introducing significant indices of concentration. Apropos of the latter, Gini made a distinction between descriptive indices and average indices. As an average index of concentration, he introduced the well-known *concen-*

tration ratio. It was clear to Gini that concentration represents a peculiar aspect of the variability of a positive transferable variate, and in connection with this, Gini introduced the concept of *relative measure of variability* on the class of all possible distributions of a positive transferable characteristic whose total amount is S, in a population of N units. This way, he was able to show that the (Pearson) *coefficient of variation* divided by $\sqrt{N-1}$ and $R(\cdot) = \Delta_1(\cdot)/2m(F)$ are measures of relative variability. Moreover, Gini [11] proved the following propositions: $R(F)$ coincides with twice the area between the segment joining (0,0), (1,1) and the Lorenz curve associated to F; $NR(F)/(N-1)$ coincides with the concentration ratio.

The methodological field where Gini and his school probably achieved their most outstanding results regards the analysis of the association between two ordered characteristics. In refs. [12], [14], [15], [16], and [17], he laid the foundations for a general and unitary theory relative to three distinct aspects of association: dissimilarity, monotone dependence, and connection. *Dissimilarity* refers to the diversity between the marginal distributions of the characteristic under examination. *Monotone dependence* concerns the propensity of a bivariate distribution to concentrate frequency around a monotone curve. In general, lack of monotone dependence—*indifference*, in Gini's terminology—and stochastic independence need not be equivalent. As a result, he introduced the term *connection* in order to designate the departure of a bivariate distribution from the situation of stochastic independence. He suggested studying monotone dependence and connection in the frame of the family of all bivariate cumulative distributions having fixed marginals. Hence, in connection with this, Gini and his followers produced some basic results in the frame of the so-called Fréchet classes; see Dall'Aglio [5], Pompilj [27], Salvemini [29]. In particular, Gini advanced some new significant indices—called indices of homophily—to measure the above-mentioned features of a bivariate distribution. At the same time, he made precise the meaning of the indices already in use, above all, the product moment correlation coefficient. In contrast with the product moment correlation coefficient, Gini's indices take into account the influence exerted on the association by dissimilarity in the marginal distributions. Their very definition reveals the inadequacy of the product moment correlation coefficient to capture monotone dependence, unless the supports of the Fréchet bounds are both included in straight lines, in which case the correlation coefficient coincides with the index of homophily of order 2. Gini also extended the above theory to grades by defining the so-called Gini's *index of cograduation*; see refs. [13] and [4]. Finally, in order to measure the discrepancy between two cumulative distribution functions, Gini proposed a suitable class of *indices of dissimilarity*. Recently, Rachev [28] has stressed the importance of these indices in connection with the theory of probability metrics.

Starting from 1939, Gini's interests in the field of statistical inference revived. As a matter of fact, his involvement in inferential problems dated back to thirty years before: his thesis for his degree was an example. In 1911 he drew inferences about the ratio of sexes in human births, according to the Bayes–Laplace paradigm, resorting to a Beta prior distribution; cf. Ref. [9]. Moreover he promoted the study of the sample distribution of *transvariability* (*transvariabilità* in Italian), which he defined as the probability of the event $\{X < Y\}$ when X and Y are random variables such that $m(X) > m(Y)$ for a fixed location measure $m(\cdot)$. Rereading Gini's papers devoted to statistical inference, one realizes his adhesion to the aforementioned paradigm combined with a rather restrictive interpretation of the domain of application of probability. In his opinion, the sole acceptable meaning of probability is that of relative frequency within a well-specified population; cf. Ref. [21]. Consequently, Gini implicitly supported de Finetti's exhortation to evaluate probabilities for observable facts only, when statistical applications are dealt with. However, unlike his fellow countryman, by identifying probability with relative frequency, he deprived probability of its innovative content with respect to the narrower meaning of relative frequency; see de Finetti [6]. Therefore, Gini was an early Bayesian who took care never to extend the Bayes–Laplace paradigm to cases in which the interpretation of the *a priori* distribution would diverge from his conception of probability. In 1911 and again in 1943, he dealt with the problem of eliciting the parameters of an initial Beta distribution, by singling out techniques which anticipated the advent of the so-called empirical Bayes approach; cf. Refs. [9] and [20]. But, if Gini's skepticism about the Bayes–Laplace paradigm stemmed from its being used outside the bounds he considered unsurmountable, his criticism of the Fisher and Neyman–Pearson approaches was quite straightforward. He resisted their spread in Italy up to the end, by stressing the logical weakness of the statistical procedures connected with those approaches. Detailed accounts of Gini's point of view on the Fisher and Neyman–Pearson theories can be found in [25] and [7].

Gini's pedagogical work began in 1909 and reached its climax in 1936 when, drawing inspiration from his unrelenting drive, the first Faculty of Statistics was eventually founded in Rome. There, he taught statistics, sociology, and biometry, and was the mastermind of a first-rate school of statisticians and demographers. Gini received honorary degrees in economics (Roman Catholic University of Milano, 1932), in sociology (University of Geneva, 1934), in sciences (Harvard, 1936) and in social sciences (University of Cordoba, 1963). In 1920 he was elected honorary fellow of the Royal Statistical Society, in 1933 vice president of the International Sociological Institute, in 1934 president of the Italian Genetics and Eugenetics Society, in 1935 president of the International Federation of Eugenics Societies in Latin-Language Countries, in 1937 president of the Italian Sociological Society, in 1939 honorary member of the International Statistical Institute, and in 1941 president of the Italian Statistical Society. In 1962 he was elected a national member of the Accademia dei Lincei. Gini was the editor of *Metron* from its foundation in 1920 up to 1965. Under Gini's leadership, this journal published important papers by outstanding statisticians. He also held many national offices and presided over the Central Institute of Statistics (1926–1932), which he organized as a center for all the official statistical services in Italy.

REFERENCES

[1] Benedetti, C. (1984). A proposito del centenario della nascita di Corrado Gini. *Metron*, **42**, 3–19. (Includes a description of Gini's contributions to applications of statistics to economics.)

[2] Boldrini, M. (1966). Corrado Gini. *J. R. Statist. Soc. A*, **29**, 148–150.

[3] Castellano, V. (1965). Corrado Gini: A memoir. *Metron*, **24**, 1–84. (Includes a comprehensive portrait of Gini's work and complete Gini's bibliography.)

[4] Cifarelli, D.M., Conti, P. L., and Regazzini, E. (1995). On the asymptotic distribution of a general measure of monotone dependence. *Ann. Statist.* to appear. (Deals with inferential applications of Gini's cograduation index.)

[5] Dall'Aglio, G. (1956). Sugli estremi delle funzioni di ripartizione doppie. *Ann. Scuola Normale Superiore*, **10**, 35–74. (Includes basic results about Gini's index of diversity.)

[6] de Finetti, B. (1966). La probabilità secondo Gini nei rapporti con la concezione soggettivista. *Metron*, **25**, 85–88. (The author stresses coincidences and differences between subjective interpretation and Gini's point of view on probability.)

[7] Forcina, A. (1982). Gini's contributions to the theory of inference. *Int. Statist. Rev.*, **50**, 65–70. (Describes Gini's contribution to statistical inference.)

[8] Gini, C. (1910). *Indici di Concentrazione e di Dipendenza*, Biblioteca dell'Economista, vol. 20. Utet, Torino.

[9] Gini, C. (1911). Considerazioni sulle probabilità a posteriori e applicazioni al rapporto dei sessi sulle nascite umane. Reprinted in *Metron*, **15**, 133–172.

[10] Gini, C. (1912). *Variabilità e Mutabilità: Contributo allo Studio delle Distribuzioni e delle Relazioni Statistiche.* Cuppini, Bologna.

[11] Gini, C. (1914). Sulla misura della concentrazione e della variabilità dei caratteri. *Atti R. Ist. Veneto Sci. Lett. e Arti*, **73**, 1203–1248.

[12] Gini, C. (1914). Di una misura della dissomiglianza fra due gruppi di quantità e applicazioni allo studio delle relazioni statistiche. *Atti R. Ist. Veneto Sci. Lett. e Arti*, **73**, 185–213.

[13] Gini, C. (1914). *Di una Misura delle Relazioni tra le Graduatorie di Due Caratteri. L'Indice di Cograduazione.* Cecchini, Roma.

[14] Gini, C. (1915). Nuovi contributi alla teoria delle relazioni statistiche. *Atti R. Ist. Veneto Sci. Lett. E Arti*, **74**, 583–610.

[15] Gini, C. (1915). Indici di omofilia e di rassomiglianza e loro relazioni col coefficiente di correlazione e con gli indici di attrazione. *Atti R. Ist. Veneto Sci. Lett. E Arti*, **74**, 1903–1942.

[16] Gini, C. (1916). Sul criterio di concordanza fra due caratteri. *Atti R. Ist. Veneto Sci. Lett. e Arti*, **75**, 309–331.

[17] Gini, C. (1916). Indici di concordanza. *Atti R. Ist. Veneto Sci. Lett. e Arti*, **75**, 1419–1461.

[18] Gini, C. (1926). The contribution of Italy to modern statistical methods. *J.R. Statist. Soc.*, **89**, 703–724.

[19] Gini, C. (1938). Di una formula comprensiva delle medie. *Metron*, **13**(2), 3–22.

[20] Gini, C. and Livada, G. (1943). Sulla probabilità inversa nel caso di grandezze intensive ed in particolare sulle sue applicazioni a collaudi per masse a mezzo di campioni. *Atti VII Riunione Scientifica della Soc. Ital. Di Statistica*, pp. 300–306.

[21] Gini, C. (1949). Concept et mesure de la probabilité. *Dialectica*, **3**, 36–54.

[22] Gini, C. (1957). *Le Medie.* Utet, Torino.

[23] Gini, C. (1966). On the characteristics of Italian statistics. *J. R. Statist. Soc. A*, **30**, 89–109.

[24] Giorgi, G. M. (1990). Bibliographic portrait of the Gini concentration ratio. *Metron*, **48**, 183–221. (Includes a bibliographical account of Gini's concentration ratio.)

[25] Herzel, A. and Leti, G. (1977). I contributi degli italiani all'inferenza statistica. *Metron*, **35**, 3–48. (Includes an account of the contributions to statistical inference by Italian statisticians.)

[26] Pietra, G. (1939). La statistica metodologica e la scuola italiana. *Suppl. Statist. Nuovi Problemi Polit. Storia ed Economia*, **5**, 125–145.

[27] Pompilj, G. (1984). *Le Variabili Casuali*. Ist. Calcolo delle Probabilità dell'Università "La Sapienca," Roma. (A comprehensive and modern treatment of the achievements of Gini's school.)

[28] Rachev, S. T. (1991). *Probability Metrics and the Stability of Stochastic Models*. Wiley, Chichester.

[29] Salvemini, T. (1939). Sugli indici di omofilia. *Suppl. Statist. Nuovi Problemi Polit. Storia ed Economia*, **5**, 105–115. (The author anticipates Frechet's bounds.)

BIBLIOGRAPHY

Gini, C. (1908). *Il Sesso dal Punto di Vista Statistico*. Sandrom, Milano.

Gini, C. (1912). *I Fattori Demografici dell'Evoluzione delle Nazioni*. Borca, Torino.

Gini, C. (1939). I pericoli della statistica. *Suppl. Statist. Nuovi Problemi Polit. Storia ed Economia*, **7**, 1–44. (Gini's first paper about the foundations of inferential methods.)

Gini, C. (1954). *Patologia Economica*. Utet, Torino.

Gini, C. (1959). *Ricchezza e Reddito*. Utet, Torino.

Gini, C. (1960). *Transvariazione*, G. Ottaviani, ed. Libreria Goliardica, Roma.

Gini, C. (1962). *L'Ammontare e la Composizione della Ricchezza delle Nazioni*. Utet, Torino.

Gini, C. (1968). *Questioni Fondamentali di Probabilità e Statistica*. Biblioteca del Metron, Roma. (Selected papers by Gini.)

EUGENIO REGAZZINI

Hansen, Morris Howard

Born: December 15, 1910, in Thermopolis, Wyoming, USA.
Died: October 9, 1990, in Washington, D.C., USA.
Contributed to: probability sampling, measurement error theory, sample survey methodology, demography.

Morris Hansen was an influential and prominent statistician in the development of survey methodology in the twentieth century. He was largely responsible for the high professional and scientific status of the U.S. Census Bureau in the period 1940 to 1970.

He was the son of Hans C. and Maud Ellen (Omstead) Hansen, and received a B.Sc. Degree from the University of Wyoming in 1934.

After a brief sojourn as a statistician with the Wyoming Relief Administration, he came to Washington in 1935 to work as a statistician for the U.S. Census Bureau. He became chief of the Statistical Research Division in 1944, statistical assistant director in 1949, assistant director for statistical standards in 1951, and associate director for research and development in 1961. He retired from the Census Bureau in 1968 and joined Westat, Inc. In Rockville, Maryland as a senior vice president, becoming chairman of the board in 1986. In the years 1945–1950 he also served as an instructor in statistics in the Graduate School of Agriculture. He obtained a M.A. in statistics from the American University in Washington, D.C. in 1940.

Many honors came Hansen's way. He received an honorary doctorate from the University of Wyoming in 1959 and a Rockefeller Public Service Award in 1962. He served as president of the Institute of Mathematical Statistics in 1953, of the American Statistical Association in 1960, and of the International Association of Survey Statisticians 1973–1977. He was a member of the U.S. National Academy of Sciences and received the rare distinction of becoming an honorary member of the International Statistical Institute.

Hansen married Mildred R. Latham in 1930, and they had two daughters and two sons. Mildred Hansen died in early 1983, and he married Eleanor Lamb in 1986.

Among Hansen's main contributions to statistical science is the two-volume work (with William N. Hurwitz and William G. Madow) *Sample Survey Methods and Theory* (Wiley, New York, 1953). This still (at the time of writing in early 1996) serves as a standard reference work in the theory and application of probability sampling. One of his early publications (ca. 1942) with W. Edward Deming—"On an important limitation to the use of data from samples"—is stored in the U.S. National Archives and Record Administration.

During his sojourn at the Census Bureau, he authored and coauthored (with W. N. Hurwitz, Leon Pritzker, and others) over a dozen publications which appeared in *Annals of Mathematical Statistics, Estadística, Journal of the American Statistical Association, Journal of Marketing* and *American Sociological Review, inter alia*. Many of these dealt with the theory of sampling from finite populations and introduced the concept of *total survey design*. This refers to incorporation of nonsampling error into the considerations influencing choices among competing survey designs. It is now known as the "Census Bureau survey error." Hansen was very active in the design of surveys at the Census Bureau, among them the Enumerative Check Census of Unemployment in 1937, the 1940 Census of Population and Housing, and the Labor Force Survey in 1943 (presently known as the Current Population Survey). He was instrumental in integrating sample and 100% (complete enumeration) data in censuses and in designing intercensal sample surveys, and was the driving force in the introduction of computing machines at the Census Bureau for statistical purposes.

Hansen's colleagues described him as a warm human being and a man of great personal integrity who welcomed collaboration with others and was punctilious in acknowledging such collaboration. He was an effective leader with an incredible capacity for hard work. He was also a popular speaker at national and international meetings, presenting a paper on "Sampling of Human Populations" at the 25th Session of the International Statistical Institute in Washington, D.C. in 1947 and coauthoring three papers (of which two were on measurement error) at the 36th Session in Sydney, Australia in 1967. He had a close association with P. C. Mahalanobis,* coordinating for a time the activities of the Census Bureau with those of the Indian Statistical Institute—noticing the considerable similarity between the activities of the two bodies.

SELECTED SOURCES

Fellegi, I. P. (1991). Obituary: Morris H. Hansen 1910–90. *J. R. Statist. Soc. A*, **154**, 352–355.

O'Brien, J. L. (1983). Interview with Morris Hansen, *Internal Rep.*, U.S. Census Bureau.

Olkin, I. (1987). A conversation with Morris Hansen. *Statist. Sci.*, **2**, 162–179.

Tepping, B. J. and Waksberg, J. (1991). Morris H. Hansen, 1910–90. *Amer. Statist.*, **45**, 2–3.

Kiaer, Anders Nicolai

Born: September 15, 1838, in Drammen, Norway.
Died: April 16, 1919, in Christiania (now Oslo), Norway.
Contributed to: survey sampling, International Statistical Institute.

Kiaer's most lasting contribution to statistical theory was his early introduction of self-conscious sampling in population surveys. He took office as Director of the Central Bureau of Statistics of Norway at its establishment in Oslo (then Christiania) in 1876 and was head of the Bureau until 1913. He developed the practice (and to a lesser extent the theory) of sampling surveys. At the Berne International Statistical Institute meeting (1895) he introduced and clarified the meaning of "*dènombrements reprèsentatives*" (representative investigations), i.e., based on individuals who have been selected according to a particular representative method. He initially used this procedure in surveys for which the sample was selected at random from the returns to the Norwegian pupulation census of 1891. For one of these surveys Kiaer linked census information with individual data collected from the income-tax offices, on the basis of which he analyzed the distribution of income and wealth in Norway. He also made significant contributions to statistical and population theory and to applications in a wide range of subject-matter fields, from population and vital statistics to national-income estimation and taxation. In addition, he was a pioneer in applying punch-card equipment for statistical data processing.

Kiaer's last work, which appeared posthumously in *J. Amer. Statist. Ass.*, **16**, 442–458, discussed determination of the birth rate in the United States. This article grew out of correspondence with W.F. Willcox* over the period 1918–1919—see also the obituary by Willcox on pp. 440—441 of the same issue.

A bibliography of Kiaer's work is included in *Den Representative Undersøgelsmethode: The Representative Method of Statistical Surveys*, Central Bureau of Statistics, Oslo, Norway (1976) (in Norwegian and English).

BIBLIOGRAPHY

Kruskal, W. H. and Mosteller, F. (1980). *Int. Statist. Rev.*, **48**, 169–195.

Seng, Y. P. (1951). *J.R. Statist. Soc. A*, **114**, 214–231.

P. J. BJERVE
W. H. KRUSKAL

Konüs, Alexandr Alexandrovich

Born: October 2, 1895, in Moscow, Russia.
Died: April 5, 1990, in Moscow, Russia.
Contributed to: index-number theory, mathematical statistics, labor value theory.

A. A. Konüs was born in Moscow and brought up by his mother, a voice teacher. In 1914 he graduated from a private high school in Moscow and volunteered for military service. He was wounded four times in World War I and was awarded the St. George's Cross for his bravery. In the fall of 1917, Konüs enrolled in Moscow University, studying physics and mathematics. In 1918–1920 he studied at the Co-operative Institute; among his teachers were the well-known Russian statisticians and economists A. V. Chayanov and N. D. Kondrat'ev. In 1923, Konüs started working at the "State of Affairs" Institute under Kondrat'ev's leadership. During this period his talent as a theoretical statistician was prominently exhibited. The Institute was liquidated in 1929, Konüs was obliged to change jobs about ten times during a decade, and finally, at the end of the thirties, he found his niche as an instructor of statistics at the Moscow Credit–Economic Institute. During the war of 1941–1945 Konüs worked in the central laboratory of a tank factory in Chelyabinsk, Siberia. His work on the application of statistical methodology to the investigation of the quality of metal production [7] served as the subject of his 1945 Ph.D. dissertation. During the years 1946–1950 Konüs was a junior scientist in economic statistics at the Academy of Sciences of the USSR. From 1950 to 1982 he worked for the USSR at the Scientific Research Institutes of Communication (1950–1956), of Labor (1957–1959), and of Economics (1960–1982). He was awarded a Doctorate of Economic Sciences in 1982 and an Honorary Doctorate in Political Economy at the University of Munich, and served for many years on the editorial board of *Metron*.

There were four directions to Konüs' scientific activities: consumer price indices and the theory of indifference; hypersurfaces; economic and mathematical statistics; and dynamic models of intra-industrial balance and problems of labor value theory using mathematical methodology. His 1924 paper "On the problem of the true index of the cost of living" brought him world renown and was reproduced (in English) in *Econometrica* in 1939. Konüs points out in this paper that no budget index—however carefully constructed—can precisely reflect the cost of living. He proposed his own index, which requires knowledge of the combinations of goods consumption leading to the same living standard. Konüs' index discloses the manner in which the monetary value of goods should be changed in order to maintain a definite level of utility of consumers' choices. The composition of these choices may differ over time, reflecting an actual economic state of affairs. The paper served as a basis for an economic-theoretical development in indexology; its ideas were further developed for example, in ref. [3].

Konüs' papers on the methods of collecting and processing statistical data related to cost indices, as carried out by the "State of Affairs" Institute in the twenties

[2, 4], are of substantial historical interest. So also is his paper on the construction of economic barometers [5], which appeared in 1933, supplemented with an extensive bibliography.

In ref. [6], Konüs proposed a method of determining the functional relationship between total consumer expenditures, the goods utilized, and a relative variation of prices over a given period in comparison with a base period.

In ref. [8] he investigated statistical methods of reducing asymmetric distribution surfaces to the form of a bivariate normal distribution. The basic idea is that when the correlation coefficient between two random variables is zero and the joint distribution can be reduced by a change of variables (by means of some functions of these variables) to a bivariate normal distribution, these variables turn out to be independent. This idea originated with the Russian statistician B. I. Sreznerskii (1857–1934).

A series of papers written during the period between the fifties and eighties (Konüs [9, 10, 12, 14–16, 18] are devoted to the construction of a price index in relation to the microeconomic theory of consumption. These works are a continuation of investigations originated in the twenties at the "State of Affairs" Institute. In particular, in ref. [12] a mathematical substantiation of the relationship between demand curves and price indices is presented, as well as calculations of demand curves by means of a formula for a "power mean."

In refs. [11, 13] an application of extended systems of equations for modelling intra-industrial relations in economics and long-range planning is considered. In the Konüs model the capital investment is viewed as production expenditures for future years, allowing for lags.

Konüs [17] provided a perceptive biography of E. E. Slutskii, which appeared in the *International Encyclopedia of Statistics*. He kept working until the very end of his long productive scientific life, publishing articles and serving as a consultant for the publication of selected works of his teacher N. D. Kondrat'ev, carried out in 1989.

REFERENCES

[1] Konüs, A. A. (1924). The problem of the true index of the cost of living (in Russian). *Econ. Bull. "State of Affairs" Inst.*, No. 9–10, pp. 64–72. English translation (1939), *Econometrica*, **7**, 10–29.

[2] Konüs, A. A. (1925). Russian price indices in 1914–1922 (in Russian). In *State of Affairs and Prices*, M. V. Ignat'ev, ed. Financial Publishing House of Peoples Commissariat of Finance USSR, Moscow, pp. 97–141.

[3] Konüs, A. A. and Byushgens, S. S. (1926). On the problem of the purchasing power of money (in Russian). *Voprosy Kon'yunktury*, **2**(1), 151–172.

[4] Konüs, A. A. (1930). A method of collecting and processing statistical data on prices. In *Dinamika Tsen Sovietskogo Khozyaistva*, Plankhozizdat, Moscow, pp. 92–123.

[5] Konüs, A. A. (1933). Economic affairs, in *Encyclopedic Dictionary of the Russian Bibliographic Institute Granat*, vol. 51, pp. 220–263.

[6] Konüs, A. A. (1939). On the theory of mean values (in Russian). *Trudy Sredneaziatskogo Gos. Univ. U*, No. 24.

[7] Konüs, A. A. (1945). Brief Instructions on a Statistical Method of Investigating Metal Production Properties Depending on the Determining Factors (in Russian). Narkomat. Tank Industry, Moscow.

[8] Konüs, A. A. (1950). Investigation of correlations reduced to the normal type and construction of correlation equations (in Russian). In *Voprosy Statisticheskogo Izmereniya Svyazei mezhdu Yavleniami* (Problems of Statistical Measurement of Relations between Occurrences). Gosplanizdat, Moscow, pp. 95–130.

[9] Konüs, A. A. (1958). Consumer price indexes and demand functions. *Rev. Inst. Int. Statist.*, **26**, 113.

[10] Konüs, A. A. (1959). Schematic analysis of the effect of price variation on consumption level (in Russian). In *Methodological Problems of the Study of the Living Standards of Workers*. Izdat. Sots.-Ekonom. Literatury, Moscow, pp. 170–197.

[11] Konüs, A. A. (1961). Extension of the system of equations of intra-industrial relations for the purposes of long-range planning (in Russian). In *Applications of Mathematics in Economic Investigations*, vol. 2. Sotsekgiz, Moscow.

[12] Konüs, A. A. (1963). Theoretical index of consumer prices and its application in planning of a solvent demand (in Russian). In *Economic-Mathematical Methods*, No. 1, Publishing House of the Academy of Sciences of the USSR, Moscow, pp. 278–291.

[13] Konüs, A.A. (1964). Long-range planning in relative numbers (in Russian). *Uchënye Zapiski Statist.*, **8**, 265–275.

[14] Konüs, A. A. (1966). Price indices of consumption budget and the theory of hypersurfaces of constant consumption level (in Russian). *Uchënye Zapiski Statist.*, **11**, 173–193.

[15] Konüs, A. A. (1968). The theory of consumer price indices and the problem of the comparison of the cost of living in time and space. In *The Social Sciences: Problems and Orientations*. UNESCO, Paris, pp. 93–107.

[16] Konüs, A. A. (1974). A method of calculating the composition of consumer budgets (in Russian). *Uchënye Zapiski Statist.*, **24** (special issue), 186–208.

[17] Konüs, A. A. (1978). Slutsky, Eugen. In *International Encyclopedia of Statistics*, W. H. Kruskal and J. M. Tanur, eds., vol. 2. Free Press, New York, pp. 1000–1001.

[18] Konüs, A. A. (1981). The method of price indices in construction of consumption models (in Russian). *Uchënye Zapiski Statist.*, **42**, 144–172.

BIBLIOGRAPHY

Diewert, W.E. (1987). Konüs, Alexandr Alexandrovich. In *The New Palgrave: A Dictionary of Economics*. London, p. 62.

Komlev, S.L. (1991). In memory of A.A. Konüs (in Russian). *Uchënye Zapiski Statist.*, **55**, 277–280.

ANTON DMITRIEV

Kőrösy, József

Born: April 20, 1844, in Pest, Hungary.
Died: June 23, 1906, in Budapest, Hungary (the cities of Pest and Buda were united in 1873).
Contributed to: official statistics; demography; fertility, birth, and death rates; mortality (life) tables; measures of association

József Kőrösy was born in Pest, in a merchant's family. His father died when József was six years old. Financial problems forced him to start work immediately after his graduation from high school. After working as an insurance clerk, he became a journalist, writing a column on economics. He was, almost entirely, self-taught, with substantial proficiency in the major European languages, in addition to knowledge in his chosen professional field.

In 1869, at age 25, he became the first Director of the newly established Capital Statistical Office in Pest. He served in this capacity for 37 years, until his death. In this period he created, nearly singlehandedly, the Statistical Department of the Municipality of Pest (up to 1867, the Austrian government's statistical services were responsible for all statistical data relating to Hungary).

It was in the Municipality of Pest that Kőrösy began to collect the eventually vast amount of detailed information upon which he based his investigation (published in English, French, and German) of birth rates and fecundity. In 1873, Kőrösy's office became the Statistical Office of the united city of Budapest. He was successful, with the collaboration of the sanitary administration, in improving substantially the hygienic conditions of the crowded population of the city.

Kőrösy was a stalwart advocate of the so-called *type studies* and was eager to extend inference from his data to other, even only marginally related, fields. He attached special importance to international uniformity in census details and to the adoption of synchronism in the enumerations carried out in all European countries.

Kőrösy was elected as a member of the Hungarian Academy of Sciences in 1879. His infectious enthusiasm, good humor, and extraordinary resourcefulness made him very popular among members of the international statistical community of his time. He was an honorary fellow of the Royal Statistical Society, and took part in their proceedings on several occasions. He was a foreign member of several academies and Doctor (Honoris Causa) of the Medical University of Philadelphia.

From 1883 onwards, in addition to his work in the Statistical Office, he was a reader in the University of Budapest, lecturing on demography. He participated vigorously in the activities of the International Statistical Institute, including presenting a paper on the theory of mortality tables at the 9th International Congress, at Budapest in 1876. In 1885, he contributed a paper "On the unification of census record tables" to the Jubilee Volume of the *Journal of the Royal Statistical Society*. His long 1896 paper in the *Philosophical Transactions of the Royal Society* (*B*, **186**, 781–875) with the title "An estimate of the degrees of legitimate natality, as derived from a table of natality compiled by the author from his observations made at Budapest" is an important methodological contribution, investigating the effects of ages of spouses and marriage duration on family size, among other topics. Körösy was a way ahead of his time in regard to the use of mathematics in statistics, introducing measures of association very similar to those developed by G. U. Yule* and Karl Pearson* some 20 years later. (See, e.g., Goodman and Kruskal [1].) His statistical studies of Budapest involved such modern indicators as corporation profits. His "Individual method" of constructing generation mortality tables was a precursor of modern cohort analysis.

As J. A. Baines wrote, in Kőrösy's obituary in the *Journal of The Royal Statistical Society* [**70** (1907), 332–333], referring to his early journalistic activities, he harnessed "his imagination to the comparative jog-trot career of urban statistics in preference to allowing it the untrammelled freedom it might have enjoyed in the record of general history from day to day."

REFERENCES

[1] Goodman, L. A. and Kruskal, W. H. (1959). Measures of association two. Further discussion and references. *J. Amer. Statist. Assoc.*, **54**, 123–163.

[2] Thirring, L. (1968). Kőrösy, József. In *International Encyclopedia of Statistics*, W. H. Kruskal and J. M. Tanur, eds. McMillan/Free Press, New York, pp. 491–492.

Lexis, Wilhelm

Born: July 17, 1837, in Eschweiler, Germany.
Died: August 24, 1914, in Göttingen, Germany.
Contributed to: theoretical statistics, economics, population studies, sociology.

The German statistician and economist W. Lexis graduated from the University of Bonn in 1859, Lexis' initial training was in mathematics and physics, but after a brief period in Bunsen's laboratory, he left for Paris and the study of the social sciences. Lexis' ensuing academic career is indicative of the range and diversity of his interests: appointed to the chair of economics at Strassburg, 1872; professor of geography, ethnology, and statistics at Dorpat, 1874; professor of economics at Freiburg, 1876, and Breslau, 1884; and professor of political science at Göttingen, 1887 (where he remained until his death).

Lexis' most important statistical contributions came during the brief period from 1876 to 1879. Dissatisfied with the uncritical and usually unsupported assumption of statistical homogeneity in sampling, often made by Quetelet* and his followers, Lexis devised a statistic Q, now called the *Lexis ratio*, to test this assumption and demonstrate its frequent invalidity. If X_{ij} $(1 \leq i \leq m,\ 1 \leq j \leq n)$ are independent Bernoulli trials with success probabilities p_{ij}, where for each i the X_{ij} represent a sample from subpopulation i, then

$$Q = \frac{\sum_{i=1}^{m}(\hat{p}_i - \bar{p})^2/m}{\bar{p}(1-\bar{p})/n},$$

where $N = mn$, $\bar{p} = \sum_{ij} p_{ij}/N$, and $\hat{p}_i = \sum_j X_{ij}/n$. It can be shown that the *theoretical* dispersion coefficient $D = \sqrt{E(Q)}$ has values $D < 1$ (subnormal dispersion), $D = 1$ (normal dispersion), or $D > 1$ (supernormal dispersion) whenever $p_{ij} \equiv p_j$ (Poissonian sampling), $p_{ij} \equiv p$ (Bernoulli sampling), or $p_{ij} \equiv p_i$ (Lexian sampling), respectively. In actual practice Q was replaced by

$$Q^* = \frac{\sum_{i=1}^{m}(\hat{p}_i - \hat{p})^2/m}{\hat{p}(1-\hat{p})/n},$$

where $\hat{p} = \sum_{ij} X_{ij}/N$.

Lexis's results were first extended by von Bortkiewicz and then placed on a sound mathematical basis by Chuprov* and Markov,* who defined a modified Lexis ratio

$$L = \frac{N-1}{n(m-1)} Q^*$$

and proved rigorously that $E(L) = 1$ [1], that $\mathrm{var}(L) \leq 2/(m-1)$ for $m \leq 5$, and that (in effect) the asymptotic distribution of $(m-1)L$ is $\chi^2(m-1)$ [8, 9].

The Lexis ratio is, up to a multiplicative constant, the variance test for homogeneity of the binomial distribution (see Snedecor* and Cochran* [13, Sec. 11.7]) and as such, is an early instance of the analysis of variance. It is a sad commentary on the gulf between continental and English statistics that it was not until 1924 that it was pointed out, by Fisher,* that mQ^* is exactly the same as Pearson's chi-square statistic; at the same time, Fisher was unaware of Markov*'s work which four years earlier had settled the degrees-of-freedom controversy between Fisher and Pearson before it had even begun. In fact, with the exception of Edgeworth* and Keynes, the work of Lexis and his school was largely ignored in England.

Literature

For details of Lexis' life and work, see the obituaries by von Bortkiewicz [15] and Klein [7], and the articles by Heiss [4] and Oldenburg [10].

The history of the Lexis ratio is ably discussed by C. C. Heyde and E. Seneta [5, Sec. 3.4]; mathematical details, including some of the results of Chuprov and Markov, are given in J. V. Uspensky's *Introduction to Mathematical Probability* [14, pp. 212–230]. *The Correspondence between A. A. Markov and A. A. Chuprov on the Theory of Probability and Mathematical Statistics* [11] is an invaluable source of information about the development of the ideas of Chuprov and Markov, and their assessment of the work of Lexis and von Bortkiewicz. Keynes's discussion of Lexis in his *Treatise on Probability* [6, Chap. 32] is also of interest. Other useful accounts of the Lexis ratio include those of J. L. Coolidge [2], H. L. Rietz [12, Chap. 6], and F. N. David,* [3, pp. 152–159].

REFERENCES

[1] Chuprov, A. A. (1916). *Izv. Akad. Nauk Pgd. VI*, **10**, 1789–1798.

[2] Coolidge, J. L. (1925). *An Introduction to Mathematical Probability.* Clarendon Press, Oxford.

[3] David, F. N. (1949). *Probability Theory for Statistical Methods.* Cambridge University Press, Cambridge.

[4] Heiss, (1978). In *International Encyclpedia of Statistics*, W. Kruskal and J. Tanur, eds. Free Press, New York.

[5] Heyde, C.C. and Seneta, E. (1977). *I.J. Bienaymé: Statistical Theory Anticipated.* Springer-Verlag, New York.

[6] Keynes, (1921). *Treatise on Probability.* Macmillan, New York.

[7] Klein, (1914). *Jahresber. Btsch. Math.-Ver.*, **23**, 314–317.

[8] Markov, A. A. (1916). *Iz. Akad. Nauk Petrograd VI*, **10**, 709–718.

[9] Markov, A. A. (1920). *Izv. Akad. Nauk Petrograd VI*, **14**, 1191–1198.

[10] Oldenburg, (1933). In *Encyclopedia of the Social Sciences*. Macmillan, New York, vol. 9.

[11] Ondar, K. O., ed. (1981). *The Correspondence between A.A. Markov and A. A. Chuprov on the Theory of Probability and Mathematical Statistics.* Springer-Verlag, New York.

[12] Rietz, H. L. (1927). *Mathematical Statistics.* Mathematical Association of America, Washington, D.C.

[13] Snedecor, G. W. and Cochran, W. G. (1980). *Statistical Methods*, 7th ed. Iowa State University Press, Ames, Iowa.

[14] Uspensky, J. V. (1937). *Introduction to Mathematical Probability.* McGraw-Hill, New York.

[15] von Bortkiewicz, L. (1915). *Bull, Int. Statist. Inst.*, **20**, 328–332.

SANDY L. ZABELL

Mahalanobis, Prasanta Chandra

Born: June 29, 1893, in Calcutta, India.
Died: on the eve of his seventy-ninth birthday, June 28, 1972, in Calcutta, India.
Contributed to: statistics, economic planning.

Prasanta Chandra Mahalanobis.

Prasanta Chandra Mahalanobis' family originally belonged to the landed aristocracy of Bengal, but had moved to Calcutta by the middle of the last century. Prasanta was educated first in Calcutta, where he received the B.Sc. honors degree in physics, and later at Cambridge University, where he passed Part I of the Mathematical Tripos in 1914 and the Natural Science Tripos, Part II, in 1915, and was elected a senior research scholar. A visit to India on a holiday turned out to be a permanent homecoming because of an opportunity to teach at the Presidency College in Calcutta and pursue his growing interest in statistical problems. The latter began to absorb his interest so completely that he abandoned his intention of pursuing a career of research in physics and decided to devote himself to statistics, although he continued to teach physics until 1948.

A man of great originality, Mahalanobis' contribution to statistical thought goes far beyond his published work. He considered as artificial any distinction between "theoretical" and "applied" statistics. His own work in statistics was always associated with, and arose out of, some field of application. His investigations on anthropometric problems led him to introduce the D^2—statistic, known later as the Mahalanobis distance, and used extensively in classification problems. This early work yielded a wealth of theoretical problems in multivariate analysis, many of which were later solved by his younger colleagues. Among other major areas in which he worked during the 1930s and 1940s are (1) meteorological statistics, (2) operations research (to give it its modern name), (3) errors in field experimentation, and (4) large-scale sample surveys.

It is characteristic of Mahalanobis' research that fundamentally new ideas are introduced in the course of studying some immediate, practical problem. Thus his concept of pilot surveys was the forerunner of sequential analysis, anticipating it by a decade. The 1944 memoir on sample surveys in the *Philosophical Transactions of the Royal Society* established the theory, gave the estimating procedures, and, at the same time, raised basic questions concerning randomness, as to what constitutes a random sample. He also introduced the concepts of optimum survey design and interpenetrating network of subsamples. His development of cost and variance functions in the design of sampling (originating in the work of J. Neyman*) may be regarded as an early use of operational research techniques.

Mahalanobis' work in field experimentation (1925), carried out in ignorance of the design of experiments introduced earlier by R. A. Fisher,* brought the two men together and started a close professional and personal friendship that lasted until Fisher's death in 1962. Mahalanobis and Fisher held similar views on the philosophy of statistics as well as its methodological aspects. Both regarded statistics as a "key technology" for "increasing the efficiency of human efforts in the widest sense."

Mahalanobis' personal contributions to statistics have, to some extent, been obscured by his two monumental achievements—the founding of the Indian Statistical Institute (ISI) and the creation of the National Sample Survey (NSS). The ISI, established in 1931, not only has produced a generation of statisticians of world stature but has played a part equal to that of western countries in making statistics the highly developed, precise science that it is today.

In 1950 Mahalanobis established the NSS as a division within the ISI. The NSS rapidly grew into an agency (for which there hardly was a parallel elsewhere in the world) noted for its use of continuing sample surveys for the collection of socioeconomic and demographic data, which covered an entire country. It came to play so vital a role during the economic five-year plans in India that the NSS was taken over by the government and now continues to function as an integral part of the Ministry of Planning.

If Professor Mahalanobis had retired in 1947, these achievements alone would have assured him an enduring place in the history of statistics. However, during the last 20 years of his life he became intensely interested in the applications of statistics to problems of planning. He came up with a two-sector and, later, a four-sector model for economic growth. (In doing so, he had no pretensions to making a contribution to economic theory.) The resources of the ISI were fully mobilized to assist the Planning Commission, of which Mahalanobis was himself a member for the years 1955–1967.

Under his leadership, the ISI prepared the draft frame of the Indian second Five-Year Plan. The importance of this aspect of Mahalanobis' activity—though far removed from the world of academic statistics—can hardly be exaggerated. It provided the government with a scientific database for industrialization as well as a cadre of highly skilled statisticians at the very outset of the Five-Year Plans. If, in the course of the last 30 years, India has been transformed from a country with virtually no industrial base into one ranking among the first 10 countries of the world with the largest industrial output, some measure of credit belongs to Professor Mahalanobis and the statistical institutions created by him.

Mahalanobis' interests were not confined to statistics and the physical sciences. He had more than a casual interest in ancient Indian Philosophy, particularly in those aspects relating to multivalued logic. Bengali literature was his second love; he had, in his younger days, been a protégé of the great Indian poet Rabindranath Tagore. In the midst of an extremely busy life he found time to take a prominent part in the activities of the Brahmo Samaj, a religious reform movement which spearheaded a renaissance in the intellectual life of Bengal in the nineteenth and early twentieth centuries.

From 1949 until his death, Professor Mahalanobis was the Statistical Advisor to the government of India. He received many awards and honors. Among the ones which he treasured most were perhaps the Fellowship of the Royal Society and one of his country's highest civilian awards, the *Padma Vibhushan*.

Professor Mahalanobis published over 200 scientific papers in addition to numerous articles in Bengali and English in nontechnical journals. A bibliography of his scientific publications and a more detailed account of his work can be found in the biographical memoir written by C. R. Rao for the Royal Society [1].

REFERENCE

[1] Rao, C. R. (1973). *Biog. Mem. Fellows R. Soc.*, **19**, 455–492.

G. KALLIANPUR

Nightingale, Florence

Born: May 12, 1820, in Florence, Italy.
Died: August 13, 1910, in London, England.
Contributed to: social reform, nursing, demography, epidemiological studies, vital statistics.

Florence Nightingale is mainly known for improving the squalid hospital conditions at Scutari during the Crimean War and for her subsequent campaigns to reform the health and living conditions of the British army, the sanitary conditions and administration of hospitals, and the nursing profession. Energy, enthusiasm, and a crusading spirit apparently drove her to physical breakdown; for the last 20 years of her working life, she directed a flow of letters and reports from a couch in the Burlington Hotel and (later) from her house in London, whither those in authority came to consult her.

To a small circle the Lady of the Lamp is known as the Passionate Statistician [1], a name linking the eloquence Florence Nightingale mustered with the massive amounts of data she compiled to convince prime ministers, viceroys, secretaries, undersecretaries, and parliamentary commissions of the truths of her cause. Thus she wrote [4, p. 249]:

> We hear with horror of the loss of 400 men on board the Birkenhead by carelessness at sea; but what should we feel, if we were told that 1,100 men are annually doomed to death in our Army at home by causes which might be prevented?

A timely combination of factors makes her place in the development of applied statistics noteworthy. The first is her flair for collecting, arranging, and presenting facts and figures, developed along with her sense of vocation against immense family opposition. Cecil Woodham-Smith writes of the early 1840s [8, Chap. 4]:

> Notebook after notebook was filled with a mass of facts, compared, indexed and tabulated. . . . In the cold dark mornings she laid the foundation of the vast and detailed knowledge of sanitary conditions which was to make her the first expert in Europe. Then the breakfast bell rang, and she came down to be the Daughter of the Home.

She became aware that mortality statistics should be age-specific and that crude death rates can be misleading [5, p. 55]:

> In comparing the deaths of one hospital with those of another, any statistics are justly considered absolutely valueless which do not give the ages, the sexes and the diseases of all the cases. . . . There can be no comparison between old men with dropsies and young women with consumptions.

Thus in 1859 she articulated the need to allow for confounding effects in observational studies. Later she was to be influenced by Adolphe Quetelet* and his approach to quantifying social behavior (see Diamond and Stone [2]).

The second factor is the contemporary state of statistics. In 1850 no scientific system of tabulating or reporting mortality or morbidity statistics could be said to exist. When Miss Nightingale arrived at Scutari in 1854, no proper records were being kept, but she introduced there a uniform system of recording mortality rates without which she would have been unable to plead her case successfully. Her campaigns for reform included proposals to standardize and improve the collection and recording of health statistics. In this she shares credit with her adviser William Farr* (1807–1883), who brought a scientific approach to vital statistics in the United Kingdom and, as registrar-general, put it on a sound footing.

She had influential friends like Sidney Herbert, the Secretary of State for War, but the opposition to reform was entrenched in the army bureaucracy and in political high places, and it was determined to defeat her. Her single most effective weapon, the presentation of sound statistical data, is the third and most important factor to be considered. No major national cause had ever been championed primarily with such a weapon. She showed, for example [3], that "those who fell before Sebastopol by disease were above seven times the number who fell by the enemy." The opposition lost because her statistics were unanswerable (see Tables 1 and 2) and their publication led to an outcry.

The monumental privately printed work [4] that constituted her main reform effort contains an enormous number of statistical tabulations and some elaborate pie charts. It demonstrated the appalling conditions in Scutari and at the front in the Crimea and the gross neglect of the health of ordinary soldiers living in barracks in peacetime. There are suggestions of attempts by army administrators such as her chief antagonist, Dr. John Hall, Chief of Medical Staff in the Crimea, to cover up the extent of the tragedy with misleading statistics (see ref. [8, Chap. 10]). Miss Nightingale explained how some of these were devised [4, Sec. X, p. XXV]:

> In constructing a Table of Mortality we take 100 men, eight die the first year, there are left 92—2 die the second year, there are left 90. The usual method of stating this mortality would be to *take the hundred over again* and strike the difference, thus—
>
> 100 + 100 = 200 8 + 2 = 10 2)10(5

TABLE 1 Florence Nightingale's Statistics of Mortality at Different Periods during the Crimean War[a]

Month	Year	Deaths per 1000 (Living) per Annum
January	1855	1173½
	1856	21½
May	1855	203
	1856	8
January–May	1855	628
	1856	11½
Crimea, May 1856		
Line at home		18.7
Guard at home		20.4

[a]Ref. [3, p. 295].

TABLE 2 Florence Nightingale's Relative Mortality Statistics of the Army at Home and of the English Male Population at Corresponding Ages[a]

Ages		Deaths Annually to 1000 Living
20–25	Englishmen	8.4
	English soldiers	17.0
25–30	Englishmen	9.2
	English soldiers	18.3
30–35	Englishmen	10.2
	English soldiers	18.4
35–40	Englishmen	11.6
	English soldiers	19.3

[a]Ref. [3, p. 253].

> Therefore, it is a mortality of 5 per cent., per annum. Now, this is manifestly wrong, and gives the Secretary of State no idea of his *accumulated* loss.

If this method is applied to the data in Table 1 [4, p. 295], the extent of the horror in the Crimea becomes glossed over. The rates show numbers of deaths per 1000 living rather than per thousand out of the pool of manpower in the forces. In modern terms, the mortality for January 1855 at the height of the catastrophe can be measured roughly as 1173.5/(1000 + 1173.5) or 54 percent. Table 1 also indicates the dramatic reduction in death rates within 18 months.

The last three rows of Table 1 suggest that soldiers in the Crimea in May 1856 had a better chance of survival than those in the insanitary barracks in England. Table 2 [4, p. 253] provides the crux of the case for army reform at home. Miss Nightingale was ahead of her time as an epidemiologist, comparing cases (soldiers in barracks) with civilian controls. In one study [8, Chap. 13] she compared these in the same boroughs, thus controlling for any "district" effect. The confounding effect in the comparisons in Table 2 worked to reinforce her conclusions [4, p. 253]:

> The Army are picked lives and the inferior lives are thus thrown back. . . .
>
> The general population includes, besides those thus rejected . . . vagrants, paupers, intemperate persons, the dregs of the race, over whose habits we have little or no control. The food, clothing, lodging, employment, and all that concerns the sanitary state of the soldier, are absolutely under our control and may be regulated to the smallest particular.
>
> Yet with all this, the mortality of the Army, from which the injured lives are *subtracted*, is double that of the whole population, to which the injured lives are added.

Of this high mortality, she wrote [4, p. 251]: "The cause is made sufficiently plain by looking at their Barrack accommodation and their mode of life."

In this age, when the collection, editing, and presentation of data are commonplace in government affairs, it is not easy to appreciate the overwhelming difficulties involved in the statistical aspect of Miss Nightingale's enterprise. Nor should we overlook the prejudices against women in Victorian England. She never appeared in a public forum; her case was always presented for her or submitted in writing. At the *International Statistical Congress* in 1860, for example, a letter from her was read and adopted; it called upon governments to publish more extensive and frequent statistical abstracts [3]. A recent study of her life and work [7] presents a negative picture of her character, but it is too early to assess its implications on her reputation as a statistician.

Florence Nightingale was instrumental in the founding of a statistical department in the army. She was ahead of her time in advocating that censuses should include data on the sick and infirm and on the housing of the population [3] and in advocating the founding of a professorship in applied statistics at a British university. For her correspondence with Francis Galton* on the latter, see Pearson [6, pp. 414–424].

In 1858 she was elected a fellow of the Royal Statistical Society, and in 1874 an honorary member of the American Statistical Association.

REFERENCES

[1] Cook, E. (1913). *The Life of Florence Nightingale*, 2 vols, MacMillan, London.

[2] Diamond, M. and Stone, M. (1981). *J. R. Statist. Soc. A*, **144**, 66–79, 176–214, 332–351.

[3] Kopf, E. W. (1916). *J. Amer. Statist. Ass.*, **15**, 388–404. (An account of Miss Nightingale's achievements as a statistician, reprinted in *Studies in the History of Statistics and Probability*, Vol. 2, M. G. Kendall and R. L. Plackett, eds.)

[4] Nightingale, F. (1858). *Notes on Matters Affecting the Health, Efficiency, and Hospital Administration of the British Army. Founded Chiefly on the Experience of the Late War.* Harrison, London, (Printed privately.)

[5] Nightingale, F. (1859). *Notes on Nursing: What It Is and What It Is Not.* Harrison, London.

[6] Pearson, K. (1924). *The Life, Letters and Labours of Francis Galton*, Vol. II. Cambridge University Press, Cambridge, England.

[7] Smith, F. B. (1982). *Florence Nightingale: Reputation and Power.* Croom Helm, London (also St. Martin's Press, New York).

[8] Woodham-Smith, C. (1951). *Florence Nightingale.* McGraw-Hill, New York. (An eminently readable and enthralling biography that barely touches on the statistical aspects of her life.)

BIBLIOGRAPHY

Cohen, I.B. (1984). *Sci. Amer.*, **250**(3), 128–137. (A fascinating and informative article on Florence Nightingale as a pioneer in the uses of social statistics. Some of the pie charts, bar charts, and tables from ref. [4] and from the Royal Commission report on sanitary conditions in the army are reproduced.)

Eyler, J. M. (1979). *Victorian Social Medicine: The Ideas and Methods of William Farr.* Johns Hopkins University Press, Baltimore.

CAMPBELL B. READ

Slutskii (Slutsky), Evgenii Evgenievich

Born: April 7, 1880, in Yaroslavl province, Russia.
Died: March 10, 1948 in Moscow, USSR.
Contributed to: mathematical economics, econometrics, mathematical statistics.

Slutskii's academic inclination on entering the Physics–Mathematics Faculty of Kiev University in his father's native Ukraine in 1899 was toward physics, and he regarded mathematics as merely a tool. Caught up in the wave of revolutionary fervor current among students in the Russian empire of the time, he was finally expelled in 1902 and forbidden to enter any Russian tertiary institution. Subsidized and encouraged by a grandmother, he spent a period over the years 1902–1905 at the Polytechnic Institute in Munich, ostensibly studying mechanical engineering, for which he showed no aptitude, but in fact deepening his knowledge of economics, until the events of 1905 made it possible for him to begin studies in political economy at Kiev University. After another turbulent period, he ultimately completed his university studies in this area in 1911, with a gold medal for a study entitled "The Theory of Limiting Utility," in which he applied mathematical methods to economic problems. This probably led to the writing of the now famous, though long overlooked, paper [8] on the theory of consumer behavior, in which ideas of F. Y. Edgeworth* and V. Pareto are developed.

His interest in statistics, and the theory of probability as a theoretical basis for it, was stimulated by a personally presented copy of the elementary book [5] by the eminent physiologist A.V. Leontovich, which exposited techniques of Gauss* and Pearson.* This led Slutskii to produce the book [6], said to be, for its time, a significant contribution to statistical literature. As a result, in 1913 Slutskii was appointed lecturer at the Kiev Commercial Institute, where he worked, rising to the rank of professor, until 1926, when he left for Moscow.

In training, continuing interests, and early career, Slutskii's development will be seen to parallel closely that of his equally eminent countryman A. A. Chuprov*; it is therefore not surprising that the two men established close academic contact that continued until the emigré Chuprov's untimely death in 1926. In Moscow Slutskii decided to pursue theoretical problems of statistics (although, as will be seen, the direction of these investigations was influenced by his interests in economics and geophysics), working at least until 1931 at the Koniunkturnyi Institute (an institute for the study of business cycles) and the Central Statistical Office. From 1931 to 1934 he worked at the Central Institute of Meteorology, from 1934 at Moscow State University, and from 1938 until his death at the Steklov Mathematical Institute of the USSR Academy of Sciences.

One of Slutskii's first papers in statistics [7] relates to fitting a function $f(x; \beta_1, \ldots, \beta_r)$ of one variable x (the β's being parameters) when there are repeated readings on response Y for each value of x considered, the system being normal and possibly heteroscedastic. If there are n_i responses for the value x_i, and their average is $\overline{Y}_i$, $i = 1, \ldots, N$, then

$$\overline{Y}_i = f(x_i;\ \beta_1, \ldots, \beta_r) + \epsilon_i, \qquad i = 1, \ldots, N,$$

where the ϵ_i are independent, $\epsilon_i \sim N(0, \sigma_i^2/n_i)$, and σ_i^2 is the variance of the normal response corresponding to setting x_i. Slutskii proposes to estimate $\boldsymbol{\beta}$ by minimizing

$$\chi^2 \sum_{i=1}^{n} \frac{n_i\{\overline{Y}_i - f(x_i;\ \beta_1, \ldots, \beta_r)\}^2}{\sigma_i^2},$$

which he recognizes as a chi-square variable, so we may (if we assume σ_i^2 known) regard this procedure as an early instance of minimum chi-square estimation. [It is also obviously maximum-likelihood estimation and (without the normality assumption) an instance of weighted least squares.]

Slutskii's views on the abstract formalization of the probability calculus [9] do not refer to any specific axiomatic system, but are of general philosophical kind as to what features a rigorous mathematical formalization of it should contain. In particular, no subjective elements should enter. There are no references to other authors, but it is likely [10] Slutskii was aware of the attempts at formalization by Bernstein* and von Mises.

The paper in ref. [10] has had fundamental influence in elucidating the notion of convergence in probability. In this Slutskii was anticipated by Cantelli* [2], but in some respects he went further, formalizing the notion of "stochastic asymptote," which generalizes that of "stochastic limit" (if for a sequence of random variables $\{X_n\}$, $n \geq 1$, $X_n - EX_n \xrightarrow{p} 0$, EX_n is said to be the stochastic asymptote of X_n) and establishing a form of what is now known as Slutsky's theorem. The paper also contains, as does subsequent work, results on the weak law of large numbers (from which the convergence-in-probability notions derive); in this direction the influence of Chuprov* and interaction with M. Watanabe are apparent.

A number of Slutskii's papers treat stationary sequences and have had significant influence on the development of time-series analysis. These contributions were stimulated by manifestations and investigations of periodicity. The most famous is ref. [11], in which the *Slutsky effect* is demonstrated: essentially that repeated filtering of even a purely random sequence of readings may in the limit produce a purely periodic sequence. More generally this paper made manifest that observed quasi-periodicity may simply be a result of statistical stationarity rather than the result of real periodic effects. Slutskii was aware of the work of G. U. Yule* in a similar direction.

The notion of stochastic limit also led him to study random functions more generally (in particular to develop the notions of stochastic continuity, differentiability, and integrability), and he may thus be regarded, with A. I. Khinchin, as one of the founders of the theory of stationary random processes. Slutskii's approach to this work was centered on the consideration of moments of the random function (under the influence of Markov* and Chuprov, it would seem) up to fixed order, which notion more recently occurs in the guise of second-order (wide-sense) stationary stochastic processes. His best-known work in the area is, perhaps, ref. [12].

REFERENCES

[1] Allen, R. G. D. (1950). The work of Eugen Slutsky. *Econometrica*, **18**, 209–216. (By an eminent mathematical economist who rediscovered ref. [8] in 1936. Includes a bibliography, pp. 214–216.)

[2] Cantelli, F. P. (1916). La tendenza ad un limite nel senzo del calcolo delle probabilità. *Rend. Circolo. Mat. Palermo*, **16**, 191–201.

[3] Gnedenko, B. V. (1960). Evgenii Evgenievich Slutskii (in Russian). In *E. E. Slutskii, Izbrannie Trudy*. Izd. AN SSSR, Moscow, USSR, pp. 5–11. (Biographical sketch and survey of his work.)

[4] Kolmogorov, A. N. (1948). Evgenii Evgenievich Slutskii (1880–1948) (in Russian). *Uspekhi Mat. Nauk*, **3**, 143–151. (Obituary.)

[5] Leontovich, A. (1911). *Elementarnoe Posobie k Primeneniu Metodov Gaussa I Pearsona pri Otsenke Oshibok v Statistike i Biologii.* Kiev, Ukraine.

[6] Slutskii, E. E. (1912). *Posobie k Izucheniu Nekotorikh Vazhneishikh Metodov Sovremennoy Statistiki.* Kiev, Ukraine.

[7] Slutsky, E. E. (1914). On the criterion of goodness of fit of the regression lines and on the best method of fitting them to data. *J. R. Statist. Soc.*, **77**, 78–84. (Reprinted in Russian in ref. [13, pp. 12–17]; commentary by N. V. Smirnov on p. 283.)

[8] Slutsky, E. E. (1915). Sulla teoria del bilancio del consumatore. *Giorn. Economisti 3*, **51**, 1–26. [In Russian: *Ekon.-mat. Metody*, **1** (1963).]

[9] Slutskii, E. E. (1922). On the problem of the logical foundations of the probability calculus (in Russian). *Vestnik Statist.*, **12**, 13–21. (A version with some changes and corrections in text appeared in 1925, in a collection of essays in statistics in memory of N. A. Kablukov. Reprinted in ref. [13, pp. 18–24].)

[10] Slutskii, E. E. (1925). Über stochastische Asymptoten und Grenzwerte. *Metron*, **5**, 3–89. (In Russian in ref. [13, pp. 25–90].)

[11] Slutskii, E. E. (1937). The summation of random causes as the source of cyclic processes. *Econometrica*, **5**, 105–146. [An earlier version appeared in Russian in *Voprosy Koniunktury*, **3**, 34–64 (1927). Reprinted in ref. [13, pp. 99–132].

[12] Slutskii, E. E. (1938). Sur les fonctions aléatoires presque périodiques et sur la décomposition des fonctions aléatoires stationnaires en composantes. *Actual. Sci. Ind. Paris*, **738**, 33–55. (In Russian in ref. [13, pp. 252–268].)

[13] Slutskii, E. E. (1960). *Izbrannie Trudy.* Izd. AN SSSR, Moscow, USSR. (Selected works. Contains commentaries by N. V. Smirnov, B. V. Gnedenko, and A. M. Yaglom on aspects of Slutsky's work, a photograph, and a complete bibliography.)

E. SENETA

Westergaard, Harald Ludvig

Born: April 19, 1853, in Copenhagen, Denmark.
Died: December 13, 1936, in Copenhagen, Denmark.
Contributed to: mathematical economics, data analysis, demography.

H. L. Westergaard

Although Harald Westergaard was best known as an economist, he made substantial contributions to the statistical assessment of demographic and economic data. It is with this latter aspect (the "most original," according to Brems [1]) of his work that we are primarily concerned.

Westergaard was born and died in Copenhagen, and worked all his life in that city, apart from studying in England and Germany in 1877–1878, after obtaining degrees in mathematics (in 1874) and economics (in 1877) from the University of Copenhagen. While he was in England, he met the British economist W. S. Jevons and became interested in the application of mathematics to problems in economics. (See ref. [2].)

In 1880–1882 he worked for a life-insurance company and developed interests in demography and statistical method. From this there came a work [3] on mortality laws. In 1883 he joined the faculty at the University of Copenhagen as a docent. He remained there for the remainder of his working life, becoming professor in 1886 and retiring in 1924. During the period 1886–1935—nearly 50 years—he was a director of the Copenhagen Fire Insurance Company.

Westergaard was in advance of his time in regard to the application of mathematics to economic theory, but somewhat behind it in appreciation of the use of tools of statistical inference—notably correlation analysis. On the other hand, he had a superb ability to perceive rational groupings and subgroupings underlying data, without the assistance of formal testing and searching procedures. "In the fine art of listening to the voice of numbers, few practitioners have had an ear as sensitive as Westergaard's" [1].

Later in life, he became interested in the historical study of "statistics," culminating in the classic book [4]. As was natural, this concentrated on historical development from *"Staatenkunde"* ("comparative description of States" was Westergaard's description) through "political arithmetic" to demography and other applications. There is, indeed, a chapter on calculus of probability, but this was treated as an interesting digression from the main discussion. The book presents the accumulated results of many years of research and thought, very clearly expressed, and even today it is a mine of information on the growth of statistical method, particularly in the context of its use in government and economics.

Westergaard married Thora Alvilda Koch in 1881. Subsequent to her death in 1891, he married Lucie Bolette Blaedel in 1892. His son, Harald Malcolm (1888–1950), had a distinguished career as an engineer, emigrating to the United States and becoming a professor at the University of Illinois in 1927, and then professor of civil engineering and dean of the graduate school of engineering at Harvard University in 1937. His daughter, Bodil, from his second marriage, was born in 1893. She married Erik Rydberg (of Swedish origin), who was later a professor of medicine in the University of Copenhagen.

Westergaard was a deeply religious person, active for many years in the evangelical wing of the Church of Denmark, and was President of the Danish Bible Society from 1927. He was very generous and helpful to those needing financial and other assistance, especially students and the elderly.

REFERENCES

[1] Brems, H. (1988). Westergaard, Harald Ludvig (1853–1936). In *The New Palgraves Dictionary of Economic Science*, vol. 4, pp. 898–899.

[2] Kaergård, N. (1995). Cooperation not opposition: Marginalism and socialism in Denmark, 1871–1924. In *Socialism and Marginalism in Economics*, I. Steedman, ed. Routledge, London and New York, pp. 87–101.

[3] Westergaard, H. (1882). *Die Lehre von der Mortalität and Morbidität.* Fischer, Jena, Germany (2nd ed., enlarged, 1901).

[4] Westergaard, H. (1932). *Contributions to the History of Statistics.* King, London. [Reprint, 1969, published by Mouton Publishers (The Hague and Paris) and S. R. Publishers (Wakefield, England).]

Willcox, Walter Francis

Born: March 22, 1861, in Reading, Massachusetts, USA.
Died: October 30, 1964, in Ithaca, New York, USA.
Contributed to: American Statistical Association, history of statistics, vital statistics.

In the course of his remarkably long life, Walter Francis Willcox was influential in the development and structuring of statistics, and especially vital statistics, as an independent subject in the general field of scientific method. He taught "an elementary course in statistical methods with special treatment of vital and moral statistics" at Cornell University in 1892–1893, soon after joining the faculty in 1891. He served at Cornell for 40 years, retiring in 1931.

However, his work was by no means confined to university teaching. He was active in working on the 12th United States Census (1900) from 1899 to 1901, and initiated and supervised a large volume of *Supplementary Analysis and Derivative Tables*. Willcox was in steady demand as a consultant on collection and analysis of demographic statistics, especially in regard to international migration. A topic attracting much of his attention was the basis of apportionment of seats in the House of Representatives of the U.S. Congress. Much interesting detail is available in ref. [1], which was produced in connection with Willcox's 100th birthday, and in the obituary [2], which contains personal reminiscences.

Willcox was president of the American Statistical Association in 1912 and of the International Statistical Institute in 1947. He was very active in the work of both of these organizations for many years, and compiled the first extensive list [3] of definitions of statistics, from G. Achenwall* in 1749 to R. A. Fisher* in 1934.

In his later years he interested himself in the history of statistics—in particular of demographic statistics—and produced a number of insightful analyses in this field.

REFERENCES

[1] Leonard, W. R. (1961). *Amer. Statist.*, **15**(1), 16–19.
[2] Rice, S. A. (1964). *Amer. Statist.*, **18**(5), 25–26.
[3] Willcox, W. F. (1935). *Int. Statist. Inst. Rev.*, **3**, 388–399.

Winkler, Wilhelm

Born: June 29, 1884 in Prague, Bohemia, Austrian Empire
Died: September 3, 1984 in Vienna, Austria
Contributed to: demography, government statistics, formation of university studies in statistics in Austria.

Wilhelm Winkler was a son of Anne and Julius Winkler, a music teacher in Prague. He had three older and four younger siblings, and at the age of 13 he had to go to work to assist the family finances. His student career at the Kleinseitner Gymnasium and the law school of the German-based Karl Friedrich University in Prague were achieved in the face of severe difficulties, though it is likely that these contributed to the later mature development of his character and talents.

Winkler's career is a remarkable conjunction of activity in two main spheres—the practical needs of government and the intellectual austerity of scientific research. After graduating from the university he spent a brief period in law practice, followed by service as a one-year volunteer in the Austrian Army. He then obtained a post, in 1909, in the Bohemian State Statistical Bureau, reflecting a growing interest in statistical, as opposed to strictly legal, matters. According to Adam (1984), his relations with his Czech colleagues were harmonious. He also enjoyed attending statistical seminars at the University. During this period, his interest in application of methods of mathematical statistics in social and economic matters continued to grow. He attended courses in higher mathematics at the Prague Technical High School to develop further his abilities in this direction.

This "calm and fruitful period" (Adams' description) was ended by the outbreak of the World War I. Winkler rejoined the Army. He fought with distinction, being twice decorated for gallantry, but was severely wounded in the Fourth Battle of the Isonzo and hospitalized in Prague for six months from November 1915. While in the hospital, a chance encounter with a former teacher—A. Spiethoff—was a turning-point in Winkler's career.

Spiethoff was in charge of a scientific committee on war economy at the Ministry of War in Vienna. He invited Winkler to join the staff of this committee. Winkler arrived in Vienna in June 1916 to take up his new duties. Unfortunately, Spiethoff had been replaced by a general with little sympathy for applications of statistical theory and methods in economic matters. However, Winkler made friends with some fellow members, that ultimately greatly influenced his career.

The committee disbanded and in 1918 Winkler became Secretary of the State Office for Military Affairs, despite opposition from establishment figures with rigidly traditionalist outlooks. In this capacity, Winkler was a member of the Austrian delegation at the Versailles Peace Conference in 1919. In 1920 he was seconded to the newlyformed Federal Statistics Office (previously Central Statistics Commission, and later Austrian Central Statistics Office). Despite some initial difficulties, he appreciated returning to work of the same kind as in his pre-war days in Prague.

In 1921 he also became a Privat-Dozent in the University of Vienna. From this time onward, his careers in government service and the university progressed in tandem. In 1925 he became chief of the Division of Population Statistics, and in 1929 he was appointed an Extraordinary Professor in the University, in charge of work in population statistics.

In this capacity he reorganized the teaching of statistics in the University, in the face of opposition similar to that he had encountered in the civil service, this time from hidebound conservative faculty. However, Winkler's work was now attracting international recognition, with his election to membership (which at that time was severely restricted) of the International Statistical Institute (ISI) in 1926. Winkler took an active part in ISI affairs until the fateful year 1938. After the Nazi invasion of Austria, Winkler was dismissed from both of his posts: in government service and at the university. [Adam (1984) ascribes this, in part, to the influence of colleagues estranged by Winkler's modernizing activities.]

There followed seven years of strife and difficulty, during which, however, Winkler found enough time and spirit to work on his book *Typenlehre der Demographie* ("*Basic Course in Demography*") which was ultimately published in 1952 by the Austrian Academy of Sciences.

At the end of World War II in 1945, Winkler was rehabilitated and appointed an Ordinary (Full) Professor in the University of Vienna. Here he pursued the reorganization of statistical courses with renewed vigor, until his retirement in 1949. After that time, he continued working as an Honorary Professor in the (then) Hochschule für Welthandel, and was Dean of the Faculty of Law and Social Sciences in the University of Vienna, 1950–1951, finally retiring from the University in 1955. Winkler's achievements received further national and international recognition, in the form of honorary membership and presidency of the ISI (1965), honorary fellowship of the Royal Statistical Society (1961), corresponding member of the Austrian Academy of Sciences (1952), and honorary Doctorates of Science from the Universities of Munich and Vienna.

In later years, Winkler saw much of his groundwork on the reorganization of statistics and the penetration of modern statistical methods in theoretical statistics, population statistics, economic statistics *et alia* come to gratifying fruition. As an example, the inception, in 1969, of the Linz Scientific Information Program at the Linz Hochschule für Sozial und Wïrtschaftwissenséhaften, lead to Statistics becoming a completely accepted subject in Austrian university study.

Winkler married Clara Deutsch in 1918. After her death in 1956, he married Franziska Haeker in 1958. He had three sons and one daughter.

Much of the information in this entry comes from the editorial introduction by Adam (1984) to a *Festschrift* on the occasion of Winkler's 100th birthday. Adam gives further details of Winkler's life, of the careers of some of his distinguished students, and in particular, some personal reminiscences of his later life. Winkler is widely regarded as the "father of Austrian university statistics." The obituary by Schmetterer (1985) contains further details and information on Winkler's personal life is derived from an entry in the *International Who's Who* (1987).

BIBLIOGRAPHY

Adam, A. (1984). Wilhelm Winkler, Vater der Österreichischen Universitätsstatistik, Leben und Wirken, In *Festschrift für Wilhelm Winkler* (ed. A. Adam) Vienna, Austria: Orac.

International Who's Who, The (1987) 51st Edition, London: Europa, pp. 1598–1599.

Schmetterer, L. (1985). "Wilhelm Winkler, 1884–1984," *J. R. Statist. Soc. Ser. A*, **148**, 67.

SECTION **6**

Applications in Medicine and Agriculture

Gosset, William Sealy ("Student")

Born: June 13, 1876, in Canterbury, England.
Died: October 16, 1937, in Beaconsfield, England.
Contributed to: sampling distributions, design of experiments, statistics in agriculture.

W. S. Gosset entered the service of the Guinness brewery business in Dublin, Ireland, in 1899, after obtaining a degree in chemistry at Oxford, with a minor in mathematics. He was asked to investigate what relations existed between the quality of materials such as barley and hops, production conditions, and the finished product. These practical problems led him to seek exact error probabilities of statistics from small samples, a hitherto unresearched area. His firm, which would later require him to write under the pseudonym "Student," arranged for him to spend the academic year 1906–1907 studying under Karl Pearson* at the Biometric Laboratory in University College, London. From Pearson, Gosset learned the theory of correlation and the Pearson system of frequency curves, both of which helped him in developing statistical techniques to analyze small-sample data.

"The study of the exact distributions of statistics commences in 1908 with Student's paper *The Probable Error of a Mean,"* wrote R. A. Fisher* [5, p. 22]. In this famous paper [11; 15, pp. 11–34], which came from his period at the Biometric Laboratory:

1. Gosset conjectured correctly the distribution of the sample variance s^2 in a normal sample.
2. Unaware of work by Abbe* and Helmert,* he proved that s^2 and the mean $\bar{x}$ of such a sample are uncorrelated and conjectured correctly that they are independent.
3. In essence he derived the t-distribution named after him, as a Pearson type VII frequency curve.

Gosset used some data which had appeared in the first volume of *Biometrika* to test the fit of the data to the theoretical distribution of $z = (\bar{x} - \mu)/s$. With the height and left-middle-finger measurements of 3000 criminals, he adopted a Monte Carlo method to generate two sets of 750 "samples" of size 4, and obtained satisfactory fits for both sets of data. The frequency curves for both sets deviated slightly from normality, but Gosset noted that "This, however, appears to make very little difference to the distribution of z."

In Gosset's notation, $s^2 = \sum (x_i - \bar{x})^2/n$ for a sample of size n. It was Fisher who realized later that sampling distributions of regression coefficients could be related to z when based on a more general definition incorporating the notion of degrees of freedom, and made the change from z to what is now called t (see Eisenhart [3]). In 1908, Gosset also correctly surmised the distribution of the sample correlation coefficient in a bivariate normal sample when the population correlation coefficient is zero [12; 15, pp. 35–42].

Gosset did not establish his results rigorously, but his achievements initiated a period of intense research into the distributions of other sampling statistics. Fisher paid tribute to his lead in this direction [4] and acknowledged the "logical revolution" in thinking effected by his approach to research, because he sought to reason inductively from observations of data to the inferences to be made from them [1, p. 240]. But, as frequently happens in scientific discovery, Gosset's work was largely unrecognized for several years; his tables of percent points of the *t*-distribution were used by few researchers other than those at Guinness and Rothamsted until the late 1920s. E. S. Pearson [8; 10, pp. 349–350] lists possible reasons for this lack of interest.

The collaboration of Gosset and Fisher in compiling, correcting, and publishing these tables lasted from 1921 to 1925, and is well documented in a collection of letters from Gosset to Fisher (see Box [2] and Gosset [6]), which also give many personal insights into the human side of Gosset's character. For example, when he was invited to referee the Master's thesis of one of Fisher's assistants, he wrote:

> I suppose they appointed me because the Thesis was about barley, so of course a brewer was required, otherwise it seems to me rather irregular. I fear that some of Miss Mackenzie's mathematics may be too 'obvious' for me. [6, letter 40]

The last refers to Fisher's use in mathematical work of the word "evidently," which for Gosset meant "two hours hard work before I can see why" [6, letter 6].

Gosset's work with the brewery led him to agricultural experiments. His work with a Dublin maltster, E. S. Beaven, and others led him to favor balanced designs. By comparing pairs of treatments in neighboring plots, he said, the correlation between them would be maximized and the error in treatment differences would be minimized. During 1936–1937 Gosset defended balanced designs such as those based on Beaven's half-drill strip method against the randomized designs of Fisher (see refs. [13], [14] and [15, pp. 192–215]). The models over which they argued may have needed randomization with some balance, but the practicality for ordinary farmers of performing and then analyzing balanced experiments was as important to Gosset as any theoretical consideration. His death in 1937 prevented them from resolving their differences.

Articles about "Student" by Fisher [4], McMullen and E. S. Pearson [7], and Pearson and Kendall [10, pp. 354–403] include a bibliography of his publications. In the latter Pearson quotes a letter of 1926 in which Gosset introduced the concept of "alternative hypothesis" in hypothesis testing for the first time, and questioned the effect of nonnormality on sampling distributions; Gosset was frequently credited with being one step ahead of his contemporaries. A number of his letters also appear in Pearson [8, 9; 10, pp. 348–351, 405–408] (see also Box [1, 2] and Eisenhart [3]).

REFERENCES

[1] Box, J. F. (1978). *R. A. Fisher. The Life of a Scientist.* Wiley, New York. (This biography contains insights into Gosset's influence on and friendship with Fisher.)

[2] Box, J. F. (1981). *Amer. Statist.*, **35**, 61–66. (A fascinating discussion of Gosset's correspondence and friendship with Fisher.)

[3] Eisenhart, C. (1979). *Amer. Statist.*, **33**, 6–10. (A mathematical discussion of the change from z to t.)

[4] Fisher, R. A. (1939). *Ann. Eugen.*, **9**, 1–19. (An appreciation of "Student" and the t-test.)

[5] Fisher, R. A. (1950). *Statistical Methods for Research Workers*, 11th ed. Oliver & Boyd, Edinburgh.

[6] Gosset, W. S. (1970). *Letters from W. S. Gosset to R. A. Fisher 1915–1936.* Arthur Guinness, Dublin. (Private circulation.)

[7] McMullen, L. And Pearson, E. S. (1939). *Biometrika*, **30**, 205–250. (Pearson's penetrating critique of "Student's" achievements is as relevant 40 years later as it was in 1939).

[8] Pearson, E. S. (1967). *Biometrika*, **54**, 341–355.

[9] Pearson, E. S. (1968). *Biometrika*, **55**, 445–457.

[10] Pearson, E. S. and Kendall, M. G., eds. (1970). *Studies in the History of Statistics and Probability.* Hafner, New York. (This collection of *Biometrika* reprints includes the preceding three references.)

[11] "Student" (1908). *Biometrika*, **6**, 1–25. (The classic paper introducing the t-distribution.)

[12] "Student" (1908). *Biometrika*, **7**, 302–309.

[13] "Student" (1936). *J. R. Statist. Soc.* (Suppl.), **3**, 115–136. (Includes an illuminating discussion of balanced vs. Randomized designs by Fisher, Wishart, Yates, and Beaven.)

[14] "Student" (1938). *Biometrika*, **29**, 363–379.

[15] "Student" (1942). *Collected Papers.* Biometrika, London.

CAMPBELL B. READ

Goulden, Cyril Harold

Born: June 2, 1897, in Bridgend, Wales.
Died: February 3, 1981, in Ottawa, Canada.
Contributed to: agricultural statistics, design and analysis of experiments, statistical education.

Cyril Goulden was one of Canada's leading agricultural scientists of the twentieth century. At the same time he was also an early Canadian pioneer in statistical methodology. The research in agriculture for which he is best known is his work on the development of rust-resistant wheat. Stem rust in wheat was a severe problem with enormous economic consequences—a severe epidemic of stem rust in 1916 in the Canadian prairies resulted in an estimated loss of 100 million bushels of wheat. In the late 1920s, early in the course of his agricultural work on wheat rust, Goulden became interested in statistical methodology. Within a short period of time Goulden developed into a very able applied statistician. He was quick to appreciate and then to use newly developed statistical methods in agricultural experiments which were of national importance to Canada. Goulden was a tireless promoter of the use of statistical methods, especially in the areas of agriculture and genetics.

Born in 1897 in Bridgend, Wales, Goulden emigrated with his family to Canada, settling near Yorkton, Saskatchewan. In 1915 he entered the University of Saskatchewan and graduated with a B.S.A. in 1921. He took his M.S.A. from Saskatchewan in 1923 and then went to the University of Minnesota for his doctoral degree. Graduating in 1925, he returned to Canada where he was appointed to the Dominion Rust Research Laboratory in Winnipeg. In the years 1925 through 1930 Goulden taught himself statistical methods and the mathematics, including calculus, basic to statistics [16].

Goulden's first use of statistics in agricultural experiments, which date from about 1925, was influenced by Gosset* and Pearson* [14, 17]. His data analyses relied heavily on the calculation in large samples of correlations between yield and other variables such as susceptibility to rust [14]. Very early in his career Goulden recognized the applicability of small-sample statistics to his work. By 1926, after reading an article published by Gosset [19] in an agronomy journal under his usual pseudonym of Student, Goulden quickly embraced Gosset's work on small-sample statistical methods and was promoting it to others in his field [5]. At about this time Goulden began corresponding with R. A. Fisher* on statistical subjects [5]. Possibly as a result of this correspondence, Goulden read Fisher's *Statistical Methods for Research Workers*. Goulden [15] applied techniques in Fisher's book to examine data on crosses of wheat varieties to determine their resistance to rust. He also began to promote Fisher's statistical methods among agronomic researchers in Canada [6].

In the summer of 1930 Goulden took a leave of absence from the Dominion Rust Research Laboratory and went to Rothamsted Experimental Station to study statistics under Fisher [2, 16]. He described what had learned in an expository article written the next year [7]. The impression that Goulden made on Fisher must have been very positive. During his trip to the United States in 1936, which involved extensive touring and lecturing, Fisher made an out-of-the-way side visit for a few days to see Goulden in Winnipeg [2].

Goulden used statistical methodology as a tool in much of his agricultural research work. In addition, Goulden wrote two papers in which statistical techniques are either developed or evaluated. At all times the motivation behind this work remained the desire to develop rust-resistant wheat. In 1936 Yates* [20] introduced lattice designs for experimental techniques. These designs were developed specifically for agricultural field trials in which large numbers of treatments are to be compared. This was exactly the situation facing Goulden. Very soon after Yates' initial paper on lattice designs using data from uniformity trials, Goulden [10] carried out an efficiency study in which lattice and other incomplete block designs were compared with randomized block designs. Later, Goulden [12] developed a method of analysis for lattice designs which included quadruple and quintuple lattices. Unfortunately for Goulden, this work was overshadowed by the publication of similar material in Cochran and Cox [4], an early version of which had been distributed in mimeograph form in 1944 [4, p. v], the year Goulden published his paper on lattice designs. Goulden received a copy of this mimeograph version, which he noted in his paper [12].

The work for which Cyril Goulden is best known in the statistical community is his book *Methods of Statistical Analysis* [11, 13]. The writing of the book resulted not only from his own research experience (several of the numerical examples are taken from his own experimental work) but also from his experience as a part-time teacher at the University of Manitoba. Goulden began offering a statistics course as early as 1931 with Fisher's *Statistical Methods for Research Workers* as the assigned textbook. In 1936 the basis of his own textbook for the course was in mimeograph form [8]. It was revised and expanded the next year, but remained in mimeograph form [9]. The first edition of the Wiley publication of the book appeared in 1939 [11]. A considerably expanded and revised second edition appeared in 1952 [13]. Goulden's main influence on statistics is through *Methods of Statistical Analysis*. It was a standard textbook in Canada and the USA and was translated into many languages.

Goulden's career as a part-time teacher at the University of Manitoba continued until 1948, when he moved to Ottawa to become chief of the Cereal Crops Division in what is now Agriculture Canada. The move to Ottawa was also another turning point in Goulden's career. He became more absorbed in administrative duties and less involved in day-to-day research activities. In 1955 Goulden was appointed director of the Experimental Farms Service of Agriculture Canada, and four years later was promoted to assistant deputy minister with responsibility for the newly formed Research Branch of Agriculture Canada. As an administrator Goulden brought about several organizational improvements to the research arm of Agriculture Canada. Goulden retired from the civil service in 1962 but continued to be active. He designed several exhibits for "Man the Provider" for the international exhibition, Expo 67, held in Montreal.

If reviews are anything to go by, the first edition of *Methods of Statistical Analysis* was generally well received. Bartlett [1] commented that the book would "prove a very useful work for those engaged in agricultural experimentation." Cochran [3] commented on the practical examples in the book, saying that "Here the author's ability and shrewd insight as a practical statistician are apparent." Comments made by Owen [18] on the second edition of the book are applicable to much of Goulden's published work. Owen wrote that "Dr. Goulden . . . writes attractively in a clear, simple and affable style. . . . Dr. Goulden is a 'tactful' writer; a light and skillful touch is in evidence throughout."

Cyril Goulden was the recipient of many honors during his career. He was an elected fellow of the Royal Society of Canada (1941), an elected fellow of the American Statistical Association (1952) and an honorary member of the Statistical Society of Canada (1981). In 1958 he served as president of the Biometrics Society. He was awarded an honorary LL.D. from the University of Saskatchewan (1954) and an honorary D.Sc. From the University of Manitoba (1964). In 1970 he received an Outstanding Achievement Award from the University of Minnesota, and ten years later was named to the Canadian Agricultural Hall of Fame.

REFERENCES

[1] Bartlett, M. S. (1939). Statistics in theory and practice. *Nature*, **144**, 799–800.

[2] Box, J.F. (1978). *R. A. Fisher: The Life of a Scientist.* Wiley, New York.

[3] Cochran, W. G. (1940). Review of *Methods of Statistical Analysis. J. R. Statist. Soc. A*, **103**, 250–251.

[4] Cochran, W. G. and Cox, G. M. (1950). *Experimental Designs.* Wiley, New York.

[5] Goulden, C. H. (1927). Some applications of biometry to agronomic experiments. *Sci. Agric.*, **7**, 365–376.

[6] Goulden, C. H. (1929). *Statistical Methods in Agronomic Research . . . a Report Presented at the Annual Conference of Plant Breeders, June, 1929.* Canadian Seed Growers' Association, Ottawa.

[7] Goulden, C. H. (1931). Modern methods of field experimentation. *Sci. Agric.*, **11**, 681–701.

[8] Goulden, C. H. (1936). *Methods of Statistical Analysis.* Burgess, Minneapolis.

[9] Goulden, C. H. (1937). *Methods of Statistical Analysis*, rev. ed. Burgess, Minneapolis.

[10] Goulden, C. H. (1937). Efficiency in field trials of pseudo-factorial incomplete randomized block methods. *Can. J. Res.*, **15**, 231–241.

[11] Goulden, C. H. (1939). *Methods of Statistical Analysis.* Wiley, New York.

[12] Goulden, C. H. (1944). A uniform method of analysis for square lattice experiments. *Sci. Agric.*, **23**, 115–136.

[13] Goulden, C. H. (1952). *Methods of Statistical Analysis, 2nd Edition.* Wiley, New York.

[14] Goulden, C. H. and Elders, A. T. (1926). A statistical study of the characters of wheat varieties influencing yield. *Sci. Agric.*, **6**, 337–345.

[15] Goulden, C. H., Neatby, K.W., and Welsh, J. N. (1928). The inheritance of resistance to *Puccinia graminis tritici* in a cross between two varieties of *Triticum vulgare. Phytopathology*, **18**, 631–658.

[16] Johnson, T. (1967). The Dominion Rust Research Laboratory, Winnipeg, Manitoba, 1925–1957. Unpublished report, seen courtesy of the Canadian Agriculture Research Station, Winnipeg.

[17] Kirk, L. E. and Goulden, C. H. (1925). Some statistical observations on a yield test of potato varieties. *Sci. Agric.*, **6**, 89–97.

[18] Owen, A. R. G. (1953). Review of *Methods of Statistical Analysis, 2nd Edition. J. R. Statist. Soc. A*, **126**, 204–205.

[19] Student (1926). Mathematics and agronomy. *J. Amer. Soc. Agron.*, **18**, 703–719.

[20] Yates, F. (1936). A new method of arranging variety trials involving a large number of varieties. *J. Agric. Sci.*, **26**, 424–455.

DAVID BELLHOUSE

Hill, Austin Bradford

Austin Bradford Hill.

Born: July 8, 1897, in London, England.
Died: April 18, 1991, near Windermere, Cumbria, England.
Contributed to: epidemiology, controlled clinical trials, statistical education in medicine.

Austin Bradford Hill (affectionately known as "Tony" to his friends and colleagues) was a son of Sir Leonard Erskine Hill, F.R.S., an eminent physiologist, and came from a distinguished family. His grandfather, Birkbeck Hill, was the editor of Boswell's *Life of Johnson*.

Hill spent his childhood in Loughton, Essex, and was educated at Chigwell School. During World War I he joined the Royal Naval Air Service, escorting battleships and searching for submarines off Eastern Scotland. He was later posted to the Aegean, providing air defense for monitors bombarding Turkish positions. He was invalided out and returned to England in 1917, very ill with pulmonary tuberculosis.

He had to abandon plans for a career in medicine, and spent the next eighteen months in bed. He was an invalid for four further years. The prominent statistician and epidemiologist, Major Greenwood—who had been his father's demonstrator—suggested that he take correspondence courses for a degree in economics from the University of London. Following this advice, Hill graduated in 1922 but had no wish to pursue a career as an economist, being strongly attracted to epidemiology.

With the help of Greenwood, who was, at that time, chairman of the Statistical Committee of the Medical Research Council (MRC), Hill obtained a grant to study the effects on health of migration from rural to urban areas. He conducted a survey covering almost every village in Essex. His findings were reported in special report No. 95 of the MRC, which became well known. He earned a Ph.D. degree from the University of London based on a dietary study, and attending Karl Pearson's* courses on statistics at University College. When Greenwood became professor of Epidemiology and Vital Statistics at London School of Hygiene and Tropical Medicine (LSHTM) in 1927, Hill joined him, becoming a reader in 1933. Up to the outbreak of World War II, he investigated the effects of environment in various industries, notably in the cotton industry.

In 1937, Hill's famous *Principles of Medical Statistics* was published, which went through eleven editions during his lifetime. During World War II, he was associated with the Ministry of Home Security and the Royal Air Force. In 1945 he succeeded Greenwood as Professor of Medical Statistics at LSHTM, and in 1947 he became honorary director of the MRC Statistical Research Section. At about this time he embarked on a series of collaborative studies on controlled clinical trials (among which work on the efficacy of streptomycin in the treatment of pulmonary tuberculosis is one of the best known).

In this connection he developed methods of general applicability and a philosophy for the use and interpretation of such trials. This philosophy was lucidly summarized in a lecture given at the National Institutes of Health in Washington, D.C., and repeated in *Statistical Methods in Clinical and Preventive Medicine* (SMCPM)—a volume of his collected works up to the year of his retirement (1961), published in 1962 by Livingstone, Edinburgh.

Hill's research, with Richard Doll, on the association between cigarette smoking and lung cancer attracted widespread international public interest (see *SMCPM*). These investigations had a considerable influence in persuading the scientific community to believe that cigarette smoking was related to onset of lung cancer and some other diseases. (The original report appeared in 1952; Hill gave up pipe smoking about two years later.)

Hill was president of the Royal Statistical Society in 1952, and was awarded the prestigious Guy Medal in Gold of that society in 1953. In 1954, he was elected a fellow of the Royal Society, and knighted in 1961. From 1955 to 1957 he was Dean of LSHTM. His most lasting practical contributions are in his successes in convincing the medical profession of the importance of scientific testing, and the introduction of sound statistical methods into medical research.

His wife of many years, Lady Florence, died in 1980, having suffered poor health for some years, during which Hill devoted his time to her at their home in Buckinghamshire. They had two sons and a daughter. His last years were spent with his daughter and son-in-law, and in nursing homes in Cumbria.

Hill was a warm-hearted and genial man, a superb after-dinner speaker. His comments on the president's addresses to the Royal Statistical Society of R. A. Fisher* (in 1953) and E. S. Pearson* (in 1956) are good examples of the flavor of his off-the-cuff remarks. Like Arthur L. Bowley,* Hill was against mathematical formalism in statistics and also had a phenomenal group of numerical information and a very well-developed intuition, allowing him to discern important aspects of data before (or sometimes in place of) formal analysis.

SELECTED SOURCES

Anon. (1991). Sir Austin Bradford Hill. *Times* (London), April 22, p. 14c., (Obituary.)

Anon. (1992). Sir Austin Hill dies. *New York Times*, April 20, A11:1. (Obituary.)

Armitage, P. (1991). Obituary: Sir Austin Bradford Hill, 1897–1991. *J. R. Statist. Soc. A*, **154**, 482–484.

Fisher, R. A. (1953). The expansion of statistics. *J. R. Statist. Soc. A*, **116**, 1–6. (Hill's comments, pp. 6–8).

Pearson, E. S. (1956). Some aspects of the geometry of statistics: the use of visual presentation in understanding the theory and application of mathematical statistics *J. R. Statist. Soc. A*, **119**, 125–146. (Hill's comments, pp. 147–148).

Nemchinov, Vasilii Sergeevich

Born: July 2, 1894, in Grabovo village, Penza region, Russia.
Died: November 5, 1964, in Moscow, USSR.
Contributed to: agricultural statistics, application of mathematical and statistical methods to the theory of sampling and to the planning of the Soviet economy.

Nemchinov graduated from the Chelyabinsk High School in 1913, studied economics at the prerevolutionary Moscow Commerce Institute from 1913 to 1917, and did some statistical work in Moscow 1915–1917. In 1919 he became director of the Statistical Bureau of the Chelyabinsk region, where in 1920 he conducted the first population census of the Russian Federation in the province.

In 1922 he became a manager in the Ural Region Statistical Administration, serving simultaneously as an instructor at the Communist University in Sverdlovsk (now Ekaterinburg). In 1928 he became head of the Statistics Department and in 1940 director of the prestigious Moscow Timiryazev Agricultural Academy. But in 1940 he was demoted after his defense of Soviet genetics and the chromosome theory of heredity against accusations of Lysenko. Refusing to dismiss the geneticist Zhebrak from the Academy, he stated: "I am not a biologist, but I am in a position to verify the theory of heredity from the viewpoint of statistics."

In 1930 he was awarded the title of professor, followed by a doctorate in economic sciences in 1935. His sole visit to the USA was in 1943, as a Soviet delegate to the Dumbarton Oaks Conference that founded the United Nations.

As early as 1939 Nemchinov established the Statistical Section of the Moscow Scientific House of the Academy of Sciences of the USSR, and served as its head until 1948. In 1945 he became chief statistician at the Institute of Economics of the Academy and was elected an Academician there in 1946, serving on its Praesidium from 1953 to 1962. He was an elected member of the International Statistical Institute, participating regularly at its sessions from 1958 onwards; he was made an Honorary Doctor of Social Sciences at the University of Birmingham, England, in 1964. His collected works of 324 titles were published in six volumes 1967–1969.

In 1957 he was the driving force in the change of approach to Soviet economic planning. His efforts resulted in a Soviet network of computer stations to gather and analyze economic statistics. The famous Central Mathematical Economics Institute was established in 1963. Nemchinov spearheaded the first all-union conference on the applications of mathematics in economic and planning studies, organizing and directing the Scientific Council devoted to these applications. In 1962–1964 he was head of the Faculty of Mathematical Methods in Economic Analysis at Moscow State University.

Nemchinov was honored with the highest awards in the USSR: the Order of the Red Banner, the Red Star, the Order of Lenin (twice), and (posthumously in 1965) the Lenin Prize. He made a very significant contribution to the consolidation of statistics in the Soviet Union. This was the establishment in 1954 of the periodical *Uchenye Zapiski po Statistike* (Learned Notes on Statistics), which was sponsored by the Academy of Sciences of the USSR. This served as an important vehicle for the dissemination of applied statistical methods for some 35 years, and it kept a close eye on Western developments in statistics.

Ideologically Nemchinov was a convinced Marxist, but a liberal one as far as the system would permit him to be. While believing that the Soviet economic system provided greater opportunities for statisticians than did the Western system, he vigorously advocated contacts between Western and Soviet scientists and strongly encouraged the translation into Russian of Western publications on econometrics and statistics.

One of Nemchinov's main contributions was the textbook *Agricultural Statistics, with Outlines of a General Theory (Sel'skokhozyaistvennaya Statistika s Osnovami Obshchei Teorii)*, 1945, which was awarded the USSR State Stalin Prize in 1946. This book propagates his philosophy of statistics as a methodological science, thus removing the barriers between statistical-economic and statistical-mathematical analysis. His significant theoretical achievement is the so-called Nemchinov method of calculating the parameters of multiple linear correlation based on Chebyshev's* orthogonal polynomials. In a number of works he tried to show that much of modern theoretical statistics can be rigorously presented on Chebyshev's polynomials as a foundation.

Several of his publications deal with problems of intra-industrial balance. In the area of location of resources he organized and participated in detailed studies of Siberia and (jointly with the Chinese) of the Amur river basin. He was active from the 1940s in the problems of statistical higher education in economics, and edited D.I. Mendeleyev's works in economics.

BIBLIOGRAPHY

The following list covers selected works by Nemchinov translated into other languages from the Russian.

Nemchinov, V. S. (1929). Les méthodes de groupement des exploitations agricoles, pratiquées par la statistique de l'USSR. (Communication presented at the XVIII Session of the International Statistical Institute, 1929; 17 pp.)

Nemchinov, V. S. (1948). *Die Situation in der biologischen Wissenschaft. Vortrag und Diskussion.* Verlag Kultur and Fortschrift GMBH, Berlin, 1948, pp. 187–188.

Nemchinov, V. S. (1952). Object and purpose of the International Economic Conference. *News*, No. 7, pp. 3–5.

Nemchinov, V. S. (1953). Die Statistik als Wissenschaft. *Sowjetwiss. Ges.-Wiss. Abt.*, No. 2, pp. 303–319.

Nemchinov, V. S. (1954). Trade as a peace factor. *News*, No. 6, pp. 3–4.

Nemchinov, V. S. (1957). The balance-sheet method in economic statistics. In *Reports by Soviet Scientists at the 30th Session of the ISI*, pp. 3–28.

Nemchinov, V. S. (1959). Some aspects of the balance-sheet method as applied in the statistics of interdependent dynamic economic systems. Report at the 31st Session of the ISI, Brussels. (1960) *Bull. Inst. Internat. Statist.*, **37**, 313–333.

Nemchinov, V. S. (1961). Les mathématiques et l'électronique au service de la planification. *Educ. Econ.*, No. 134, pp. 1–12.

Sources about Nemchinov in English:

Davies, R. W. and Barker, G. R. (1965). Obituary: Vasilii Sergeevich Nemchinov. *J. R. Statist. Soc. A,* ***128***, 614.

Kotz, S. (1965). Statistics in the USSR. *Survey*, No. 57. Pp. 132–141.

Snedecor, George Waddel

Born: October 20, 1881, in Memphis, Tennessee, USA.
Died: February 15, 1974, in Amherst, Massachusetts, USA.
Contributed to: analysis of variance and covariance, applied sampling, data analysis, design of experiments, statistical methods, world-wide use of statistical methods.

Through his many years of experience in statistical consulting with research workers, particularly in the biological and agricultural sciences, through his equally many years of teaching statistical methods using real experimental data, usually drawn from such consultations, and through the seven editions of *Statistical Methods*, George Waddel Snedecor is among the greatest pioneers in improving the quality of scientific methods insofar as it concerns the use of analytical statistical methodologies. These applications using real data were based primarily on new statistical methodologies developed by R. A. Fisher* and other English statisticians.

The Snedecor family traces its genealogy from a Dutch settler in New Amsterdam (New York) in 1639, with descendants moving south after the Revolutionary War, reaching Alabama in 1818 to enter an aristocratic plantation period, and later becoming involved in law, education, and public service. The oldest of eight children, George Snedecor was born in Tennessee but grew up mainly in smaller towns and rural areas of Florida and Alabama; his lawyer father became a minister and finally an educator working with young blacks.

Snedecor earned two degrees in mathematics and physics in Alabama and Michigan and had taught for eight years before coming to Iowa State College (now University [ISU]) in 1913 as an assistant professor of mathematics. Quickly promoted, he began teaching the first formal statistics courses in 1915.

In 1924 Snedecor assisted with the historically important Saturday seminars conducted by Henry A. Wallace, during the spring quarter, on statistics, including machine calculation of correlation coefficient, partial correlation, and calculation of regression lines. Wallace and C. F. Sarle borrowed some card-handling equipment from an insurance company in Des Moines, Iowa, to bring to Ames on certain Saturdays to illustrate their use. Thus business machines (including IBM punch-card tabulation equipment) were used at this early time for research computations. These seminars led to the ISU bulletin "Correlation and Machine Calculations" by Snedecor and Wallace (1925), which attained world-wide distribution. See Jay L. Lush, "Early Statistics at Iowa State University," in Bancroft [1]; also see ref. [3].

In 1927 Snedecor and A. E. Brandt became directors of a newly created Mathematics Statistical Service in the Department of Mathematics to provide a campus-wide statistical consulting and computation service. It was the forerunner of the Statistical Laboratory, organized in 1933, under the ISU President's Office, with Snedecor as the first director.

While special courses in statistics were taught in several departments, degree programs were with mathematics, and in 1931 the first degree in statistics at ISU, an M.S., was awarded to Snedecor's student Gertrude M. Cox.*

Snedecor was instrumental in bringing R. A. Fisher* to Ames as a visiting professor twice, in 1931 and 1936. Since the combination of the service functions of the Statistical Laboratory and the degree programs in statistics through Mathematics (both provided primarily by part-time service from the same faculty members) was unique among universities, many outstanding statisticians visited ISU, some to lecture and some just to observe.

In order to strengthen the graduate program in statistics at the Ph.D. level, Snedecor brought C. P. Winsor from Harvard as a faculty member in 1938. W. G. Cochran* came from Rothamsted in 1938 as a visiting professor and continued as a regular professor in 1939. In recognition of the Statistical Laboratory's leadership role, contractual and cooperative projects were initiated in the 1940s by the U.S. Weather Bureau and the Bureau of the Census.

Snedecor became both a dedicated research worker and teacher and an able administrator, yet retained a personal humility. Tall, lean, and vigorous, with a direct, unpretentious, patient, and kind manner, Snedecor was warmly regarded by colleagues and students.

In recognition of the importance of his contributions, he was given many honors and awards by ISU, other universities (U.S. and foreign), and the statistical societies. At ISU: establishment of the Snedecor Ph.D. Student Award, faculty citation, honorary D.Sc., dedication of Snedecor Hall, and designation as professor emeritus. At other universities: honorary D.Sc., visiting lecturer and/or professor appointments in the 1950s and 1960s at North Carolina state universities, Virginia Polytechnic Institute and State University, University of Florida, Alabama Polytechnic Institute, and, for the Rockefeller Foundation, the University at São Paulo, Brazil. The statistical societies: president of the American Statistical Association, honorary membership in the Royal Statistical Society, Samuel S. Wilks Memorial Medal award, establishment of a Snedecor Award by the American Statistical Association and Iowa State University for the best publication in biometrics.

Snedecor's world-wide leadership among research workers has been based on the seven editions of *Statistical Methods*, beginning with his first in 1937 and extending to the last, in 1980, which was coauthored with Cochran. Note that the early editions' full titles were *Statistical Methods Applied to Experiments in Biology and Agriculture*. As of April 1984, over 207,000 copies of the seven editions had been published in English. In addition, these editions had been translated and published abroad in nine languages, including Spanish, French, Roumanian, Japanese, and Hindi. There were over 3000 entries in the 1981 volume of *Science Citation Index* for *Snedecor's Statistical Methods*, establishing it as among the most cited publications (see David [6]). Snedecor also wrote numerous papers, including noteworthy expository works on data analysis, statistical methods, design of experiments, sampling, analysis of variance and covariance, biometry, and scientific method.

Snedecor retired as director of the Statistical Laboratory in 1947; however, he remained active both at ISU and on off-campus visiting appointments until 1958. During this period he advised on the establishment, in 1947, of a separate Department of Statistics at Iowa State, initiated and prepared a text for a pioneering, introductory course in statistics [8, 9], and conducted a special seminar, often using queries submitted to *Biometrics*, for graduate students majoring in statistics. After 1958 Snedecor spent another five years as consultant at the U.S. Naval Electronics Laboratory in San Diego—to complete an exceptionally long and vigorous career.

At the age of 92, George Snedecor died. Brief but insightful appraisals [2, 5, 7] of his character and contributions are given by Cox and Homeyer and in obituaries in statistical journals. Apart from his books on statistical methods, his great contribution was his vision in propagating the role of statistics in quantitative studies.

The life of G. W. Snedecor is encapsulated by the statement of W. G. Cochran*[4]:

> Our profession owes Snedecor a great debt for his vision in foreseeing the contributions that statistical methods can make in quantitative studies, for his book made these methods available to workers with little mathematical training, and his administrative skill in building a major training center and in attracting leaders like Fisher, Yates, Mahalanobis, Kendall and Neyman to Ames as visitors.

REFERENCES

[1] Bancroft, T. A., ed. (Assisted by S. A. Brown) (1972). *Statistical Papers in Honor of George W. Snedecor.* Iowa State University Press, Ames.

[2] Bancroft, T. A. (1974). *Amer. Statist.*, **28**, 108–109.

[3] Bancroft, T. A. (1982). *Iowa State J. Res.*, **17**, 3–10.

[4] Cochran, W. G. (1974). *J. R. Statist. Soc. A*, **137**, 456–457.

[5] Cox, G. M. and Homeyer, P. G. (1975). *Biometrics*, **31**, 265–301. (Contains a bibliography.)

[6] David, H.A. (1984). In *Statistics: An Appraisal*, H. A. David and H. T. David, eds. Iowa State University Press, Ames, pp. 3–18.

[7] Kempthorne, O. (1974). *Int. Statist. Rev.*, **42**, 319–321. (Followed by a bibliography, pp. 321–323.)

[8] Snedecor, G. W. (1948). *J. Amer. Statist. Ass.*, **43**, 53–60.

[9] Snedecor, G. W. (1950). *Everyday Statistics—Facts and Fallacies*, 1st ed. William C. Brown Co., Dubuque, Ia. [Limited 1st ed.; limited 2nd ed. (1951).]

T. A. BANCROFT

Spearman, Charles Edward

Born: September 10, 1863 in London, England
Died: September 17, 1945 in London, England
Contributed to: psychometrics, correlation, factor analysis.

Charles Edward Spearman

Charles Spearman was a third son in a minor aristocratic family and, like many young men of similar rank, chose a military career. In 1897, however, 34 years old and after some fifteen years of service mainly in India, Spearman resigned his commission to study for a Ph.D. in experimental psychology at Leipzig. His studies were interrupted, however, by recall during the South African War to serve on the Army General Staff in Guernsey, where he met and married the daughter of a local doctor. The Spearmans had four daughters and a son; the latter killed in action during World War II.

Having eventually returned to Leipzig, Spearman obtained his degree in 1906 and the following year accepted a post at University College, London, where he stayed until retiring as Emeritus Professor in 1931. Under Spearman's leadership there emerged the so-called "London School" of psychology, distinguished by its rigorous statistical and psychometric approach.

Spearman first came to prominence with two articles published in 1904. By refining existing correlational methods to correct for measurement error (and thereby precipitating a life-long feud with Pearson), Spearman had demonstrated that the pattern of intercorrelations for scores on various tests of mental ability was consistent with the existence of a *quantifiable factor* (later called *g*) common to every mental activity, and a second factor specific to the particular task. With this mathematical formulation of a *two-factor theory of intelligence*, Spearman had laid the foundations of factor analysis. Stemming also from this seminal work is the so-called Spearman–Brown prophesy formula for measuring the reliability of psychological tests and the familiar rank correlation measure (although, ironically, the version bearing his name is not the formula that he advocated; see Lovie, 1995).

For almost three decades, Spearman was locked in a continual battle defending the two-factor theory against its many detractors and, with the help of volunteers and conscripts alike, toiled to strengthen its statistical foundations. This work reached its pinnacle with the publication of *The Abilities of Man* in 1927. Although multiple factor theories, bolstered by a sophisticated statistical methodology well beyond Spearman's capabilities, had gradually gained ascendance by the early 1930s, this in no way lessens his achievement as the architect of factor analysis. Moreover, in an academic career spanning almost 40 years, Spearman published more than 100 articles and six books, the last of which (still defending the two-factor theory) appeared posthumously.

Spearman was a formidable academic opponent, adept at recruiting allies to his cause, with a meticulously ordered academic life, yet he was notoriously absent-minded. On a personal level, colleagues found him courteous and sociable; he was also generous in acknowledging the contributions of his many helpers. He enjoyed playing tennis and travelling, especially to the United States. Spearman died at 82, after falling from a window of the hospital where he was undergoing treatment for pneumonia.

In view of his association with factor analysis, test reliability and, most enduringly of all, with rank correlation, it is perhaps surprising to discover that Spearman himself considered his statistical and psychometric work (for which he was elected Fellow of the Royal Society) as subordinate to his quest for the fundamental laws of psychology. A longer account of Spearman's life and work can be found in Lovie and Lovie (1996).

REFERENCES

Lovie, A. D. (1995). 'Who discovered Spearman's rank correlation?' *Br. J. Math. Statis. Psy.*, **48**, 255–69.

Lovie, P. and Lovie, A. D. (1996). Charles Edward Spearman, F. R. S. (1863–1945), *Notes and Records of the Royal Society of London*, **50**, 1–14.

Spearman, C. (1904). The proof and measurement of association between two things, *Amer. J. Psy.*, **15**, 72–101.

Spearman, C. (1904). General intelligence objectively determined and measured, *Amer. Psy.*, **15**, 202–293.

Spearman, C. (1927). *The Abilities of Man, their Nature and Measurement*, Macmillan, London.

PATRICIA LOVIE

Wilson, Edwin Bidwell

Born: April 25, 1879, in Hartford, Connecticut, USA.
Died: December 28, 1964, in Brookline, Massachusetts, USA.
Contributed to: mathematics, physics, statistical inference, biostatistics.

Edwin Bidwell Wilson was a scientific generalist. As a mathematician, he published two of the most influential advanced texts of the early part of this century, and he criticized David Hilbert on the foundations of geometry. As a physicist, he did path-breaking work on the mathematics of aerodynamics, and he criticized Albert Einstein on relativity. As a statistician, he anticipated Neyman* on confidence intervals, he devised one of the earliest normalizing transformations, he founded the biostatistics program at the Harvard School of Public Health, and he criticized Ronald Fisher* on inference.

Wilson's lifelong commitment to science and education was presumably instilled in him at a very early age: his father was a teacher and superintendent of schools in Middleton, Connecticut. Wilson graduated from Harvard College in 1899, majoring in mathematics, and he then went to Yale to continue his mathematical study as a student of J. Willard Gibbs. In 1901, the same year he received his Ph.D., Wilson published *Vector Analysis, Founded upon the Lectures of J. Willard Gibbs*, a text that, with Wilson's later *Advanced Calculus* (1912), provided a significant portion of the upper-level mathematics curriculum in America for the first third of the century.

From 1900 to 1907, Wilson taught at Yale, with a year off in 1902–1903 for study in Paris. In 1907 he moved from Yale to a faculty position at Massachusetts Institute of Technology (M.I.T.), where his interests evolved to mathematical physics and academic administration. From 1920 to 1922 he served as one of a committee of three, functioning collectively as interim president of M.I.T. In 1922 he moved to Harvard as professor and head of Vital Statistics in the Harvard School of Public Health, where he remained until his retirement in 1945. Wilson was managing editor of the *Proceedings of the National Academy of Sciences* for fifty years, from its first issue in January 1915 until his death in December 1964. He was active on nearly every national committee involved with social science over the last half of his life. A student of his, Paul Samuelson, has described him as "the only intelligent man I ever knew who liked committee meetings." [6]

WORK IN STATISTICS

Wilson's most important contribution to statistics was arguably as an institution builder, founding a program that still flourishes under the title of Biostatistics in Harvard's School of Public Health. He also brought his sharp critical intelligence and knowledge of quantitative social science to bear on methodological issues on many national committees and in a far-ranging national correspondence (see, e.g., ref. [7]). He had a keen sense of data analysis, as indicated in his reinvestigation with Margaret Hilferty of C.S. Peirce's extensive data on reaction times [9], a study still cited in the literature on robustness. But he also made a number of important technical contributions. In work with Jane Worcester on quantal response [15], he advanced the study of the estimation of the median lethal dose, which they called "LD 50." Wilson's 1931 paper with Margaret Hilferty [10] introduced what has become known as the Wilson–Hilferty transformation, a device that allowed the use of the normal approximation for chi-square probabilities over a wide range of degrees of freedom.

In 1927 Wilson published a short note that anticipated (albeit for a very limited class of problems) the concept behind Neyman's confidence intervals. Wilson clearly described in that paper how the confidence-interval idea could be invoked for inference about the binomial parameter, explaining the difference between the confidence idea and the common use of standard errors as an approximate way of doing inverse probability (or Bayesian) inference [8, 12].

In 1941 Wilson published a note in *Science* that contradicted R. A. Fisher on the analysis of two-by-two tables [11]. Fisher's polite reply [2] elicited a published retreat by Wilson [13, 14], and the two enjoyed a long correspondence on statistical issues [1].

Wilson wrote several general articles on statistical inference and scientific methodology; articles that show an acute sensitivity to both similarities and differences between measurement problems in the social and physical sciences [3]. He also made important contributions to mathematical utility theory, showing that Pareto's derivation of the law of demand held under more general conditions than had been previously believed [5]. Wilson's work on statistics, whether on bioassay, contingency tables, factor analysis, population growth, or the foundations of inference, showed a keen and acutely perceptive intelligence that was unusual among writers at that time. Although he only rarely achieved originality of concept, his knowledge and critical assessment of contemporary advances was far ahead of most of his contemporaries'.

REFERENCES

[1] Bennett, J. H., ed. (1990). *Statistical Inference and Analysis: Selected Correspondence of R.A. Fisher.* Clarendon Press, Oxford.

[2] Fisher, R. A. (1941). The interpretation of experimental four-fold tables. *Science*, **94**, 210–211. (Reprinted in vol. 4 of Fisher's *Collected Papers* as No. 183.)

[3] Hunsaker, J. and Mac Lane, S. (1973). Edwin Bidwell Wilson 1979–1964. *Biographical Mem.*, **43** 285–320. (Includes a complete bibliography of Wilson's work.)

[4] Irwin, J. O. and Worcester, J. (1965). Edwin Bidwell Wilson, 1879–1964. *J.R. Statist. Soc A*, **128**, 616–618.

[5] Milgate, M. (1987). Wilson, Edwin Bidwell. In *The New Palgrave Dictionary of Economics*. Macmillan, London, vol. 4, pp. 922–923.

[6] Samuelson, P. (1989). Gibbs in economics. In *Proceedings of the Gibbs Symposium*, Amer. Math. Soc., pp. 255–267.

[7] Stigler, S. M. (1994). Some correspondence on methodology between Milton Friedman and Edwin B. Wilson; November–December 1946. *J. Econ. Literature*, **32**. 1197–1203.

[8] Wilson, E. B. (1927). Probable inference, the law of succession, and statistical inference. *J. Amer. Statist. Ass.*, **22**, 209–212.

[9] Wilson, E. B. and Hilferty, M. M. (1929). Note on C. S. Peirce's experimental discussion of the law of errors. *Proc. Nat. Acad. Sci. U.S.A.*, **15**, 120–125.

[10] Wilson, E. B. and Hilferty, M. M. (1931). The distribution of chi-square. *Proc. Nat. Acad. Sci. U.S.A.*, **17**, 684–688.

[11] Wilson, E. B. (1941). The controlled experiment and the four-fold table. *Science*, **93**, 557–560.

[12] Wilson, E. B. (1942). On confidence intervals. *Proc. Nat. Acad. Sci. U.S.A.*, **28**, 88–93.

[13] Wilson, E. B. (1942). On contingency tables. *Proc. Nat. Acad. Sci. U.S.A.*, **28**, 94–100.

[14] Wilson, E. B. and Worcester, J. (1942). Contingency tables. *Proc. Nat. Acad. Sci. U.S.A.*, **28**, 378–384.

[15] Wilson, E. B. and Worcester, J. (1943). The determination of Lethal Dose 50 and its sampling error in bio-assay. *Proc. Nat. Acad. Sci. U.S.A.*, **29**, 79–85, 114–120, 257–262.

STEPHEN M. STIGLER

Yates, Frank

Born: May 12, 1902, in Manchester, U.K.
Died: June 17, 1994, in Harpenden, U.K.
Contributed to: design and analysis of experiments, survey sampling, statistical computing.

Frank Yates (as he was universally called) went to school first in Manchester, where he grew up, then to Clifton College, and to university at St. John's College, Cambridge, from which he graduated with first-class honors in 1924. His first important job (1927–1931) was as research officer and mathematical adviser to the Geodetic Survey of the Gold Coast (now Ghana). Here he became acquainted with least squares, the applications of which he greatly extended when he went to Rothamsted in 1931 as assistant to R.A. (later Sir Ronald) Fisher.* Within two years Fisher had left Rothamstead and Yates became head of the Statistics Department, a post which he held for 35 years, until he retired in 1968.

Yates' career belongs to Rothamsted, and he built upon Fisher's legacy to make the Statistics Department there a continuing source of important ideas, which although originating in agriculture, spread to many other fields. Fisher had founded the subject of design of experiments, and propagated the ideas of replication, randomization, and blocking (to reduce error) together with factorial designs, where several treatment factors are varied together, rather than one at a time. Yates clarified the ideas of orthogonality, confounding, and balance [2], and in his famous TC35 monograph, *The Design and Analysis of Factorial Experiments* (1937) [6], he dealt with factorial experiments having factors with two, three, or four levels, and the use of split plots with both complete and partial confounding. In TC35 will be found the Yates algorithm for calculating in two-level factorial experiments the main effects and interactions from the original yields. The computing trick here is the same as was later to underlie the fast Fourier transform. The inverse algorithm for calculating fitted values from a subset of the effects is also given. The other major new idea in design was Yates' introduction of balanced incomplete block designs, whereby only a subset of treatments occurs in each block, but treatments are balanced in such a way that any two occur together in an equal number of blocks [4]. He also realized that where the number of blocks is substantial, information about treatment effects can also be extracted from block totals, the so-called interblock information. For variety trials with many varieties he introduced the lattice designs, where varietal effects can be mapped on to a pseudofactorial structure [5]; this greatly simplified the analysis. There is now a huge combinatorial literature, spreading as far as coding theory, that derives from this work. Curiously, Yates is probably most widely known for his continuity correction to chi-square in tables of counts [3], something which has become of much less importance now that the exact test statistic can be easily computed.

By 1939, when World War II began, Yates had become an authority on sampling schemes, and this knowledge was put to work in assessing the effect of bombing. This work, in close collaboration with Zuckerman, showed, for example, that the bombing of vital points of the railway system during the Sicily campaign of 1943 had been more effective than attacks on cities and factories. Here we see the introduction of quantitative methods that led in time to operational research as a subject in its own right. At the same time Yates continued his work at Rothamsted, where, in a major study with E. M. Crowther [7], he drew together all the past work on the effect of fertilizers on crops; this was to form the basis of government policy on fertilizer imports during the war years. This work is an example of the combination of information from many diverse experiments, and exhibits a skill that many statisticians seem only recently to have appreciated the need for.

In 1950, the United Nations was responsible for a survey of world agriculture, and Yates was commissioned to write a sampling manual for it. *Sampling Methods for Censuses and Surveys* [8] was indeed a manual and not a textbook. It described, with numerous examples, different sampling schemes, and how estimates should be made of population values, and their uncertainties calculated. The fourth edition is still in print. In Yates and Grundy [9] he introduced sampling without replacement with probability proportional to size, an early example of what is now known as rejective sampling. He also gave the rule for calculating estimates of sampling error for multistage designs of any degree of complexity.

The first analyses of surveys were done with Hollerith sorters and tabulators. Yates managed to invert symmetric matrices using Hollerith cards! However, in 1954 the world changed, and Yates secured for Rothamsted one of the first commercial electronic computers, the Elliott 401. It had 512 words of memory on a rotating disk, paper tape for I/O, no floating-point arithmetic, and not even integer division. Frank Yates and his colleagues showed remarkable ingenuity in doing useful work with this machine. What must have been the first routines for operations on multiway tables were among the products of this work, and Yates himself began optimizing program speed by the positioning of instructions on the disk. The 401 was succeeded by the 402 and then by the Ferranti Orion, perhaps the first modern computer in the series. Program development went on steadily under Yates' direction, resulting, for example, in Genfac, for the analysis of factorial experiments [10], and RGSP, the Rothamstead General Survey Program [12], a particular interest of his.

Yates never lost touch with agriculture; he was a member of the committee that supervised all research work on the experimental farm at Rothamsted for 35 years, being chairman for 13. He was appointed a deputy director in 1958, and after retirement he kept a room at Rothamsted and completed 60 years of work there in 1993. His published work began in 1929, with the last paper appearing in 1988. The statistical tables [1] that he published with Fisher (first edition, 1936) were widely used throughout the world, and in 1970 there appeared a book of selected papers on experimental design [11] with his additional comments. He was elected a fellow of the Royal Society in 1948 and gained the Society's Royal Medal in 1966. He was president of the British Computer Society in 1960–1961 and of the Royal Statistical Society in 1967–1968. In 1953 he was awarded the Weldon Memorial Prize of Oxford University for work in biometrical science.

I believe that Yates regarded himself as a scientist whose subject was statistics.

He believed, like Fisher, that the interesting problems came from the real world, and that their solution would aid the processes of science and technology.

REFERENCES

[1] Fisher, R. A. and Yates, F. (1936). *Statistical Tables for Biological, Agricultural and Medical Research.* Oliver and Boyd, Edinburgh. (Subsequent editions 1942, 1948, 1953, 1957, 1963. The introduction contains much of interest historically.)

[2] Yates, F. (1933). The principles of orthogonality and confounding in replicated experiments. *J. Agric. Sci.*, **23**, 108–145.

[3] Yates, F. (1934). Contingency tables involving small numbers and the chi-squared test. *J. R. Statist. Soc. Suppl.*, 1, 217–235.

[4] Yates, F. (1936). Incomplete randomised blocks. *Ann. Eugen.*, **7**, 121–140.

[5] Yates, F. (1936). A new method of arranging variety trials involving a large number of varieties. *J. Agric. Sci.*, 26, 424–455.

[6] Yates, F. (1937). *The Design and Analysis of Factorial Experiments.* Commonwealth Bureau of Soil Science, Technical Communication No. 35. (The classical exposition of factorial experiments.)

[7] Yates, F. and Crowther, E. M. (1941). Fertiliser policy in war-time. The fertilizer requirements of arable crops. *Emp. J. Exp. Agric.*, **9**, 77–97.

[8] Yates, F. (1949). *Sampling Methods for Censuses and Surveys.* Griffin, London. (Subsequent editions 1953, 1960, 1981.)

[9] Yates, F. and Grundy, P. M. (1953). Selection without replacement from within strata with probability proportional to size. *J. R. Statist. Soc. B*, **15**, 253–261.

[10] Yates, F. and Anderson, A. J. B. (1966). A general computer programme for the analysis of factorial experiments. *Biometrics*, **22**, 503–524.

[11] Yates, F. (1970). *Experimental Design: Selected Papers.* Griffin, London.

[12] Yates, F. (1973). The analysis of surveys on computers—features of the Rothamsted Survey Program. *Appl. Statist.*, **22**, 161–171.

BIBLIOGRAPHY

Dyke, G. V. (1995). Obituary: Frank Yates. *J. R. Statist. Soc. A*, **158**, 333–338. (Evokes Rothamsted background and personality well.)

Healy, M. J. R. (1995). Obituary: Frank Yates, 1902–1994. *Biometrics*, **51**, 389–391. (Another well-written and enlightening account.)

JOHN NELDER

SECTION **7**

Applications in Science and Engineering

Boltzmann, Ludwig Edward

Born: February 20, 1844, in Vienna, Austria.
Died: September 5, 1906, in Duino, near Trieste (now Italy).
Contributed to: physics (especially statistical mechanics).

Boltzmann's statistical contributions are associated with his efforts, following on from J.C. Maxwell,* to explain the thermodynamics of gases on the basis of kinetic theory: i.e., to view gases as particles undergoing movement at different velocities and collisions according to the principles of mechanics. Maxwell, considering an equilibrium situation, had arrived at the probability density function $f(v) = 4\alpha^{-3}\pi^{-1/2}v^2 \exp(-v^2/\alpha^2)$, $v \geq 0$, for the (root-mean-square) velocity V of a particle in three dimensions, with $V = (X^2 + Y^2 + Z^2)^{1/2}$, where X, Y, Z are independent, identically distributed (i.i.d.) zero-mean normal random variables, corresponding to velocity components along the coordinate axes. The density of kinetic energy, U (of a randomly chosen particle), since it is proportional to V^2, is in this situation of the form const $\cdot\, u^{1/2} \exp(-hu)$, $u \geq 0$, which is essentially the density of a chi-square distribution with 3 degrees of freedom. Boltzmann [1] in 1881 actually wrote down the density of the equilibrium kinetic energy distribution in the case of movement in n dimensions, apparently unaware of the work of earlier authors (Abbe,* Bienaymé,* and Helmert*), who had obtained expressions for the chi-square density of the sum of squares of n i.i.d. $N(0,1)$ random variables.

His major contribution in the kinetic theory of gases, however, is in connection not with the equilibrium state itself, but with approach to equilibrium and consequently with the relation between the evolution of thermodynamic entropy and the probability distribution of velocities. If $f(x, y, z; t)$ is the joint density of velocity components X, Y, Z at time t, he asserted that, as t increases, the entropy

$$E(t) = -\iiint f(x, y, z; t) \times \log f(x, y, z; t)\, dx\, dy\, dz$$

is nondecreasing to the value given by $f(x, y, z; t) = \text{const} \cdot \exp[-h(x^2 + y^2 + z^2)]$: i.e., the equilibrium-state velocity distribution. This is essentially Boltzmann's H-theorem, with $E(t)$ corresponding to modern probabilistic usage of entropy in connection with information. Boltzmann's derivation was not rigorous, and the fact that in classical mechanics collisions between particles are reversible, anticipating recurrence of any overall configuration, seemed inconsistent with the overall irreversibility predicted by increasing entropy. As a result of the ensuing controversy, the celebrated paper [2], which introduced the Ehrenfest urn model as an illustrative example, suggested a probabilistic formulation in Markov-chain terms. The appropriate Markov chain $\{\mathbf{X}_n\}$, $n \geq 0$, describes the evolution of the relative-frequency table of velocities assumed by N particles at a specific time, where the states of $\mathbf{X}_n$ are vectors describing all such possible frequency tables, it being assumed that each particle must have one of r velocity values (such subdivision into quanta, leading to combinatorial treatment, is characteristic of Boltzmann's work). The chain $\{\mathbf{X}_n\}$ is such that each state will recur with probability 1, but a suitable scalar function $\phi(\mathscr{E}\mathbf{X}_n)$ ("entropy") is nondecreasing in n. The probability distribution, $\mathscr{E}\mathbf{X}_n$, reflects the density $f(v, t)$ of root-mean-square velocity at time t. The limiting stationary distribution of $N\mathbf{X}_n$ is the multinomial with general term

$$\frac{N!}{x_1! \ldots x_r!} r^{-N},$$

now identified with "Maxwell–Boltzmann statistics" in physics, and manifests the ultimate (equilibrium) tendency of any one particle to have any of the possible velocities equiprobably.

We owe to Boltzmann also the word "ergodic," although its meaning has evolved considerably up to the present.

Additional information on Boltzmann's contributions to statistics can be found in refs. [3–6].

REFERENCES

[1] Boltzmann, L. (1909). *Wissenschaftliche Abhandlungen*, F. Hasenöhrl, ed., 3 vols. Leipzig. (Boltzmann's collected technical papers; the expression appears in Vol. 2, p. 576.)

[2] Ehrenfest, P. and T. (1907). *Phys. Z.*, **8**, 311–314.

[3] Gnedenko, B. V. and Sheynin, O. B. (1978). In *Matematika XIX Veka [Mathematics of the 19th Century*]. Izd. Nauka, Moscow, pp. 184–240. (Pages 229–232 contain a picture of Boltzmann and an assessment of his statistical contributions.)

[4] Kac, M. (1959). *Probability and Related Topics in Physical Sciences.* Interscience, London.

[5] Moran, P. A. P. (1961). *Proc. Cambridge Phil. Soc.*, **57**, 833–842. (A modern exploration of the theorem in a Markovian setting.)

[6] Sheynin, O. B. (1971). *Biometrika*, **58**, 234–236. (Historical manifestations of the chi-square distribution.)

E. SENETA

Deming, William Edwards

William Edwards Deming.

Born: October 14, 1900, in Sioux City, Iowa, USA.
Died: December 20, 1993, in Washington, D.C., USA.
Contributed to: sampling theory, sampling practice, design of statistical studies, distinction between enumerative studies and analytic problems, frame and universe, half-open interval, equal complete coverage, operational definitions, quality, management, Deming system of profound knowledge.

Dr. Deming's earliest recollections were of running barefoot across the grassy hills of Iowa while ducks swam on the pond, and listening to his mother give music lessons. His father had studied law, and was good in Latin, arithmetic, and algebra. The Deming family moved to Wyoming when he was seven. The family had obtained forty acres of land to work, and lived for four or five years in a tar-paper shack. Deming got a job filling the boiler at Mrs. Judson's Hotel before and after school, and bringing in kindling and coal; at about age 14 he took on the job of lighting the new gasoline lamps in town.

While at the University of Wyoming Deming sang in the choir and played in the band. He was a bass soloist in adulthood; he learned to play the piano and other instruments along the way, and composed liturgical and other music. Since the "Star Spangled Banner" has too great a range for most people to sing, he wrote a version that reduced the range.

In 1921 he obtained a Bachelor of Science degree in electrical engineering at the University of Wyoming. For two years he was assistant professor of physics at the Colorado School of Mines while he studied toward the masters degree that he completed at the University of Colorado in 1924. He received the Doctor of Philosophy degree in mathematical physics in 1928 from Yale University.

FAMILY

In 1922 he married Agnes Bell. She died eight years later, less than a year after they had adopted an infant.

In 1932 Deming married Lola E. Shupe, who had coauthored a number of articles with him on the Beattie–Bridgeman equation of state and on the physical properties of compressed gases. She continued to work at the Fixed Nitrogen Laboratory until 1942, worked for a few years in Selective Service, and retired in 1960 after about fifteen years with the National Bureau of Standards. Over the years she often helped Deming with calculations for his studies. They had two daughters.

AFFILIATIONS AND WRITINGS

In 1927 Deming began a twelve-year association with the U.S. Department of Agriculture, working as a mathematical physicist in the Bureau of Chemistry and Soils. He went to London in the spring of 1936 to study under Sir Ronald A. Fisher,* Egon Pearson,* and Jerzy Neyman.* He was so impressed with the lectures that he arranged to bring Neyman to Washington to lecture in 1937, and edited Neyman's lectures. The Graduate School of the Department of Agriculture (GSDA) sponsored the Neyman lectures and published the text. American statisticians had not heard of Neyman's work until Deming brought it to the USA through the lectures and the text; it influenced substantially the direction of American statistical thought. Later, Deming brought Fisher to Washington, as well.

Deming similarly brought the work of Walter Shewhart to the attention of American statisticians. He invited Shewhart to lecture at the GSDA and then made sure that the text of the lectures was published [9]. Interest in Shewhart's views on variation soared.

In 1938 the book *Statistical Adjustment of Data* [1] was published. Its principal aim was to help practitioners view a method of statistical adjustment as a "way of arriving at a figure that can be used for a given purpose—in other words for action," rather than as an absolute or "true" value. The methods of this book were developed over a period of sixteen years while he was in the government service. It was synthesized from notes from his practice and from classes he taught at the Graduate School of the Department of Agriculture. The influence on Deming's statistical thinking of the writings of the philosopher Clarence I. Lewis [8] is nowhere more evident than in this book. This work brought some solutions to unsolved statistical problems associated with population studies. The U.S. Census of Population in 1940 successfully introduced sampling techniques for the first time. Deming once said that Frederik Stephan, Phil Hauser, and Morris Hansen* furnished the brains for the census, and that he just carried out the work. His colleagues gave him much more credit.

Deming enjoyed work with other statisticians, often inviting help from them. He carried on lively communications with many well-known statisticians, who included Frederik Stephan, Walter Shewhart, Churchill Eisenhart, R. A. Fisher, and Jerzy Neyman.

Deming recognized the importance of the Rev. Thomas Bayes'* work and attributed the controversy surrounding it to lack of reading Bayes' original article. To give wider publicity to Bayes' essay on inverse probability and to bring attention to a little-known note by the same author on the summation of divergent series, Deming prepared in 1940 a small book containing these works [2].

After the U.S. entered World War II, Deming recommended to Stanford University that it sponsor a series of short courses in statistical methods for industry representatives who were producing products for the war effort. Under the direction of Holbrook Working, the first of dozens of these courses was held in 1942. Attendance was free and by invitation.

As part of the war effort Deming was on loan from the Bureau of the Census to the Ordnance Department, where he was an adviser in sampling from 1939 to 1946. Deming was adviser in sampling techniques to the Supreme Command of the Allied Powers, Tokyo (1947 and 1950), and member of the United Nations Sub-commission on Statistical Sampling (1947–1952).

In 1946 Deming became a professor of statistics at the Graduate School of Business Administration, New York University, where for 47 years he taught sampling and design of statistical studies, as well as his ideas on management. He commuted weekly to New York City from Washington, D.C., often by air, sometimes by train, and maintained a small apartment there for convenience when working with New York area clients or teaching.

The year 1946 also marked the beginning of his extensive and varied consulting practice. His client list included large and small companies, service and manufacturing, private and public, government and industry. He conducted hundreds of studies for railroad companies and hundreds more for motor freight carriers. The work he carried out was thorough and exacting. Statisticians around the world use his studies as models for their own work. It was not unusual for the opposing attorney in a legal case to be so impressed with Deming's statistical work that the attorney would hire Deming for the next client he represented. Deming's expert testimony was in continual demand. His international practice included activities in Canada, Greece, India, Japan, Germany, Mexico, Turkey, England, China, and Argentina.

To meet the needs of practicing statisticians and to fill the gaps between theory and practice, Deming wrote *Some Theory of Sampling* in 1950 [3]. It is a textbook of statistical theory that focuses on the planning of surveys, understanding errors, biases, and variation, to enable the lowering of costs and improved reliability of studies. The book has many ideas on the structure and analysis of statistical studies that Deming developed as a result of solving problems that he encountered in his own work. His solutions proved enormously helpful to statistical practice in general. The concept of a half-open interval to accommodate the ever-present miscounting, blanks, or skipping of numbers during the enumeration of a frame, and the distinction between the universe and the frame, are particularly noteworthy. Ten years later another book [4] introduced a simplification of the theory and practice of Mahalanobis'* interpenetrating samples for replication that maintained efficiency, yet facilitated estimation of the standard error and mathematical bias. The concepts of equal complete coverage and operational definitions of accuracy and bias were new. A plan for establishment of field procedures, statistical control as part of a sampling plan for nonsampling errors, standards of professional statistical practice, and delineation of responsibilities of the statistician and of the client brought order and new understanding to how to conduct a study. The treatment of uncertainties not attributable to sampling was a major contribution.

Deming first visited Japan in early 1947. In 1948 an invitation came from Kenichi Koyanagi, the manager and director of the Union of Japanese Scientists and Engineers, for Deming to come to teach quality control in Japan. He went for ten weeks in the summer of 1950 to give 35 lectures to engineers and to top management. Six months later, he went again, and six months after that yet again. He made over twenty-five trips to Japan. The results of his teachings are known the world over. Japan became a leader in quality products. The Japanese give much of the credit for their success to Deming.

In June 1980 NBC aired a provocative white paper "If Japan Can, Why Can't We?" documenting what happened in Japan in the 1950s with Deming's help. This challenge to U.S. companies created overnight a tremendous demand for Deming's services. With a huge demand and limited supply (one man) for Deming's knowledge, others looked to fill the need. Would-be consultants looked at Deming's writings to see what they had that they might offer. Some offered teambuilding, others offered statistical process control, some promised leadership skills, others consumer research, each offering something they knew was at least a part of what Deming was talking about. So now they could say, "We help your company implement Deming." It didn't take long for executives to realize that consultants were teaching different things. Then came new names and acronyms: Total Quality Management (TQM), Continuous Quality Improvement (CQI), Total Quality Control, Statistical Process Control (SPC), etc.

Deming disassociated himself from these efforts. He taught a four-day seminar to upwards of 20,000 people a year. His theory of management was radically different from what existed in most companies; it required fundamental changes in the way executives thought about business. Enterprises throughout the world studied his management theory and began the transformation that Deming talked about, knowing it would take ten to twenty years to accomplish. His 1982 book [5], coupled with videotapes produced by MIT Center for Advanced Engineering Study, brought more detail about his management theory to U.S. executives. His last book [6] explained his management theory as a system of profound knowledge that was the synthesis of ideas from four areas: psychology, theory of knowledge, appreciation for a system, and theory of variation.

Deming published close to 200 articles and ten books. Additional studies and papers written for clients number in the hundreds. His first 22 journal papers, written between 1928 and 1934, focused on issues and questions of physics, principally connected with compressed gases. In 1934 his first article on least squares was published. His 1940s articles focused on census issues, Shewhart's methods of quality control, the design of surveys, and sampling methods. From the 1950s onward Deming's articles brought statistical thinking into problems and issues of management.

At age 93 Deming still followed a schedule that exhausted many a younger person who tried to keep up with him. Almost every month saw him consulting with clients; teaching students; writing studies, articles, and books; giving speeches, keynote addresses, and press interviews; and conducting several four-day seminars.

AWARDS AND HONORS

The Deming Prize is awarded to a Japanese statistician each year by the Union of Japanese Scientists and Engineers for contributions to statistical theory; the Deming

Prize for Application is awarded to a company for improved use of statistical theory. In 1955 Deming received the Shewhart Medal from the American Society for Quality Control. In 1960 he received the Second Order Medal of the Sacred Treasure from the Emperor of Japan, for improvement of quality and of the Japanese economy, through the statistical control of quality.

Recognition of Deming in the USA came primarily in the last ten years of his life. In 1983 he received the Taylor Key Award from the American Management Association, and was elected to the National Academy of Engineering. He was enshrined in the Science and Technology Hall of Fame, Dayton, in 1986. President Ronald Reagan awarded him the National Medal of Technology in 1987. Deming was the recipient of 20 honorary doctorates, was an honorary member or fellow of numerous societies, associations, and committees, and received hundreds of awards and honors.

THE W. EDWARDS DEMING INSTITUTE

In November 1993 Dr. Deming formed the W. Edwards Deming Institute as a nonprofit Washington, D.C. corporation. It has as its aim to foster understanding of The Deming System of Profound Knowledge to advance commerce, prosperity, and peace.

COLLECTION OF HIS WORKS

The Library of Congress in Washington, D.C. has a large collection of Deming's papers, personal correspondence and trip journals, work for clients, and statistical studies. The collection includes correspondence with R. T. Birge, Harold Dodge, Churchill Eisenhart, Gregory Lidstone, Jerzy Neyman, John Tukey, W. Allen Wallis, Peter Drucker, Joseph Juran, Ichiro Ishikawa, and Kenichi Koyanagi, as well as many of his studies for railway and motor freight companies. It includes Deming's handwritten notes from the Neyman–Fisher–Pearson lectures in London in 1936 and many notes from other lectures and classes that he attended.

Reference [7], written by his secretary of 38 years, is a rich source of anecdotes from Deming's early years and glimpses of his personal life and his trips to Japan. This book also contains a list of his principal papers, academic and honorary degrees, positions held, and other honors.

REFERENCES

[1] Deming, W. E. (1938). *Statistical Adjustment of Data.* (Reprinted, Dover, Mineola, New York 1943.)

[2] Deming, W. E. (1940). *Two Papers by Bayes.* Noble, New York. (Reprinted, Hafner, New York, 1963.)

[3] Deming, W. E. (1950). *Some Theory of Sampling.* Wiley, New York. (The first five chapters provide a primer of things to consider before embarking on a statistical study. They may be read by anybody. Later chapters, which develop specific survey designs, require some knowledge of statistics.)

[4] Deming, W. E. (1960). *Sample Design in Business Research.* Wiley, New York.

[5] Deming, W. E. (1982). *Quality, Productivity and Competitive Position.* MIT Center for Advanced Engineering Study. (Revised and reissued in 1986 as Out of the Crisis.)

[6] Deming, W. E. (1993). *The New Economics.* MIT Center for Advanced Engineering Study.

[7] Killian, C. S. (1992). *The World of W. Edwards Deming.* SPC Press. Knoxville, Tennessee.

[8] Lewis, C. I. (1929). *Mind and the World Order.* Scribner's, New York.

[9] Shewhart, W. A. (1939). *Statistical Method from the Viewpoint of Quality Control.* Graduate School of the Department of Agriculture, Washington.

JOYCE NILSSON ORSINI

Gauss, Carl Friedrich

Born: April 30, 1777, in Brunswick, Germany.
Died: February 23, 1855, in Göttingen, Germany.
Contributed to: statistics, mathematical physics, astronomy.

Carl Friedrich Gauss

Carl Friedrich Gauss was born into a humble family in Brunswick, Germany; he became famous as an astronomer and mathematician before he was 25 years old, being educated at the Universities of Göttingen and Helmstedt, where he received his doctorate in 1799. For a time he was supported by a stipend from the Duke of Brunswick, but in 1807 he moved to Göttingen to become director of the observatory, and remained there for the rest of his life.

Kenneth O. May [11] described Gauss as "one of the greatest scientific virtuosos of all time"; his penetrating research and prolific output bear witness to that. At times his results were produced more rapidly than he could set them down and publish them. One example of this was his accurate prediction in 1801 of the location in the heavens of a supposed new distant planet, Ceres, which for a time had been lost. Gauss was heaped with praise on account of this achievement; but he did not set down until 1809 the processes by which he had made his prediction, namely, the refinement of orbit theory and the method of least squares.

The range of Gauss' influence in science, particularly in mathematical physics, has been enormous. He made strides in celestial mechanics, geodesy, number theory, optics, electromagnetism, real analysis, theoretical physics, and astronomy, as well as in statistics. It is surprising that he found no collaborators in mathematics, worked alone for most of his life, and never visited Paris (perhaps because he was strongly opposed to political revolutions). May [11] contains a selective but comprehensive bibliography of Gauss, including translations and reprints.]

Gauss' principal contributions to statistics are in the theory of estimation, and are known as *least squares*. The problem is to estimate k unknown parameters $\theta_1, \theta_2, \ldots, \theta_k$ on the basis of $n > k$ observations $y_1, y_2, \ldots, y_n$, where $y_i = \xi_i(\theta_1, \theta_2, \ldots, \theta_k) + e_i$. The quantities e_i are observational errors, assumed to be random and free from systematic error. In matrix notation this is

$$\mathbf{Y} = \boldsymbol{\xi} + \mathbf{e}, \tag{1}$$

where $\mathbf{Y}, \boldsymbol{\xi}, \mathbf{e}$ are the respective column vectors of the y_i's, ξ_i's, and e_i's. This problem is historically referred to as the *combination of observations*, and was regarded by Gauss as one of the most important in natural philosophy; he attached great importance to his contributions to this field.

For clarity, it is useful to distinguish between the *principle* of least squares and the *theory* of least squares.

PRINCIPLE OF LEAST SQUARES

The principle of least squares chooses θ to minimize $Q = \sum(y_i - \xi_i)^2$. Thus θ is a solution of $\partial Q/\partial\theta_j = 0, j = 1, 2, \ldots, k$. These are the least-squares equations

$$\sum(y_i - \xi_i)\frac{\partial\xi_i}{\partial\theta_j} = 0. \tag{2}$$

In the special but widespread case where the ξ_i are linear in the θ's, $\xi_i = \sum_j x_{ij}\theta_j$, or in matrix notation, $\boldsymbol{\xi} = \mathbf{X\theta}$, the x's being known constants, (1) is

$$\mathbf{Y} = \mathbf{X\theta} + \mathbf{e}, \tag{3a}$$

the Gauss linear model discussed in statistics texts as linear regression. The least-squares equations are now $\mathbf{X'X} = \mathbf{X'Y}$, and the least-squares estimate is

$$\hat{\boldsymbol{\theta}} + (\mathbf{X'X})^{-1}\,\mathbf{X'Y}. \tag{3b}$$

There has been much discussion surrounding priority in the development of least squares, since Legendre also considered (1) and (3a) and essentially developed (2) and (3b). Whereas Gauss published in 1809 [5], Legendre published in 1805 [10] and was responsible for the name "least squares" (moindre carrés). Of relevance to this issue of priority is the fact that Legendre confined himself to the *principle* of least squares outlined above, whereas Gauss developed in addition the statistical theory of least squares as outlined below.

STATISTICAL THEORY

Gauss' First Approach [5]

Assuming that $e_i = y_i - \xi_i$ are independent random variables with distribution $f(e_i)$, the joint distribution of the observational errors is $\Omega = \Pi f(e_i) = \Pi f(y_i - \xi_i)$. Assuming that all values of the θ's are equally probable, the distribution of the θ's given the observed values $\mathbf{y}$ is by Bayes' theorem proportional to Ω. Gauss chose the most probable value $\hat{\boldsymbol{\theta}}$ (i.e., the mode of Ω) as an estimate of $\boldsymbol{\theta}$. This is obtained as a root of

$$\frac{\sum \partial \log f(y_i - \xi_i)}{\partial\,\xi_i}\frac{\partial\xi_i}{\partial\theta_j} = 0, \qquad j = 1, 2, \ldots, k. \tag{4}$$

To proceed further, the mathematical form of f must be known. To this end, Gauss assumed that for the special case of (1) in which $y_i = \theta_1 + e_i$ for all i (so that there is only one parameter θ_1), the least-squares estimate should be the arithmetic mean, $\hat{\theta}_1 = \bar{y}$. It follows that f must be the normal distribution $(\sqrt{2\pi}\,\sigma)^{-1}\exp(-e^2/2\sigma^2)$. The distribution of the θ's is then proportional to

$$\left(\frac{1}{\sqrt{2\pi}\,\sigma}\right)^n \exp\left(-\frac{Q^2}{2\sigma^2}\right),$$

where $Q = \sum(y_i - \xi_i)^2$. This probability is maximized by minimizing Q, which is the principle of least squares described above. Gauss also considered the case of unequal variances σ_i^2, leading to weighted least squares $Q = \sum(y_i - \xi_i)^2/\sigma_i^2$. He then went through arguments, now standard in statistical texts, to show that $\boldsymbol{\theta}$ has the multivariate normal distribution about $\hat{\boldsymbol{\theta}}$ with covariance matrix $(\mathbf{X}'\mathbf{X})^{-1}\sigma^2$ (see Seal [12] for details).

Gauss' Second Approach [7]

The essential feature of this second approach is the assumption that when the true value θ is estimated by $\hat{\theta}$, an error $\theta - \hat{\theta}$ is committed, which entails a loss. The estimate $\hat{\theta}$ is then chosen to minimize the expected loss. He took a *convenient*, although admittedly *arbitrary* loss which is proportional to the squared error $(\theta - \hat{\theta})^2$. Then $\hat{\theta}$ is chosen to minimize the mean squared error (MSE), $E(\theta - \hat{\theta})^2$.

Gauss assumed that the errors e_i were sufficiently small that their squares and higher powers could be ignored, and thus restricted attention to *linear* estimates $\hat{\boldsymbol{\theta}} = \mathbf{CY}$ such that $\mathbf{CX} = \mathbf{I}$, the $k \times k$ identity matrix. He then showed that among all such estimates, the least-squares estimate (3b) minimizes the MSE. The resulting MSE is $(\mathbf{X}'\mathbf{X})^{-1}{}_{\sigma}{}^{2}$, thus reproducing the results of the first approach (with the exception, of course, of the normality of $\boldsymbol{\theta}$). In addition, the following results were obtained:

The least-squares estimate of any linear parametric function $\alpha = \sum g_i\theta_i = \mathbf{g}'\boldsymbol{\theta}$ is given by $\hat{\alpha} = \mathbf{g}'\hat{\boldsymbol{\theta}}$, with standard error

$$\sigma_\alpha = \{\mathbf{g}'(\mathbf{X}'\mathbf{X})^{-1}\mathbf{g}\}^{1/2}\sigma.$$

The minimum of Q is $Q_m = \mathbf{Y}'(\mathbf{Y} - \mathbf{X}\hat{\boldsymbol{\theta}})$.

If $\alpha = \mathbf{g}'\boldsymbol{\theta}$ is held fixed and the θ_i's otherwise allowed to vary, then Q can take on a relative minimum Q_r such that $Q_r - Q_m \leq C^2$ implies that $|\hat{\alpha} - \alpha| \leq C\sigma_\alpha/\sigma$.

An additional observation y with corresponding x-values

$$\mathbf{x}' = (x_{n+1,1}, x_{n+2,2}, \ldots, x_{n+1,k})$$

can be incorporated into the original least-squares estimate $\hat{\boldsymbol{\theta}}$ to form the updated estimate $\boldsymbol{\theta}^*$ via

$$\boldsymbol{\theta}^* = \hat{\boldsymbol{\theta}} - \mathbf{M}\,\frac{\mathbf{x}'\hat{\boldsymbol{\theta}} - y}{1 + w},$$

where $\mathbf{M} = (\mathbf{X}'\mathbf{X})^{-1}\mathbf{x}$, and $w = \mathbf{x}'(\mathbf{X}'\mathbf{X})^{-1}\mathbf{x} = \mathbf{x}'\mathbf{M}$ (which is a scalar). The covariance matrix of $\boldsymbol{\theta}^*$ is $[(\mathbf{X}'\mathbf{X})^{-1} - \mathbf{MM}'/(1 + w)]\sigma^2$, and the new minimum value of Q is $Q^*_m = Q_m + (\mathbf{x}'\hat{\boldsymbol{\theta}} - y)^2/(1 + w)$. This allows continuous updating of the least-squares estimate in the light of further observations obtained sequentially, without having to invert the new $\mathbf{X'X}$ matrix each time, and may be called "recursive least squares."

$E(Q_m) = (n - k)\sigma^2$; thus σ should be estimated by $\hat{\sigma} = \sqrt{Q_m/(n - k)}$, and not by $\sqrt{Q_m/n}$. Gauss also obtained the standard error of the estimate and noted that, when the e_i's are standard normal, this standard error becomes the standard error of the sum of $n - k$ independent errors e_i.

If $\mathbf{X} = \mathbf{I}_n$, so that there are n parameters $\xi_i = \theta_i$ and n observations $y_i = \theta_i + e_i$, and if there are r linear restrictions $\mathbf{F\theta} = \mathbf{0}$ where $\mathbf{F} = (f_{ij})$, $i = 1, \ldots, r$, then among linear functions the estimate of $\alpha = \mathbf{g'\theta}$ that minimize the MSE is $\hat{\sigma} = \mathbf{g'\theta}$,* where $\mathbf{\theta}^*$ is the least-squares estimate of $\mathbf{\theta}$ subject to $\mathbf{F\theta} = \mathbf{0}$. Its standard error is $\hat{\sigma} = \sqrt{Q_m/r}$.

Discussion

Gauss not only developed the foregoing theory, but also *applied* it to the analysis of observational data, much of which he himself collected. Indeed, the development of the theory was undoubtedly a response to the problems posed by his astronomical and geodetic observations. For example, at the end of *Theoria Motus Corporum Coelestium* [5], Gauss states that he has used this principle since 1795. See also numerical examples in Gauss [4, 7 (1826), 8, 9].

The difference in generality between the first and second approaches should be noted. The first approach allows $\hat{\mathbf{\theta}}$ to be any function of the observations, but requires the errors of observation e_i to be normally distributed about a zero mean. The second approach restricts the estimate $\hat{\mathbf{\theta}}$ to linear functions of the observations but allows the e_i to have any distribution with a mean of zero and finite variance.

The maximizing of Ω leading to (4) is equivalent to the method of estimation known today as maximum likelihood, the most useful general method of estimation. The properties of the maximum-likelihood estimate were obtained by Fisher* (1922 [3] and later).

The only way in which modern textbooks add to the foregoing theory of Gauss is in the explicit setting out of tests of linear hypotheses $\mathbf{A\theta} = \mathbf{0}$ and in the concomitant exact distributional theory associated with normally distributed observational errors, e.g., the *t*, *F*, and χ^2 distributions.

An unfortunate feature of the treatment in modern textbooks is their reference to the foregoing as the Gauss–Markov theory, and in their insistence that Gauss was seeking unbiased estimates, $E(\hat{\mathbf{\theta}}) = \mathbf{\theta}$. Regarding the former, there seems no justification for associating the name of Markov with the foregoing theory. Regarding the latter, it is true that $\mathbf{CX} = \mathbf{I}_k$ implies that $\hat{\mathbf{\theta}}$ of (3b) is unbiased. But the *requirement* of unbiasedness is unreasonable and can lead to absurd results. That Gauss did not insist on unbiased estimates is evident from his estimate of σ as $\sqrt{Q/(n - k)}$, which is biased. The condition $\mathbf{CX} = \mathbf{I}_k$ is in fact a "consistency" criterion, the purpose of which is to specify what is being estimated. In the present case, the requirement is that if the observational errors are all $e_i \equiv 0$, then the equations $\mathbf{Y} = \mathbf{Y\theta}$ must be consistent and the estimate must be identical to the true value, $\hat{\mathbf{\theta}} \equiv \theta$. As stated by Bertrand [1, p. 255], the official translator of Gauss's statistical work into French, "car, sans cela, toutes les mesures étant supposées exact, la valeur qu'on en déduit ne le serait pas."

Also, modern texts do not restrict the domain of application of linear estimates, as did Gauss, by assuming the observational errors to be sufficiently small that their squares and higher powers can be ignored. Similarly, many modern texts overemphasize the criterion of MSE, regarding it as the foundation on which to base a theory of estimation. In this way they ignore the qualifications imposed by Gauss.

There are other statistical contributions of Gauss. A notable one [6] is the demonstration that for a normal distribution, the most precise estimate of σ^2 among estimates depending on $S_k = \sum |e_i|^k$ is obtained when $k = 2$. His calculations were independently obtained by Fisher [2], who went on to isolate the property of sufficiency. In another noteworthy paper Gauss [9] exemplifies the use of weighted least squares in determining longitude by the use of a chronometer. He presents there a discussion of the structure of the observation errors e_i obtained in the use of a chronometer, and takes the standard deviation σ_i to be proportional to the square root of the time elapsed between observations. All of the above demonstrate a mixture of theory and of *application* that seems all too rare today.

A more detailed account and discussion of the matters raised in this article appear in Sprott [13].

REFERENCES

[1] Bertrand, J. (1888). *Calcul des Probabilitiés.* (2nd ed., 1972, Chelsea, New York.)

[2] Fisher, R. A. (1920). *Monthly Notices R. Astron. Soc.*, **80**, 758–770.

[3] Fisher, R. A. (1922). *Phil. Trans. R. Soc. A*, **222**, 309–368.

[4] Gauss, C. F. (1803–1809). Disquisitiones de Elementis ellipticis pallidis. *Werke*, **6**, 1–24.

[5] Gauss, C. F. (1809). *Theoria Motus Corporum Coelestium. Werke*, **7**. (English transl.: C. H. Davis. Dover, New York, 1963.)

[6] Gauss, C. F. (1816). Bestimmung der Genauigkeit der Beobachtungen. *Werke*, **4**, 109–117.

[7] Gauss, C. F. (1821, 1823, 1826). Theoria combinationis erroribus minimis obnoxiae, parts 1, 2 and suppl. *Werke*, **4**, 1–108.

[8] Gauss, C. F. (1823). Anwendungen der Wahrscheinlichkeitsrechnung auf eine Aufgabe der praktischen Geometrie. *Werke*, **9**, 231–237.

[9] Gauss, C. F. (1824). Chronometrische Langenbestimmungen. *Astron. Nachr.*, **5**, 227.

[10] Legendre, A. M. (1805). *Nouvelles Méthodes pour la Determination des Orbits des Cométes.* (Appendix: Sur la méthode des moindre carrés.)

[11] May, K. O. (1972). In *Dictionary of Scientific Biography*, Vol. 5. Scribners, New York, pp. 298–315.

[12] Seal, H. L. (1967). *Biometrika*, **54**, 1–24.

[13] Sprott, D. A. (1978). *Historia Math.*, **5**, 183–203.

BIBLIOGRAPHY

Gauss, C. F. (1957). Gauss's Work (1803–1826). On The Theory of Least Squares. Mimeograph, Dept. Of Science and Technology, Firestone Library, Princeton University. (English translations of refs. [7, 6, 8, 9] and the statistical parts of refs. [5] and [4], by H. F. Trotter, based on the French translation by J. Bertrand published in 1855 and authorized by Gauss.)

D. A. SPROTT

Maxwell, James Clerk

Born: June 13, 1831, in Edinburgh, Scotland.
Died: November 5, 1879, in Cambridge, England.
Contributed to: mathematical physics (including statistical mechanics).

James Clerk Maxwell (1831–1879) was a native Scot, educated at Edinburgh Academy and at the Universities of Edinburgh and Cambridge. He held senior professorial positions in natural philosophy and experimental physics at Aberdeen University, King's College in London, and at the University of Cambridge, where he supervised the planning and early administration of the Cavendish Laboratory.

Max Planck [9] asserted that among Maxwell's many contributions to diverse areas in mathematical physics his work materially influenced two main areas, those dealing with the physics of continuous media such as his electromagnetic theory, and with the physics of particles. The latter includes Maxwell's study of the kinetic theory of gases [5, 6], which initiated a new period in physics by describing physical processes in terms of a statistical function rather than a mechanical or deterministic one.

DYNAMICAL SYSTEMS

Maxwell became acquainted with probabilistic arguments while in Edinburgh, under his friend and teacher James Forbes, and through a review by Sir John Herschel of Quetelet's* treatise on probability in the *Edinburgh Review* in 1850; see Everitt [3]. He was also aware that Rudolf Clausius [2] had used statistical arguments in 1859 to show that, under the assumption that all molecules have equal velocity, the collisions of any given particle occur in what we would describe as a Poisson process.

In his first paper on the dynamical theory of gases [5] in 1860, Maxwell used geometry to determine the distributions $f(x)$, $f(y)$, and $f(z)$ of mutually orthogonal components of velocity. If N particles start from the origin together, then the number of particles in a "box" of volume $dx\,dy\,dz$ after a large number of collisions has occurred is

$$Nf(x)f(y)f(z)\,dx\,dy\,dz;$$

the independence of the densities $f(x)$, $f(y)$, $f(z)$ follows from the orthogonality of the coordinate axes. "But the directions of the coordinates are perfectly arbitrary, and therefore . . .

$$f(x)f(y)f(z) = \phi(x^2 + y^2 + z^2)."$$

This functional equation solves to give the normal law in the form

$$f(x) = \frac{1}{\alpha\sqrt{\pi}}\, e^{-x^2/\alpha^2}.$$

"It appears from this proposition," Maxwell wrote, "that the velocities are distributed among the particles according to the same law as the errors are distributed among the observations in the theory of the method of least squares," a reference to Gauss'* contributions to statistics.

The velocity $v = \sqrt{(x^2 + y^2 + z^2)}$ has the *Maxwell distribution* with density

$$g(v) = \frac{4}{\alpha^3\sqrt{\pi}} v^2 e^{-v^2/\alpha^2}, \qquad v > 0, \tag{2}$$

with mean velocity $2\alpha/\sqrt{\pi}$ and variance $[3/2 - (4/\pi)]\alpha^2$. It would then have been easy to show that the molecular kinetic energy has a chi-square distribution with three degrees of freedom. In 1867 Maxwell [6] derived similar results when the collisions of the particles are incorporated; the above distributions are then stationary. His approach attracted the keen attention of Ludwig Boltzmann,* who built upon it the foundations of statistical mechanics.

THERMODYNAMICS AND MAXWELL'S DEMON

Maxwell had a penetrating mind, which, as his friend P.G. Tait wrote, "could never bear to pass by any phenomenon without satisfying itself of at least its general nature and causes" [10]. Having resolved the crux of the phenomenon, he could then leave others to work out the details. Nowhere does this mark of his character appear more forcefully than in his insistence that the second law of thermodynamics is a statistical one; at the same time others, such as Clausius and Boltzmann,* were trying to explain it as a mechanical one, to the effect that heat can never pass from a colder to a warmer body without some change (in the form of external work) occurring at the same time. Challenging the universality of this assertion, Maxwell [7] considered gas in a compartmentalized container, with a diaphragm between a section A with hotter gas and a section *B* with colder gas; he postulated "a finite being who knows the paths and velocities of all the molecules by simple inspection but who can do no work except open and close a hole in the diaphragm by means of a slide without mass." Labeled "Maxwell's demon" by William Thomson (see ref. [4] for a full discussion), this being would allow molecules from *A* with sufficiently low velocities to pass into *B*, and would allow molecules from *B* with sufficiently high velocities to pass into *A*, carrying out these procedures alternately. The statistical distribution (2) guarantees the availability of such molecules; the gas in *A* would become hotter and that in *B* colder, no work would be done, and the second law would be violated. "Only we can't, not being clever enough" [7].

It was again to be Ludwig Boltzmann who would give explicit form to Maxwell's insight, developing in his *H*-theorem the relationship between entropy and probability [1]. Maxwell returned to his theme in 1878, while reviewing the second edition of Tait's *Sketch of Thermodynamics* [8].

The law is continually violated, he says,

> . . . in any sufficiently small group of molecules belonging to a real body. As the number of molecules in the group is increased, the deviations from the mean of the whole become smaller and less frequent; and when the number is increased till the group includes a sensible portion of the body, the probability of a measurable variation from the mean occurring in a finite number of years becomes so small that it may be regarded as practically an impossibility.
>
> This calculation belongs of course to molecular theory and not to pure thermodynamics, but it shows that we have reason for believing the truth to the second law to be of the nature of a strong probability, which, though it falls short of certainty by less than any assignable quantity, is not an absolute certainty.

Maxwell stood apart from the controversy between Tait and the German school over credit and priorities in thermodynamics. Quantitative data in support of the statistical molecular theory which he and Boltzmann had initiated did not appear until after 1900. Klein [4] gives a very readable detailed narrative and discussion, and includes an extensive bibliography.

REFERENCES

[1] Boltzmann, L. E. (1877). *Wiss. Abh.*, **2**, 164–223.

[2] Clausius, R. (1859). *Phil. Mag. 4th Ser.*, **17**, 81–91.

[3] Everitt, C. W. F. (1975). *James Clerk Maxwell, Physicist and Natural Philosopher.* Scribners, New York.

[4] Klein, M. J. (1970). *Amer. Scientist*, **58**, 84–97.

[5] Maxwell, J. C. (1860). *Phil. Mag. 4th Ser.*, **19**, 19–32; *ibid.*, **20**, 21–37.

[6] Maxwell, J. C. (1866). *Phil. Mag. 4th Ser.*, **32**, 390–393.

[7] Maxwell, J. C. (1867). Letter to P. G. Tait. [See Knott, C. B. (1911), *Life and Scientific Work of Peter Guthrie Tait*, Cambridge University Press, Cambridge, England, pp. 213–214.]

[8] Maxwell, J. C. (1878). *Nature*, **17**, 278–280.

[9] Planck, M. (1931). In *James Clerk Maxwell: A Commemoration Volume, 1831–1931.* Cambridge, England.

[10] Tait, P. G. (1880). *Nature*, **21**, 321.

BIBLIOGRAPHY

Campbell, L. and Garnett, W. (1883). *Life of James Clerk Maxwell.* MacMillan, London.

Domb, C. (1980/81). *Notes and Records R. S. London*, **35**, 67–103. (An account of Maxwell's life and work in London and Aberdeen, 1860–1871.)

CAMPBELL B. READ

Scott, Elizabeth Leonard

Born: November 23, 1917, in Fort Sills, Oklahoma, USA.
Died: December 20, 1988, in Berkeley, California, USA.
Contributed to: weather modification, correlation medical statistics, design of experiments, astronomy, distribution theory.

Elizabeth Leonard Scott

Elizabeth Leonard Scott's father was an officer in the U.S. Army, and she often referred to herself, jokingly, as an "army brat." When she was only four years old, her father retired to study law, and the family moved to Berkeley to be close to the university. From that time on, her life centered around the University of California, Berkeley. Her degrees, both in astronomy, were from there, and her first publications were in the astronomy field. All her academic life was spent at Berkeley and, after its inception, in the statistics department.

During World War II she worked in the Statistical Laboratory that Jerzy Neyman* had just started at Berkeley. The work was mostly concerned with improving the precision of air bombing. By 1948, when she published her first statistical paper, she had already coauthored 13 papers in astronomy. Neyman had led her to statistics, and she led him to astronomy; the two of them wrote a long series of papers on the spatial distribution of galaxies.

In 1951 Scott was appointed assistant professor of mathematics at U.C. Berkeley, and she became an associate professor of statistics in 1957 and a professor of statistics in 1962. She worked on a variety of statistical studies. One such study, jointly with Neyman, was on weather modification (cloud seeding). That started in 1952. (She was still working on a booklet on the subject at the time of her hospitalization in 1988.) Also with Neyman, she published a series of papers on stochastic models for carcinogenesis.

In the early seventies Scott worked to promote the status of women in academia. She published several papers in which she compared salaries of faculty women with those of faculty men and found a substantial difference at equal qualifications nationally and locally. She was a role model for many aspiring young female scientists and took an active interest in their careers even after her retirement in 1987.

Scott had been president (1977–1978) of the Institute of Mathematical Statistics (IMS), vice-president (1981–1983) of the International Statistical Institute (ISI), vice-president of the American Statistical Association (ASA), and president of the Bernoulli Society. (She was a main force behind the organization of the World Congress of the Bernoulli Society in Tashkent in 1986.) She received many honors, including election as honorary fellow of the Royal Statistical Society and as fellow of the American Association for the Advancement of Science (AAAS).

Scott's weakness was in trying to do too much. She did not know how to say no, and in consequence was sometimes behind in her commitments, which necessitated long hours of work when she should have been resting and relaxing.

BIBLIOGRAPHY

Le Cam, L. (1989). Obituary, Elizabeth Leonard Scott: 1917–1988. *IMS Bull.*, **18**(1), 80.

Billard, L. and Ferber, M. (1991). Elizabeth Scott: Scholar, teacher, administrator. *Statist. Sci.*, **6**(2), 206–216.

Wilcoxon, Frank

Born: September 2, 1892, in County Cork, Ireland.
Died: November 18, 1965, in Tallahassee, Florida, USA.
Contributed to: (statistics) rank tests, multiple comparisons, sequential ranking, factorial design; (chemistry) fungicidal action, synthesis of plant growth substances, insecticides research (pyrethrins, parathion, Malathion).

Frank Wilcoxon was born in Glengarriffe Castle, near Cork, Ireland, to wealthy American parents on September 2, 1892. His father was a poet, outdoorsman, and hunter. Wilcoxon spent his boyhood at Catskill, New York and developed his lasting love of nature and water there. Adolescence seems to have been difficult, with a runaway period during which he was briefly a merchant seaman in New York harbor, a gas-pumping station attendant in an isolated area of West Virginia, and a tree surgeon. As this period ended, he was enrolled in Pennsylvania Military College and, although the school's system did not agree with his ideas of personal freedom, he received the B.S. degree in 1917.

Wilcoxon entered Rutgers University in 1920 after a World War I position with the Atlas Powder Company at Houghton, Michigan and received the M.S. degree in chemistry in 1921. He continued his education at Cornell University and received the Ph.D. degree in physical chemistry in 1924. At Cornell, he met Frederica Facius, and they were married on May 27, 1926. Frank and Freddie later became well known and loved in the statistical community, particularly through their regular participation in the Gordon Research Conference on Statistics in Chemistry and Chemical Engineering.

From 1924 to 1950, Frank Wilcoxon was engaged in research related to chemistry. In 1925 he went to the Boyce Thompson Institute for Plant Research under a Crop Protection Institute fellowship sponsored by the Nichols Copper Company and worked on colloid copper fungicides until 1927. He then (1928–1929) worked with the sponsoring company in Maspeth in Queens. In 1929 he returned to the Boyce Thompson Institute and remained there until 1941, working on the chemistry and mode of action of fungicides and insecticides. With a leave of absence from 1941 to 1943, he designed and directed the Control Laboratory of the Ravenna Ordnance Plant operated by the Atlas Powder Company. Wilcoxon joined the American Cyanamid Company in 1943 and continued with that company until his retirement in 1957, first with the Stamford Research Laboratories as head of a group developing insecticides and fungicides, and then as head of the statistics group of the Lederle Division in Pearl River, New York. He served as a consultant to various organizations, including the Boyce Thompson Institute, until 1960, when he joined the faculty of the new Department of Statistics at Florida State University. There he was active in research and teaching and contributed to the development of the department until his death in 1965.

Frank Wilcoxon made many contributions to chemistry and biochemistry with some 40 publications in the field. His first paper [13], on acidimetry and alkalimetry, was published in 1923 in *Industrial and Engineering Chemistry* with van der

Meulen. A series of papers with S. E. A. McCallan was written on the mode of action of sulphur and copper fungicides. A series of papers with Albert Hartzell dealt with extracts of pyrethrum flowers and resulted in a mercury reduction method for the determination of pyrethrin I. Also, in his work with the Boyce Thompson Institute, he synthesized a number of plant growth substances, including alpha-naphthaleneacetic acid, which were studied with P.W. Zimmerman and led to a series of papers on the action of these growth substances on plants. In his work with the American Cyanamid Company he led the research group that studied the insecticide parathion and which developed the less toxic Malathion.

Wilcoxon's interest in statistics began in 1925 with a study of R. A. Fisher's* book, *Statistical Methods for Research Workers*. This study was done in a small reading group in which his colleague at the Boyce Thompson Institute, W.J. Youden, participated also. Wilcoxon became increasingly interested in the application of statistics in experimentation, and this is apparent in a number of his papers on spore germination tests. His first publication in a statistics journal [14] dealt with uses of statistics in plant pathology. His research contributions in statistics were to range over rank tests, multiple comparison, sequential ranking, factorial experiments, and bioassay. Throughout his research, he sought statistical methods that were numerically simple and easily understood and applied.

Wilcoxon introduced his two rank tests, the rank-sum test for the two-sample problem and the signed-rank test for the paired-samples problem, in his 1945 paper [15]. This paper and a contemporary one by Mann and Whitney [10] led the way to the extensive development of nonparametric statistics, including Wilcoxon's own further contributions. There is no doubt that this paper was Wilcoxon's most important contribution. Although there were independent proposals of the two-sample statistic (see Kruskal [9] for historical notes), the paper became a major inspiration for the development of nonparametric methods. In addition, the statistical methodology introduced has had a major impact in applied statistics, particularly for applications in the social sciences, becoming one of the most popular of statistical tools.

Research on Wilcoxon test theory continued. He himself provided additional tables [16] in 1947 and again, with Katti and Wilcox [20], in 1963. Serfling [12] studied the properties of the two-sample rank-sum test in a setting where dependence is allowed within samples. Hollander et al. [8] studied the properties of the rank-sum test in a model where the assumption of independence between samples is relaxed. Hettmansperger [7] provides a modern treatment of nonparametric inference based on ranks.

Wilcoxon was interested in extensions of his basic rank procedures to new situations. He was an initiator of research in nonparametric sequential methods. Since nonparametric techniques were so successful in fixed-sample-size situations, he felt that their good properties would naturally carry over to the sequential setting. This idea led to a number of sequential rank procedures developed under the leadership of Wilcoxon and Bradley [2, 3, 17, 18]. Other researchers continued the development of nonparametric sequential methods after this early work by Bradley and Wilcoxon, and a comprehensive development of the theory is given by Sen [11]. Wilcoxon was interested also in the problem of testing whether two p-variate populations, $p \geq 2$, are equivalent, and two of his proposals for this problem motivated refs. [1] and [4].

In experiments in the natural sciences and in the behavioral sciences, typically there are more than just one or two conditions (treatments, etc.) under investigation. Often multiple hypotheses need to be tested. In such settings, it is important to control the overall or experimentwise error rate. Wilcoxon recognized this and had a strong interest in multiple comparisons. In particular, his 1964 revision of the booklet [19] (joint with Roberta A. Wilcox) *Some Rapid Approximate Statistical Procedures*, features multiple comparison procedures based on Wilcoxon rank sums for the one-way layout and multiple comparison procedures based on Friedman rank sums for the two-way layout. The booklet played a significant role in the (now) widespread use of nonparametric multiple comparison procedures. In the early 1960s Wilcoxon also suggested and largely directed a dissertation on multiple comparisons by Peter Dunn-Rankin, part of which was published as a joint paper [6].

Other areas of Wilcoxon research seem less well known. From his research on insecticides and fungicides, he, with J. T. Litchfield, Jr. and K. Nolen, developed an interest in and a series of papers on a simplified method of evaluating dose–effect experiments. Daniel and Wilcoxon [5] devised fractional designs robust against linear and quadratic trends, anticipating to some degree the concept of trend-free block designs.

While Frank Wilcoxon was not an academician for much of his career, he was a teacher and a student throughout his life. His enthusiasm for statistics and his encouragement of others led many to more intensive study of the subject. It was typical of the man that, prior to visits to Russia in 1934 and 1935, he undertook a study of the language and retained a remarkably proficient reading knowledge throughout his life.

Frank Wilcoxon was recognized by his associates. He was a Fellow of the American Statistical Association and of the American Association for the Advancement of Science. He was a leader in the development of the Gordon Research Conference on Statistics in Chemistry and Chemical Engineering and a past chairman of that conference.

Karas and Savage list the publications of Wilcoxon. This and other material are listed in the bibliography.

REFERENCES

[1] Bradley, R. A. (1967). *Proc. Fifth Berkeley Symp.*, Vol. 1, L. LeCam and J. Neyman, eds. Univ. Of Calif. Press, Berkeley, pp. 593–605.

[2] Bradley, R. A., Martin, D. C., and Wilcoxon, F. (1965). *Technometrics*, **7**, 463–483.

[3] Bradley, R. A., Merchant, S. D., and Wilcoxon, F. (1966). *Technometrics*, **8**, 615–623.

[4] Bradley, R. A., Patel, K. M., and Wackerly, D. D. (1971). *Biometrics*, **27**, 515–530.

[5] Daniel, C. and Wilcoxon, F. (1966). *Technometrics*, **8**, 259–278.

[6] Dunn-Rankin, P. And Wilcoxon, F. (1966). *Psychometrika*, **31**, 573–580.

[7] Hettmansperger, T. P. (1984). *Statistical Inference Based on Ranks.* Wiley, New York.
[8] Hollander, M., Pledger, G., and Lin, P. (1974). *Ann. Statist.*, **2**, 177–181.
[9] Kruskal, W. H. (1957). *J. Amer. Statist. Ass.*, **52**, 356–360.
[10] Mann, H. and Whitney, D. R. (1947). *Ann. Math. Statist.*, **18**, 50–60.
[11] Sen, P. K. (1981). *Sequential Nonparametrics.* Wiley, New York.
[12] Serfling, R. J. (1968). *Ann. Math. Statist.*, **39**, 1202–1209.
[13] van der Meulen, P. A. and Wilcoxon, F. (1923). *Ind. And Eng. Chem.*, **15**, 62–63.
[14] Wilcoxon, F. (1945). *Biometrics Bull.*, **1**, 41–45.
[15] Wilcoxon, F. (1945). *Biometrics Bull.*, **1**, 80–83.
[16] Wilcoxon, F. (1947). *Biometrics*, **3**, 119–122.
[17] Wilcoxon, F. and Bradley, R. A. (1964). *Biometrics*, **20**, 892–895.
[18] Wilcoxon, F., Rhodes, L. J., and Bradley, R. A. (1963). *Biometrics*, **19**, 58–84.
[19] Wilcoxon, F. and Wilcox, R. A. (1964). *Some Rapid Approximate Statistical Procedures.* Stamford Research Laboratories, Pearl River, New York. [Revision of a 1947 (revised 1949) booklet by Wilcoxon.]
[20] Wilcoxon, F., Katti, S. K., and Wilcox, R. A. (1970). *Selected Tables in Mathematical Statistics*, vol. 1, H. L. Harter and D. B. Owen, eds. Markham, Chicago, pp. 171–259. (Originally prepared and distributed in 1963, revised 1968, by Lederle Laboratories, Pearl River, New York and the Department of Statistics, Florida State University, Tallahassee, Florida.)

BIBLIOGRAPHY

Anon. (1965). *New York Times*, Nov. 19, p. 39, col. 2.
Bradley, R. A. (1966). *Biometrics*, **22**, 192–194.
Bradley, R. A. (1966). *Amer. Statist.*, **20**, 32–33.
Bradley, R. A. and Hollander, M. (1978). In *International Encyclopedia of Statistics*, vol. 2, W.H. Kruskal and J.M. Tanur, eds. The Free Press, New York, pp. 1245–1250.
Dunnett, C. W. (1966). *Technometrics*, **8**, 195–196.
Karas, J. and Savage, I. R. (1967). *Biometrics*, **23**, 1–11.
McCallan, S. E. A. (1966). Boyce Thompson Institute for Plant Research, *Contributions*, **23**, 143–145.

RALPH A. BRADLEY
MYLES HOLLANDER

APPENDIX

A Statistical Analysis of Lifelengths of Leading Statistical Personalities

The following simple analyses may assist in forming a demographic picture of the group of individuals (some 115 names) who were chosen as the prominent statisticians. They were, of course, selected entirely on the basis of our perceived value of their work as pioneers in the development and application of statistical methods, and also their personalities (although geographical diversity has been taken into account). We had no prior ideas on the nature of the collection to which we were led—and, quite possibly, other persons would certainly have made other choices (hopefully not too drastically different). The majority of personalities are males of the European origin.

In Table 1, the available data on dates of birth and death, and derived (approximate) age at death, are set out.

Table 2 gives distributions of calendar months of birth and death. Chi-square tests of uniformity (with approximate null hypothesis of a probability of 1/12 for each month) yield nonsignificant results for births and only 5–10% significance for deaths. (The lack of homogeneity for the latter is most pronounced by high values for March and December, but low values for January and May.) In Table 3, we give a small part of the distribution of length of period from nearest birthday to date of death. Although it has been suggested by Phillips (1972) that persons tend to die soon after attaining a birthday, the data in this table strongly suggests quite the opposite.

Note that, in Phillips' calculations, he considered only *months* of births and deaths and concentrated on the "death dip"—the relative lack of deaths in the calendar month immediately preceding the calendar month of death. Thus if a person was born on May 26 and died on May 3, he would classify as "same month." Our classification, on the other hand used number of *days* from birth to death dates, so we would classify this as "*minus* 23 days".

Figure 1 exhibits grouped frequencies for year of birth and age at death, with group widths 25 years and 5 years respectively, and Table 1a provides values of lower quartile, median, upper quartile, maximum and arithmetic mean for the marginal distributions of Table 1. Values for the mode are also included, but these are of somewhat doubtful accuracy.

Table 4 compares grouped frequency distributions of age last birthday at death (5-year age groups) of LSP data and a sample of 75 from obituary notices from the *Washington Post*, with a few comments, during the period 10/25/96–11/2/96.

TABLE 1

	Date of		Approximate Age at Death
	Birth	Death	
Name	Mo./Day/Yr.	Mo./Day/Yr.	Years
Abbe	1/23/1840	1/14/1905	69.95
Achenwall	10/20/1719	5/1/1772	52.6
Aitken	4/1/1895	11/3/1967	72.05
Anderson	8/2/1887	2/12/1960	72.5
Arbuthnot	4/29/1667	2/27/1735	77.7
Bayes	/1701	4/7/1761	(60)
Bernstein	3/5/1880	10/26/1968	88.5
Bienaymé	8/28/1796	10/19/1878	81.15
Birnbaum	5/27/1923	7/1/1976	53.25
Bol'shev	3/16/1922	8/29/1978	55.45
Boltzmann	2/20/1844	9/5/1906	62.55
Bonferroni	1/28/1892	8/18/1960	67.55
Bortkiewicz	8/7/1868	7/15/1931	62.9
Boscovich	5/18/1711	2/13/1787	76.95
Bose	6/19/1901	10/30/1987	86.4
Bowley	11/6/1869	1/21/1957	87.1
Cantelli	12/20/1875	7/21/1966	90.6
Cauchy	8/21/1789	5/22/1857	67.75
Chebyshev	5/26/1821	12/8/1894	73.6
Chuprov	2/18/1874	4/19/1926	52.15
Cochran	7/15/1909	3/29/1980	70.6
*Cox	1/13/1900	10/17/1978	78.75
Cramér	9/25/1893	10/5/1985	92.05
*David	8/23/1909	7/18/1993	83.95
De Finetti	6/13/1906	7/20/1985	79.1
De Moivre	5/26/1667	11/27/1754	87.5

*Indicates female.

TABLE 1 (*continued*)

	Date of		Approximate Age at Death
	Birth	Death	
Name	Mo./Day/Yr.	Mo./Day/Yr.	Years
De Montmort	10/27/1678	10/7/1719	40.95
De Witt	9/25/1625	8/20/1672	46.9
Deming	10/14/1900	12/20/1993	93.2
Déparcieux	10/18/1703	9/2/1768	64.85
Edgeworth	2/8/1845	2/13/1926	81.0
Elfving	6/28/1908	3/25/1984	75.75
Engel	3/26/1821	12/8/1896	75.7
Farr	11/30/1807	4/14/1883	75.5
Feller	7/7/1906	1/14/1970	63.5
Fisher, I.	2/27/1867	4/29/1947	80.15
Fisher, R. A.	2/17/1890	7/29/1962	72.45
Franscini	10/23/1796†	7/19/1857	60.75
Galton	2/16/1822	1/17/1911	88.9
Gauss	4/30/1777	2/23/1855	77.8
Gini	5/23/1884	3/13/1965	80.8
Gnedenko	1/1/1912	12/30/1955	84.0
Gosset	6/13/1876	10/16/1937	61.35
Goulden	6/2/1897	2/3/1981	83.65
Graunt	4/24/1620	4/18/1674	53.95
Gumbel	7/18/1891	9/10/1966	75.15
Guttman	2/10/1916	10/25/1987	71.7
Hájek	2/4/1926	6/10/1974	48.3
Hannan	1/29/1921	1/7/1994	72.95
Hansen	12/15/1910	10/9/1990	79.8
Hartley	4/13/1912	12/30/1980	68.7
Helmert	7/31/1843	6/15/1917	73.9
Hill	7/8/1897	4/18/1991	93.75
Hoeffding	6/12/1914	2/25/1991	76.95
Hotelling	9/29/1895	12/26/1973	78.25
Hsu	9/1/1910	12/18/1970	60.3
Huygens	4/14/1629	6/8/1695	66.15
Jeffreys	4/22/1891	3/18/1989	97.93
Jordan	12/16/1871	12/24/1959	88.25
Kendall	9/6/1907	3/29/1989	75.55
Kiaer	9/15/1838	4/16/1919	80.6
Kitagawa	10/3/1909	3/13/1993	83.45
Kolmogorov	4/25/1903	10/20/1987	84.5
Konüs	10/2/1895	4/5/1990	94.5
Kőrösy	4/20/1844	6/23/1906	62.1
Lambert	/1728	/1777	(49)
Langevin	1/23/1872	12/19/1946	74.9

†Month and day of birth not available in time to be included in analysis

TABLE 1 (*continued*)

Name	Date of Birth Mo./Day/Yr.	Date of Death Mo./Day/Yr.	Approximate Age at Death Years
Laplace	3/23/1749	3/5/1827	77.95
Lévy	9/1/1886	12/15/1971	85.8
Lexis	7/17/1837	8/24/1914	77.1
Liapunov	6/7/1857	3/11/1918	61.75
Linnik	1/8/1915	6/30/1972	57.45
Lüroth	2/18/1844	9/14/1910	66.6
Mahalanobis	6/29/1893	6/28/1972	79.0
Markov	6/2/1856	7/20/1922	66.1
Maxwell	6/13/1831	11/5/1879	48.1
Nemchinov	7/2/1894	11/5/1964	70.25
Newton	12/25/1642	3/20/1727	84.3
Neyman	4/16/1894	8/5/1981	87.25
*Nightingale	5/12/1820	8/13/1910	90.25
Pascal	6/19/1623	8/9/1662	39.15
Pearson, E. S.	8/11/1895	6/12/1980	84.85
Pearson, K.	3/27/1857	4/27/1936	79.15
Petty	5/26/1623	12/16/1687	64.55
Pitman	10/27/1897	7/21/1993	95.7
Quetelet	2/22/1796	2/17/1874	72.0
Ramsey	2/22/1903	1/19/1930	26.95
Rényi	3/20/1921	2/1/1970	48.9
Savage	11/20/1917	11/1/1971	53.95
*Scott	11/23/1917	12/20/1988	71.1
'sGravesande	/1688	/1742	(54)
Sinclair	5/10/1754	12/21/1835	81.6
Slutskii	4/7/1880	3/10/1948	67.9
Smirnov	10/17/1900	6/2/1966	65.65
Snedecor	10/20/1881	2/15/1974	93.3
Süssmilch	9/3/1707	3/22/1767	59.55
Sverdrup	2/23/1917	3/15/1944	77.05
Thiele	12/24/1838	9/26/1910	71.75
Wald	10/31/1902	12/13/1950	48.1
Wargentin	9/11/1717	12/13/1783	66.25
Westergaard	4/19/1853	12/13/1936	83.65
Wiener	11/26/1894	3/19/1964	69.3
Wilcoxon	9/2/1892	11/18/1965	73.3
Wilks	6/17/1906	3/7/1964	57.7
Willcox	3/22/1861	10/30/1964	103.7
Wilson	4/25/1879	12/28/1964	85.65
Winkler	6/29/1884	9/3/1984	100.15
Wold	12/25/1908	2/16/1992	83.15
Wolfowitz	3/19/1910	7/16/1981	71.3

*Indicates female.

TABLE 1 (*continued*)

Name	Date of Birth Mo./Day/Yr.	Date of Death Mo./Day/Yr.	Approximate Age at Death Years
Yanson	11/5/1835	1/31/1892	56.15
Yastremskii	5/9/1877	11/28/1962	85.55
Yates	5/12/1902	6/17/1944	92.1
Yule	2/18/1871	6/26/1951	80.3

Remarks:

1. Neither James (1654–1705) nor Daniel Bernoulli (1700–1782) were included in this table. Their life lengths (50 and 82 years, respectively) would not have affected the conclusions.
2. The data for the four women from the original selection were, however, incorporated.
3. Some odd points:
   (a) *Two* successive identical month/day of birth (February 22)—Quetelet and Ramsay.
   (b) *Three* successive identical month/day of death (December 13)—Wald, Wargentin, Westergaard.
   (c) *Eight* successive days of birth in 20s from K. Pearson to E. Scott (namely K. Pearson, Petty, Pitman, Quetelet, Ramsey, Rényi, Savage, and Scott were born on 27, 26, 27, 22, 22, 20, 20, 23 days of a month, respectively.)
4. Newton and Wold were both born on Christmas Day; Gnedenko was born on New Year's Day; but no deaths are recorded on Christmas Day or New Year's Day.

TABLE 1a

	Age at Death Years	Year of Birth
Minimum	26.95	1620
Lower quartile	61.75	1792
Median	70.9	1867
Upper quartile	81.0	1893
Maximum	103.7	1861
Arithmetic mean	72.35	1845
(Mode)	(75)	(1902)

(*'Box–Whisker' Plots*)

AGE AT DEATH

20 25 30 35 40 45 50 55 60 65 70 75 80 85 90 95 100 105

AGE AT DEATH (YEARS)

YEAR OF BIRTH

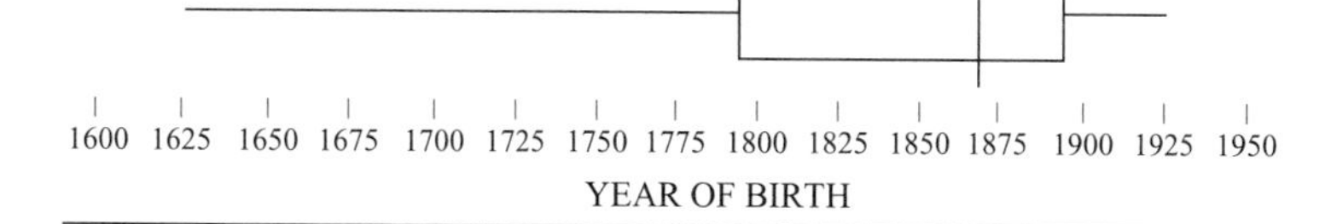

TABLE 2 Births and Deaths by Calendar Months

Month	Number of Births	Number of Deaths
1	7	7
2	13	11
3	8	14
4	13	9
5	10	2
6	13	10
7	7	10
8	6	7
9	10	6
10	10	11
11	6	7
12	6	17
	$N = 109$	111
$\chi^2_{11,0.90} = 17.275$; $\chi^2_{11,0.95} = 19.675$		
Calculated	$\chi^2 = 9.6$	$\chi^2 = 18.2$

TABLE 3

Date of Death Minus Date on Nearest Birthday	Frequency	Approximate Expected Value
–40 to –30	3	3
–30 to Zero	13	9
Zero to 30	4	9
30 to 40	3	3

(Among 109 individuals if $p = 1/12$; expected value $= 109/12 \doteq 9$ Standard deviation (S. D.) $= \sqrt{109 - 11 \times 144} \doteq 3.46$) ($-5/3.46 \doteq -1.445$)

CENTURIES	AGE AT DEATH (YEARS) LBD 25–29	30–34	35–39	40–44	45–49	50–54	55–59	60–64	65–69	70–74	75–79	80–84	85–89	90–94	95–99	100–105	YEAR OF BIRTH	
			1			1		1									1600–1624	3
					1				1			1					1625–1649	3
17th											1		1				1650–1674	2
				1		1											1675–1699	2
						1	2	1	1		1						1700–1724	6
					1						1						1725–1749	2
18th										1		1					1750–1774	2
								1	1	1	1	1					1775–1799	5
										1	2		1	1			1800–1824	5
					1		1	2	2	2	1	2					1825–1849	11
						1		2	1			3	2	1		1	1850–1874	11
19th								1	4	6	2	2	5	5	2	1	1875–1899	28
	1				2	2	3	2	2	5	7	5	1	2			1900–1924	32
20th					1												1925–1949	1
	1	–	1	1	6	6	6	10	12	16	16	15	10	9	2	2		113

FIGURE 1

TABLE 4 Comparison Between LSP Data and Sample Newspaper Obituary Data

Age at Death						
	LSP Data (for Deaths After 1950 only)			*Washington Post* 10/25/96–11/2/96 Sample Data (Men only)		
Age LBD (years)	Frequency	Cumulative Frequency	Proportion (EDF)	Frequency	Cumulative Frequency	Proportion (EDF)
25–29	1 (–)	1 (–)	0.009 (–)	– (–)	– (–)	0.000 (–)
30–34	– (–)	1 (–)	0.008 (–)	1 (–)	1 (–)	0.013 (–)
35–39	1 (–)	2 (–)	0.018 (–)	2 (2)	3 (2)	0.040 (0.045)
40–44	1 (–)	3 (–)	0.027 (–)	1 (1)	4 (3)	0.053 (0.068)
45–49	6 (3)	9 (3)	0.080 (0.048)	4 (2)	8 (5)	0.107 (0.114)
50–54	6 (2)	15 (5)	0.133 (0.079)	5 (4)	13 (9)	0.173 (0.205)
55–59	6 (4)	21 (9)	0.186 (0.143)	4 (2)	17 (11)	0.227 (0.250)
60–64	10 (2)	31 (11)	0.274 (0.175)	3 (2)	20 (13)	0.267 (0.295)
65–69	12 (4)	43 (15)	0.381 (0.238)	6 (4)	26 (7)	0.347 (0.386)
70–74	16 (10)	59 (25)	0.522 (0.397)	12 (6)	38 (23)	0.507 (0.523)
75–79	16 (10)	75 (35)	0.664 (0.556)	9 (3)	47 (26)	0.627 (0.591)
80–84	15 (9)	90 (44)	0.796 (0.698)	4 (4)	51 (30)	0.680 (0.681)
85–89	10 (8)	100 (52)	0.885 (0.825)	11 (9)	62 (39)	0.827 (0.886)
90–94	9 (7)	109 (59)	0.965 (0.937)	10 (5)	72 (44)	0.960 (1.005)
95–99	2 (2)	111 (61)	0.982 (0.968)	2 (–)	74 (44)	0.987 –
100–104	2 (2)	113 (63)	1.000 (1.000)	1 (–)	75 (44)	1.300 –

Remarks:

1. There seems to be little difference between the two EDFs. Sample data have a somewhat higher EDF between ages 50 and 60.
2. The two sets of data—although collected not from really comparable sources—do share the feature that some prominence is (presumably) needed for inclusion.

REFERENCE

Phillips, David (1972) Deathday and birthday: An unexpected connection, in J. M. Tanur and W. H. Kruskal (eds): *Statistics: A Guide to the Unknown*, Holden-Day, pp. 52–65.

Index of Names

Subject Index

WILEY SERIES IN PROBABILITY AND STATISTICS

Probability and Statistics

ANDERSON · An Introduction to Multivariate Statistical Analysis, *Second Edition*
*ANDERSON · The Statistical Analysis of Time Series
ARNOLD, BALAKRISHNAN, and NAGARAJA · A First Course in Order Statistics
BACCELLI, COHEN, OLSDER, and QUADRAT · Synchronization and Linearity: An Algebra for Discrete Event Systems
BARTOSZYNSKI and NIEWIADOMSKA-BUGAJ · Probability and Statistical Inference
BERNARDO and SMITH · Bayesian Statistical Concepts and Theory
BHATTACHARYYA and JOHNSON · Statistical Concepts and Methods
BILLINGSLEY · Convergence of Probability Measures
BILLINGSLEY · Probability and Measure, *Second Edition*
BOROVKOV · Asymptotic Methods in Queuing Theory
BRANDT, FRANKEN, and LISEK · Stationary Stochastic Models
CAINES · Linear Stochastic Systems
CAIROLI and DALANG · Sequential Stochastic Optimization
CHEN · Recursive Estimation and Control for Stochastic Systems
CONSTANTINE · Combinatorial Theory and Statistical Design
COOK and WEISBERG · An Introduction to Regression Graphics
COVER and THOMAS · Elements of Information Theory
CSÖRGŐ and HORVÁTH · Weighted Approximations in Probability Statistics
*DOOB · Stochastic Processes
DUDEWICZ and MISHRA · Modern Mathematical Statistics
DUPUIS and ELLIS · A Weak Convergence Approach to the Theory of Large Deviations
ETHIER and KURTZ · Markov Processes: Characterization and Convergence
FELLER · An Introduction to Probability Theory and Its Applications, Volume 1, *Third Edition,* Revised; Volume II, *Second Edition*
FREEMAN and SMITH · Aspects of Uncertainty: A Tribute to D. V. Lindley
FULLER · Introduction to Statistical Time Series, *Second Edition*
FULLER · Measurement Error Models
GHOSH, MUKHOPADHYAY, and SEN · Sequential Estimation
GIFI · Nonlinear Multivariate Analysis
GUTTORP · Statistical Inference for Branching Processes
HALD · A History of Probability and Statistics and Their Applications before 1750
HALL · Introduction to the Theory of Coverage Processes
HANNAN and DEISTLER · The Statistical Theory of Linear Systems
HEDAYAT and SINHA · Design and Inference in Finite Population Sampling
HOEL · Introduction to Mathematical Statistics, *Fifth Edition*
HUBER · Robust Statistics
IMAN and CONOVER · A Modern Approach to Statistics
JOHNSON and KOTZ · Leading Personalities in Statistical Sciences: From the Seventeenth Century to the Present
JUREK and MASON · Operator-Limit Distributions in Probability Theory
KASS and VOS · Geometrical Foundations of Asymptotic Inference: Curved Exponential Families and Beyond

*Now available in a lower priced paperback edition in the Wiley Classics Library.

Probability and Statistics (Continued)

KAUFMAN and ROUSSEEUW · Finding Groups in Data: An Introduction to Cluster Analysis
LAMPERTI · Probability: A Survey of the Mathematical Theory, *Second Edition*
LARSON · Introduction to Probability Theory and Statistical Inference, *Third Edition*
LESSLER and KALSBEEK · Nonsampling Error in Surveys
LINDVALL · Lectures on the Coupling Method
MANTON, WOODBURY, and TOLLEY · Statistical Applications Using Fuzzy Sets
MARDIA · The Art of Statistical Science: A Tribute to G. S. Watson
MORGENTHALER and TUKEY · Configural Polysampling: A Route to Practical Robustness
MUIRHEAD · Aspects of Multivariate Statistical Theory
OLIVER and SMITH · Influence Diagrams, Belief Nets and Decision Analysis
*PARZEN · Modern Probability Theory and Its Applications
PRESS · Bayesian Statistics: Principles, Models, and Applications
PUKELSHEIM · Optimal Experimental Design
PURI and SEN · Nonparametric Methods in General Linear Models
PURI, VILAPLANA, and WERTZ · New Perspectives in Theoretical and Applied Statistics
RAO · Asymptotic Theory of Statistical Inference
RAO · Linear Statistical Inference and Its Applications, *Second Edition*
*RAO and SHANBHAG · Choquet-Deny Type Functional Equations with Applications to Stochastic Models
RENCHER · Methods of Multivariate Analysis
ROBERTSON, WRIGHT, and DYKSTRA · Order Restricted Statistical Inference
ROGERS and WILLIAMS · Diffusions, Markov Processes, and Martingales, Volume I: Foundations, *Second Edition;* Volume II: Îto Calculus
ROHATGI · An Introduction to Probability Theory and Mathematical Statistics
ROSS · Stochastic Processes
RUBINSTEIN · Simulation and the Monte Carlo Method
RUBINSTEIN and SHAPIRO · Discrete Event Systems: Sensitivity Analysis and Stochastic Optimization by the Score Function Method
RUZSA and SZEKELY · Algebraic Probability Theory
SCHEFFE · The Analysis of Variance
SEBER · Linear Regression Analysis
SEBER · Multivariate Observations
SEBER and WILD · Nonlinear Regression
SERFLING · Approximation Theorems of Mathematical Statistics
SHORACK and WELLNER · Empirical Processes with Applications to Statistics
SMALL and McLEISH · Hilbert Space Methods in Probability and Statistical Inference
STAPLETON · Linear Statistical Models
STAUDTE and SHEATHER · Robust Estimation and Testing
STOYANOV · Counterexamples in Probability
STYAN · The Collected Papers of T. W. Anderson: 1943–1985
TANAKA · Time Series Analysis: Nonstationary and Noninvertible Distribution Theory
THOMPSON and SEBER · Adaptive Sampling
WELSH · Aspects of Statistical Inference
WHITTAKER · Graphical Models in Applied Multivariate Statistics
YANG · The Construction Theory of Denumerable Markov Processes

Applied Probability and Statistics

ABRAHAM and LEDOLTER · Statistical Methods for Forecasting
AGRESTI · Analysis of Ordinal Categorical Data

*Now available in a lower priced paperback edition in the Wiley Classics Library.

Applied Probability and Statistics (Continued)

AGRESTI · Categorical Data Analysis
AGRESTI · An Introduction to Categorical Data Analysis
ANDERSON and LOYNES · The Teaching of Practical Statistics
ANDERSON, AUQUIER, HAUCK, OAKES, VANDAELE, and WEISBERG · Statistical Methods for Comparative Studies
ARMITAGE and DAVID (editors) · Advances in Biometry
*ARTHANARI and DODGE · Mathematical Programming in Statistics
ASMUSSEN · Applied Probability and Queues
*BAILEY · The Elements of Stochastic Processes with Applications to the Natural Sciences
BARNETT and LEWIS · Outliers in Statistical Data, *Second Edition*
BARTHOLOMEW, FORBES, and McLEAN · Statistical Techniques for Manpower Planning, *Second Edition*
BATES and WATTS · Nonlinear Regression Analysis and Its Applications
BECHHOFER, SANTNER, and GOLDSMAN · Design and Analysis of Experiments for Statistical Selection, Screening, and Multiple Comparisons
BELSLEY · Conditioning Diagnostics: Collinearity and Weak Data in Regression
BELSLEY, KUH, and WELSCH · Regression Diagnostics: Identifying Influential Data and Sources of Collinearity
BERRY · Bayesian Analysis in Statistics and Econometrics: Essays in Honor of Arnold Zellner
BERRY, CHALONER, and GEWEKE · Bayesian Analysis in Statistics and Econometrics: Essays in Honor of Arnold Zellner
BHAT · Elements of Applied Stochastic Processes, *Second Edition*
BHATTACHARYA and WAYMIRE · Stochastic Processes with Applications
BIEMER, GROVES, LYBERG, MATHIOWETZ, and SUDMAN · Measurement Errors in Surveys
BIRKES and DODGE · Alternative Methods of Regression
BLOOMFIELD · Fourier Analysis of Time Series: An Introduction
BOLLEN · Structural Equations with Latent Variables
BOULEAU · Numerical Methods for Stochastic Processes
BOX · R. A. Fisher, the Life of a Scientist
BOX and DRAPER · Empirical Model-Building and Response Surfaces
BOX and DRAPER · Evolutionary Operation: A Statistical Method for Process Improvement
BOX, HUNTER, and HUNTER · Statistics for Experimenters: An Introduction to Design, Data Analysis, and Model Building
BOX and LUCENO · Statistical Control by Monitoring and Feedback Adjustment
BROWN and HOLLANDER · Statistics: A Biomedical Introduction
BUCKLEW · Large Deviation Techniques in Decision, Simulation, and Estimation
BUNKE and BUNKE · Nonlinear Regression, Functional Relations and Robust Methods: Statistical Methods of Model Building
CHATTERJEE and HADI · Sensitivity Analysis in Linear Regression
CHATTERJEE and PRICE · Regression Analysis by Example, *Second Edition*
CLARKE and DISNEY · Probability and Random Processes: A First Course with Applications, *Second Edition*
COCHRAN · Sampling Techniques, *Third Edition*
*COCHRAN and COX · Experimental Designs, *Second Edition*
CONOVER · Practical Nonparametric Statistics, *Second Edition*
CONOVER and IMAN · Introduction to Modern Business Statistics
CORNELL · Experiments with Mixtures, Designs, Models, and the Analysis of Mixture Data, *Second Edition*
COX · A Handbook of Introductory Statistical Methods

*Now available in a lower priced paperback edition in the Wiley Classics Library.

*COX · Planning of Experiments
COX, BINDER, CHINNAPPA, CHRISTIANSON, COLLEDGE, and KOTT · Business Survey Methods
CRESSIE · Statistics for Spatial Data, *Revised Edition*
DANIEL · Applications of Statistics to Industrial Experimentation
DANIEL · Biostatistics: A Foundation for Analysis in the Health Sciences, *Sixth Edition*
DAVID · Order Statistics, *Second Edition*
*DEGROOT, FIENBERG, and KADANE · Statistics and the Law
*DEMING · Sample Design in Business Research
DETTE and STUDDEN · Theory of Canonical Moments with Applications in Statistics and Probability
DILLON and GOLDSTEIN · Multivariate Analysis: Methods and Applications
DODGE and ROMIG · Sampling Inspection Tables, *Second Edition*
DOWDY and WEARDEN · Statistics for Research, *Second Edition*
DRAPER and SMITH · Applied Regression Analysis, *Second Edition*
DUNN · Basic Statistics: A Primer for the Biomedical Sciences, *Second Edition*
DUNN and CLARK · Applied Statistics: Analysis of Variance and Regression, *Second Edition*
ELANDT-JOHNSON and JOHNSON · Survival Models and Data Analysis
EVANS, PEACOCK, and HASTINGS · Statistical Distributions, *Second Edition*
FISHER and VAN BELLE · Biostatistics: A Methodology for the Health Sciences
FLEISS · The Design and Analysis of Clinical Experiments
FLEISS · Statistical Methods for Rates and Proportions, *Second Edition*
FLEMING and HARRINGTON · Counting Processes and Survival Analysis
FLURY · Common Principal Components and Related Multivariate Models
GALLANT · Nonlinear Statistical Models
GLASSERMAN and YAO · Monotone Structure in Discrete-Event Systems
GNANADESIKAN · Methods for Statistical Data Analysis of Multivariate Observations, *Second Edition*
GREENWOOD and NIKULIN · A Guide to Chi-Squared Testing
GROSS and HARRIS · Fundamentals of Queueing Theory, *Second Edition*
GROVES · Survey Errors and Survey Costs
GROVES, BIEMER, LYBERG, MASSEY, NICHOLLS, and WAKSBERG · Telephone Survey Methodology
HAHN and MEEKER · Statistical Intervals: A Guide for Practitioners
HAND · Discrimination and Classification
*HANSEN, HURWITZ, and MADOW · Sample Survey Methods and Theory, Volume 1: Methods and Applications
*HANSEN, HURWITZ, and MADOW · Sample Survey Methods and Theory, Volume II: Theory
HEIBERGER · Computation for the Analysis of Designed Experiments
HELLER · MACSYMA for Statisticians
HINKELMAN and KEMPTHORNE: · Design and Analysis of Experiments, Volume 1: Introduction to Experimental Design
HOAGLIN, MOSTELLER, and TUKEY · Exploratory Approach to Analysis of Variance
HOAGLIN, MOSTELLER, and TUKEY · Exploring Data Tables, Trends and Shapes
HOAGLIN, MOSTELLER, and TUKEY · Understanding Robust and Exploratory Data Analysis
HOCHBERG and TAMHANE · Multiple Comparison Procedures
HOCKING · Methods and Applications of Linear Models: Regression and the Analysis of Variables
HOEL · Elementary Statistics, *Fifth Edition*
HOGG and KLUGMAN · Loss Distributions
HOLLANDER and WOLFE · Nonparametric Statistical Methods

*Now available in a lower priced paperback edition in the Wiley Classics Library.

Applied Probability and Statistics (Continued)

HOSMER and LEMESHOW · Applied Logistic Regression
HØYLAND and RAUSAND · System Reliability Theory: Models and Statistical Methods
HUBERTY · Applied Discriminant Analysis
IMAN and CONOVER · Modern Business Statistics
JACKSON · A User's Guide to Principle Components
JOHN · Statistical Methods in Engineering and Quality Assurance
JOHNSON · Multivariate Statistical Simulation
JOHNSON and BALAKRISHNAN · Advances in the Theory and Practice of Statistics: A Volume in Honor of Samuel Kotz
JOHNSON and KOTZ · Distributions in Statistics
Continuous Multivariate Distributions
JOHNSON, KOTZ, and BALAKRISHNAN · Continuous Univariate Distributions, Volume 1, *Second Edition*
JOHNSON, KOTZ, and BALAKRISHNAN · Continuous Univariate Distributions, Volume 2, *Second Edition*
JOHNSON, KOTZ, and BALAKRISHNAN · Discrete Multivariate Distributions
JOHNSON, KOTZ, and KEMP · Univariate Discrete Distributions, *Second Edition*
JUDGE, GRIFFITHS, HILL, LÜTKEPOHL, and LEE · The Theory and Practice of Econometrics, *Second Edition*
JUDGE, HILL, GRIFFITHS, LÜTKEPOHL, and LEE · Introduction to the Theory and Practice of Econometrics, *Second Edition*
JUREČKOVÁ and SEN · Robust Statistical Procedures: Aymptotics and Interrelations
KADANE · Bayesian Methods and Ethics in a Clinical Trial Design
KADANE AND SCHUM · A Probabilistic Analysis of the Sacco and Vanzetti Evidence
KALBFLEISCH and PRENTICE · The Statistical Analysis of Failure Time Data
KASPRZYK, DUNCAN, KALTON, and SINGH · Panel Surveys
KISH · Statistical Design for Research
*KISH · Survey Sampling
LAD · Operational Subjective Statistical Methods: A Mathematical, Philosophical, and Historical Introduction
LANGE, RYAN, BILLARD, BRILLINGER, CONQUEST, and GREENHOUSE · Case Studies in Biometry
LAWLESS · Statistical Models and Methods for Lifetime Data
LEBART, MORINEAU., and WARWICK · Multivariate Descriptive Statistical Analysis: Correspondence Analysis and Related Techniques for Large Matrices
LEE · Statistical Methods for Survival Data Analysis, *Second Edition*
LEPAGE and BILLARD · Exploring the Limits of Bootstrap
LEVY and LEMESHOW · Sampling of Populations: Methods and Applications
LINHART and ZUCCHINI · Model Selection
LITTLE and RUBIN · Statistical Analysis with Missing Data
LYBERG · Survey Measurement
MAGNUS and NEUDECKER · Matrix Differential Calculus with Applications in Statistics and Econometrics
MAINDONALD · Statistical Computation
MALLOWS · Design, Data, and Analysis by Some Friends of Cuthbert Daniel
MANN, SCHAFER, and SINGPURWALLA · Methods for Statistical Analysis of Reliability and Life Data
MASON, GUNST, and HESS · Statistical Design and Analysis of Experiments with Applications to Engineering and Science
McLACHLAN and KRISHNAN · The EM Algorithm and Extensions
McLACHLAN · Discriminant Analysis and Statistical Pattern Recognition
McNEIL · Epidemiological Research Methods
MILLER · Survival Analysis

*Now available in a lower priced paperback edition in the Wiley Classics Library.

Applied Probability and Statistics (Continued)

MONTGOMERY and MYERS · Response Surface Methodology: Process and Product in Optimization Using Designed Experiments
MONTGOMERY and PECK · Introduction to Linear Regression Analysis, *Second Edition*
NELSON · Accelerated Testing, Statistical Models, Test Plans, and Data Analyses
NELSON · Applied Life Data Analysis
OCHI · Applied Probability and Stochastic Processes in Engineering and Physical Sciences
OKABE, BOOTS, and SUGIHARA · Spatial Tesselations: Concepts and Applications of Voronoi Diagrams
OSBORNE · Finite Algorithms in Optimization and Data Analysis
PANKRATZ · Forecasting with Dynamic Regression Models
PANKRATZ · Forecasting with Univariate Box-Jenkins Models: Concepts and Cases
PIANTADOSI · Clinical Trials: A Methodologic Perspective
PORT · Theoretical Probability for Applications
PUTERMAN · Markov Decision Processes: Discrete Stochastic Dynamic Programming
RACHEV · Probability Metrics and the Stability of Stochastic Models
RÉNYI · A Diary on Information Theory
RIPLEY · Spatial Statistics
RIPLEY · Stochastic Simulation
ROSS · Introduction to Probability and Statistics for Engineers and Scientists
ROUSSEEUW and LEROY · Robust Regression and Outlier Detection
RUBIN · Multiple Imputation for Nonresponse in Surveys
RYAN · Modern Regression Methods
RYAN · Statistical Methods for Quality Improvement
SCHOTT · Matrix Analysis for Statistics
SCHUSS · Theory and Applications of Stochastic Differential Equations
SCOTT · Multivariate Density Estimation: Theory, Practice, and Visualization
*SEARLE · Linear Models
SEARLE · Linear Models for Unbalanced Data
SEARLE · Matrix Algebra Useful for Statistics
SEARLE, CASELLA, and McCULLOCH · Variance Components
SKINNER, HOLT, and SMITH · Analysis of Complex Surveys
STOYAN, KENDALL, and MECKE · Stochastic Geometry and Its Applications, *Second Edition*
STOYAN and STOYAN · Fractals, Random Shapes and Point Fields: Methods of Geometrical Statistics
THOMPSON · Empirical Model Building
THOMPSON · Sampling
TIERNEY · LISP-STAT: An Object-Oriented Environment for Statistical Computing and Dynamic Graphics
TIJMS · Stochastic Modeling and Analysis: A Computational Approach
TITTERINGTON, SMITH, and MAKOV · Statistical Analysis of Finite Mixture Distributions
UPTON and FINGLETON · Spatial Data Analysis by Example, Volume 1: Point Pattern and Quantitative Data
UPTON and FINGLETON · Spatial Data Analysis by Example, Volume II: Categorical and Directional Data
VAN RIJCKEVORSEL and DE LEEUW · Component and Correspondence Analysis
WEISBERG · Applied Linear Regression, *Second Edition*
WESTFALL and YOUNG · Resampling-Based Multiple Testing: Examples and Methods for *p*-Value Adjustment
WHITTLE · Optimization Over Time: Dynamic Programming and Stochastic Control, Volume I and Volume II
WHITTLE · Systems in Stochastic Equilibrium

*Now available in a lower priced paperback edition in the Wiley Classics Library.

Applied Probability and Statistics (Continued)

WONNACOTT and WONNACOTT · Econometrics, *Second Edition*
WONNACOTT and WONNACOTT · Introductory Statistics, *Fifth Edition*
WONNACOTT and WONNACOTT · Introductory Statistics for Business and Economics, *Fourth Edition*
WOODING · Planning Pharmaceutical Clinical Trials: Basic Statistical Principles
WOOLSON · Statistical Methods for the Analysis of Biomedical Data
*ZELLNER · An Introduction to Bayesian Inference in Econometrics

Tracts on Probability and Statistics

BILLINGSLEY · Convergence of Probability Measures
KELLY · Reversability and Stochastic Networks
TOUTENBURG · Prior Information in Linear Models

*Now available in a lower priced paperback edition in the Wiley Classics Library.